Statistical Inference in Linear Models

Statistical Inference in Linear Models

Editor

Sandra Ferreira

Basel • Beijing • Wuhan • Barcelona • Belgrade • Novi Sad • Cluj • Manchester

Editor
Sandra Ferreira
University of Beira Interior
Covilhã
Portugal

Editorial Office
MDPI
St. Alban-Anlage 66
4052 Basel, Switzerland

This is a reprint of articles from the Special Issue published online in the open access journal *Mathematical and Computational Applications* (ISSN 2297-8747) (available at: https://www.mdpi.com/journal/mca/special_issues/1853X8M81G).

For citation purposes, cite each article independently as indicated on the article page online and as indicated below:

Lastname, A.A.; Lastname, B.B. Article Title. *Journal Name* **Year**, *Volume Number*, Page Range.

ISBN 978-3-7258-0257-9 (Hbk)
ISBN 978-3-7258-0258-6 (PDF)
doi.org/10.3390/books978-3-7258-0258-6

Contents

About the Editor

Sandra Ferreira

Sandra Ferreira is an assistant professor at the Department of Mathematics, University of Beira Interior. She received a master's degree in Applied Mathematics from Évora University (2000) and a Ph.D. in Mathematics at the University of Beira Interior (2006). She is a researcher at the Center of Mathematics and Applications, in the Physics and Mathematical Modeling research group. She is also a researcher (external collaborator) at the Center for Mathematics and Applications (CMA), NOVA School of Science and Technology — FCT NOVA, in the Statistics and Risk Management research group. Her research interests include mathematical statistics; statistical analysis, modeling, and inference; applied and computational statistics; R statistical package; sample size; mixed, linear, and linear mixed models; data analysis models; F-tests; applied mathematics; applied probability and probability theory; random sampling; confidence intervals; linear regression; multivariate, descriptive, and nonparametric statistics; and regression analysis. Her publications and current research interests focus on statistical inference for estimable functions and variance components in linear mixed models with a commutative orthogonal block structure (COBS). Her research group is currently working on a project titled "Estimation in Mixed Linear Models—MLM". She has authored 78 scientific papers, receiving more than 340 citations (H-index 10). Moreover, she was awarded the "Mathematical sciences sponsorship fund" (2018) by Elsevier.

Preface

Linear models are statistical models that play a crucial role in several fields of science and are of practical importance in statistics. Their most typical type is the linear regression model. Many phenomena, such as those in biology, medicine, economics, management, geology, meteorology, agriculture, and industry, can be approximately described with linear models. Moreover, the further research and development of linear models is still a hot research topic.

This reprint encompasses the study and practical implementation of linear models in diverse scientific domains through a collection of several articles. The text furnishes a comprehensive examination of various classifications of linear models, such as linear regression, ANOVA, and MANOVA, elucidating their respective applications. Additionally, it delves into the presumptions and limitations inherent in these models, employing practical instances to exemplify their pragmatic utility. This compilation of articles on linear models assumes a significant role as it equips analysts with a potent instrument to scrutinize and model intricate data, enabling them to expound upon the relationship between variables, conduct hypothesis testing, and make predictions. The ongoing research endeavors to devise novel linear models and data analysis methodologies hold tremendous potential in advancing our comprehension and enhancing our predictive capabilities. A comprehensive overview of linear models and their applications is presented herein. This compilation was completed as research in this field is witnessing a surge due to the mounting availability of data and the consequent requirement for robust analytical tools. These models furnish a versatile and formidable framework for unraveling complex data patterns and facilitating prediction. The pursuit of novel linear models and data analysis techniques remains a vibrant field of research, harboring the promise of furthering our understanding and improving our predictive prowess. This publication will undoubtedly captivate researchers and students specializing in statistics, data science, and allied fields, who seek to broaden their knowledge of linear models and their practical implementations. It also caters to professionals who aspire to comprehend and apply these models to real-world scenarios and their own research endeavors.

I would like to avail myself of this occasion to express my gratitude for the entirety of the assistance extended by this scholarly publication, which extended an invitation to me to act as a guest editor for this distinctive edition, as well as for the remarkable efforts exerted by all the contributors and reviewers throughout the process of editing.

The paper titled "A Quantile Functions-Based Investigation on the Characteristics of Southern African Solar Irradiation Data" investigates the climatic attributes of solar irradiation data in the region of Southern Africa, thus offering crucial information for planners, designers, and investors involved in the solar power generation sector. Moreover, it scrutinizes the seasonal changes in these data, including the highest daily solar irradiation. Within this paper, a suggestion is made to employ quantile functions to model solar irradiation data. This approach is grounded in the idea that quantile functions can reveal insights into data skewness, outliers, and tail behaviors. Quantile functions can reveal insights into data skewness, outliers, and tail behaviors. Furthermore, their use can be extended to various probability distributions, broadening their applicability in the analysis of solar irradiation data. Future investigations can explore the use of quantile functions in different probability distributions. Such advancements would simplify the approximation process and enhance the accuracy of quantile functions in the modeling of solar irradiation data.

The paper "A New Sine Family of Generalized Distributions: Statistical Inference with Applications" introduces a new family of distributions called the alpha-sine-G family, which is derived from the trigonometric function and includes an additional parameter. The statistical properties of this new distribution are investigated, and several families of trigonometric distributions are developed. Two real data sets were analyzed to help illustrate the suggested model. The performance was presented based on various criteria measures, such as the Akaike information criteria, the consistent Akaike information criteria, the Bayesian information criteria, the Hannan–Quinn information criteria, box plots, TTT plots, and PP plots to support the assertion that the AS-W model provides the best fit among its competitors.

The paper "Computation of the Distribution of the Sum of Independent Negative Binomial Random Variables" compares different methods to estimate the distribution of the sum of negative binomial random variables. The authors compare different methods for estimating this distribution, including a finite-sum exact expression, a series expression via convolution, a normalized saddle point approximation, and normal and single distribution negative binomial approximations. This paper highlights the limitations and advantages of each method, providing insights for applied practitioners in terms of memory usage, computing time, and precision of the estimates. Exact series expressions, for estimating the distribution of the sum of negative binomial random variables, are deemed not practical for high numbers of random variables due to their high memory usage. Thus, it is commendable that the authors propose the exploration of alternative approximation methods or techniques for estimating the distribution of the sum of negative binomial random variables in future studies.

In the paper titled "Prediction Interval for Compound Conway–Maxwell–Poisson Regression Model with Application to Vehicle Insurance Claim Data", the authors introduce a regression model utilizing a compound Conway–Maxwell–Poisson distribution—a succinct Poisson distribution with just two parameters—that is adept at accommodating under-dispersed count data, which render it versatile across numerous domains. A two-step approach involves estimating the parameters of the compound Conway–Maxwell–Poisson regression model within the context of a generalized linear model. Additionally, a novel technique is outlined for deriving prediction intervals for the response variable in this compound regression model. The practicality of this methodology is illustrated through an application to actual vehicle insurance claims data, showcasing its efficacy in parameter estimation and prediction interval generation.

The paper "The Arctan Power Distribution: Properties, Quantile and Modal Regressions with Applications to Biomedical Data" tests the arctan power distribution, which is a versatile distribution capable of representing data with left-skewedness, right-skewness, and J and reversed-J shapes. The authors proceed to develop a bivariate extension of the arctan power distribution to represent the interdependence of two random variables or pairs of data. They further elucidate that the parameters of the arctan power distribution can be accurately estimated through the utilization of a Bayesian approach. Additionally, quantile and modal regression models based on the arctan power distribution are presented by the authors. It is discovered that these models yield superior fits to biomedical data in comparison to other existing regression models.

The paper "Forecasting Financial and Macroeconomic Variables Using an Adaptive Parameter VAR-KF Model" examines the demand for robust economic and financial forecasting tools, a necessity driven by growing uncertainty in these sectors. Various prediction methods have been developed covering time series models. This research spotlights the hybrid vector autoregressive and Kalman filter approach for economic and financial trend projection, incorporating updated coefficients via the unified state-parameter Kalman filter process. Findings from simulated experiments, involving Thailand and Indonesia's main stock exchange index, real effective exchange rate, and consumer price index, spanning from January 1997 to May 2021, demonstrate the general superiority of the adaptive parameter vector autoregressive and Kalman filter model in predictive accuracy compared to other models. The model's success is attributed to the two-step Kalman filter process that gleans valuable insights from training data, combined with adaptive parameters that enhance hybrid model performance. This study recognizes that the vector autoregressive lag 1 assumption may present limitations and recommends exploring higher lag orders and introducing additional macroeconomic variables in future research.

In this research paper entitled "Pseudo-Poisson Distributions with Concomitant Variables", the impact of accompanying variables on the parameters of the bivariate pseudo-Poisson distribution is thoroughly examined, focusing on distributional and inferential aspects. This study covers a simulation analysis and references the work of Kokonendji and Puig, introducing the concept of the bivariate Fisher dispersion index. Furthermore, the authors investigate the null hypothesis and the parameter space to evaluate how concomitant variables affect the dependence structure of the pseudo-Poisson model. The concept of concomitant variables that influence the parameters of the considered distribution is also introduced. Additionally, it explores the distributional characteristics and inferential dimensions of models enriched by these variables. The inclusion of a real-world application highlights the pragmatic utility of augmented models in analyzing bivariate data. Augmented models provide a valuable framework for integrating these variables and produce valuable insights into the distributional characteristics and inferential capabilities of the model.

"New Lifetime Distribution for Modeling Data on the Unit Interval: Properties, Applications and Quantile Regression". This paper presents a new distribution, the Cauchy-bounded truncated exponential power distribution, designed to model unit interval data sets with versatile formats and hazard rate functions. It introduces a bivariate extension of this distribution to capture dependencies within datasets. The effectiveness of this distribution is demonstrated through its successful application to COVID-19 data, particularly in modeling mortality and recovery rates. The manuscript incorporates a quantile regression model that presents a reasonable fit to the data, validated through residual analysis. The authors of this manuscript intend to conduct future research that will be dedicated to further exploring the properties of the bivariate distribution, parameter estimation, and their potential applications.

The article "Flexible Parametric Accelerated Hazard Model: Simulation and Application to Censored Lifetime Data with Crossing Survival Curves" introduces a flexible, fully parametric regression model for censored time-to-event data with crossing survival curves, known as the accelerated hazard model. This model offers the ability to analyze various types of survival data with crossover survival curves. The proposed accelerated hazard model explores the use of a versatile parametric baseline distribution, specifically the generalized log-logistic distribution. This choice allows for a more flexible representation of the baseline hazard and enables different hazard rate shapes to be captured. The article showcases that both Bayesian and classical likelihood inference can be performed using the proposed accelerated hazard model and a package in the R programming language. This addresses the limited utilization of the semi-parametric accelerated hazard model, which has been hindered by a lack of efficient and reliable estimation methods. Further examinations can be conducted to assess the performance of the proposed accelerated hazard model with various types of baseline distributions and to compare it against existing models. The authors suggest that there is room for additional research to examine the performance of the suggested model with various baseline distributions and to make comparisons with existing models. They also propose that the application of the AH parametric model can be extended to encompass various forms of survival data characterized by the intersection of survival curves, including applications in areas such as medical research and engineering.

The investigation presented in the paper titled "Bivariate Generalized Half-Logistic Distribution:

Properties and Its Application in Household Financial Affordability in KSA" examines the study of adaptive life distributions, which exhibit various characteristics of probability density and risk rate. This area of research is particularly relevant in the field of reliability analysis. Within this study, the authors introduce the bivariate generalized semi-logistic distribution (BGHLD), referred to as FGMBGHLD. This distribution is well-suited for describing bivariate life data sets in which variables show weak correlations. The primary focus of this manuscript is to conduct a comparative analysis with a contemporary bivariate Weibull distribution, highlighting the competitive performance of the proposed model. The authors also mention that future research perspectives should involve the exploration of this distribution's applications in different types of bivariate data, developing its multivariate counterpart and expanding its use to various types of regression models.

The present study, titled "Odd Exponential-Logarithmic Family of Distributions: Features and Modeling", showcases a wide range of distributions called the "odd exponential logarithmic family", which can be effectively employed in practical applications. This family encompasses distinct members that have versatile shape properties, allowing for great flexibility in data modeling. A comprehensive mathematical analysis of this family is provided in this article. It also demonstrates how the model's parameters can be estimated using the maximum likelihood method and the observed information matrix, which are discussed in detail within. Moreover, a comparative analysis is carried out, revealing the superior data fitting performance of the odd exponential logarithmic model when compared to two competitors—an analysis carried out using three sets of practical data. The findings included in this study strongly indicate that the odd exponential logarithmic family adds significant value to the existing knowledge in the field of distribution modeling and that it is particularly relevant in several disciplines where it has already demonstrated its importance.

Sandra Ferreira
Editor

Mathematical and Computational Applications

Article

Odd Exponential-Logarithmic Family of Distributions: Features and Modeling

Christophe Chesneau [1,*], Lishamol Tomy [2], Meenu Jose [3] and Kuttappan Vallikkattil Jayamol [4]

[1] Department of Mathematics, LMNO, CNRS-Université de Caen, Campus II, Science 3, CEDEX, 14032 Caen, France
[2] Department of Statistics, Deva Matha College, Kuravilangad 686633, Kerala, India
[3] Department of Statistics, Carmel College Mala, Thrissur 680732, Kerala, India
[4] Department of Statistics, Maharajas College Ernakulam, Ernakulam 682011, Kerala, India
[*] Correspondence: christophe.chesneau@gmail.com

Abstract: This paper introduces a general family of continuous distributions, based on the exponential-logarithmic distribution and the odd transformation. It is called the "odd exponential logarithmic family". We intend to create novel distributions with desired qualities for practical applications, using the unique properties of the exponential-logarithmic distribution as an initial inspiration. Thus, we present some special members of this family that stand out for the versatile shape properties of their corresponding functions. Then, a comprehensive mathematical treatment of the family is provided, including some asymptotic properties, the determination of the quantile function, a useful sum expression of the probability density function, tractable series expressions for the moments, moment generating function, Rényi entropy and Shannon entropy, as well as results on order statistics and stochastic ordering. We estimate the model parameters quite efficiently by the method of maximum likelihood, with discussions on the observed information matrix and a complete simulation study. As a major interest, the odd exponential logarithmic models reveal how to successfully accommodate various kinds of data. This aspect is demonstrated by using three practical data sets, showing that an odd exponential logarithmic model outperforms two strong competitors in terms of data fitting.

Keywords: exponential-logarithmic distribution; T-X transformation; moments; entropy; maximum likelihood estimation; simulation; data sciences

Citation: Chesneau, C.; Tomy, L.; Jose, M.; Jayamol, K.V. Odd Exponential-Logarithmic Family of Distributions: Features and Modeling. *Math. Comput. Appl.* **2022**, 27, 68. https://doi.org/10.3390/mca27040068

Academic Editor: Sandra Ferreira

Received: 5 July 2022
Accepted: 4 August 2022
Published: 8 August 2022

Publisher's Note: MDPI stays neutral with regard to jurisdictional claims in published maps and institutional affiliations.

1. Introduction

There has been a growing interest in defining new flexible distributions in the modern age, which has been submerged by the volume of data arriving from all disciplines. To define such mathematical objects, "thoroughly changing" a baseline (continuous) distribution is a straightforward and fast method. The addition of parameters has been shown to be useful in investigating tail properties as well as increasing the goodness-of-fit of the related models. Among the proposed distributions, the T-X family of continuous distributions (focds) by [1] is the most popular one. An exhaustive review of it can be found in [2]. Also, one of the most useful transformers for the T-X focds is the following odd transformation: $W[G(x;\Im)] = G(x;\Im)/[1 - G(x;\Im)]$, where $G(x;\Im)$ denotes the cumulative density function (cdf) and $\Im$ the parameters of the cdf. That is, the focds defined by $G(x;\Im)$ is modified, defining a new focd based on a transformed cdf through the use of $W[G(x;\Im)]$. Such a transformed focds is generally called an "odd family" of distributions. Some odd families available in the modern literature are the odd log-logistic (OLL) focds by [3], odd-gamma generated type 3 (OGGT3) focds by [4], odd exponentiated generated (odd exp-G) focds by [5], odd Weibull generated (OW-G) focds by [6], odd generalized exponential (OGE) focds by [7], odd generalized exponential log-logistic (OGELL) focds by [8], odd log-logistic normal (OLLN) focds by [9], new generalized odd log-logistic (NGOLL) focds by [10], odd Fréchet generated (OF-G) focds by [11], generalized odd gamma generated (GOG-G) focds by [12], generalized odd Lindley generated (GOL-G) focds by [13], Marshall-Olkin odd Lindley generated (MOOL-G) focds by [14], extended odd generated (EO-G) focds by [15], generalized odd inverted exponential generated (GOIE-G) focds by [16], odd flexible Weibull-H (OFW-H) family by [17], transmuted odd Fréchet generated (TOF-G) focds by [18], odd generalized gamma generated (OGG-G) focds by [19], modified odd Weibull generated (MOW-G) focds by [20], Topp-Leone odd Fréchet generated (TLOF-G) focds by [21], weighted odd Weibull generated (WOW-G) focds by [22], additive odd (AO) focds by [23], exponentiated odd Chen-G (EOC-G) focds by [24], generalized odd linear exponential (GOLE) focds by [25], and sine extended odd Fréchet generated (SEOF-G) focds by [26].

The new idea in this paper is centered around the notorious exponential-logarithmic (EL) distribution introduced by [27]. The EL distribution plays a fundamental role in reliability in several disciplines such as manufacturing, finance, biological sciences, and engineering. It is mathematically defined as follows. Let $p \in (0,1)$ and $\beta > 0$. Then, the EL distribution with parameters p and β is defined by the following cdf:

$$F_*(x; p, \beta) = 1 - \frac{1}{\log(p)} \log\left[1 - (1-p)e^{-\beta x}\right], \quad x > 0. \tag{1}$$

Thus, it has the feature of combining exponential and logarithmic functions. The related probability density function (pdf) is given by

$$f_*(x; p, \beta) = \left(\frac{1}{-\log(p)}\right) \frac{\beta(1-p)e^{-\beta x}}{1 - (1-p)e^{-\beta x}}, \quad x > 0.$$

This pdf has the following notable properties: it is strictly decreasing with respect to x, it tends to zero as $x \to +\infty$, it is unimodal with a modal value at $x = 0$ and it is reduced to the pdf of the exponential distribution with rate parameter β as $p \to 1$. Also, as a complementary key function, the corresponding hazard rate function (hrf) is given by

$$h_*(x; p, \beta) = \frac{-\beta(1-p)e^{-\beta x}}{[1 - (1-p)e^{-\beta x}] \log[1 - (1-p)e^{-\beta x}]}, \quad x > 0.$$

It is proved to be decreasing (contrary to the former exponential distribution having a constant hrf). As an advantage for statistical analysis, the quantile function (qf) of the EL distribution has a closed-form expression; it is given by

$$Q_*(u; p, \beta) = \frac{1}{\beta} \log\left(\frac{1-p}{1-p^{1-u}}\right), \quad u \in [0,1).$$

Also, the EL distribution has a solid physical interpretation. Indeed, consider $T = (T_n)_{n \in \mathbb{N}^*}$ to be a sequence of independent and identically distributed random variables with an exponential distribution and a common parameter, β. Let N be a random variable following the discrete logarithmic distribution with parameter $1 - p$, also independent of T. Then, the random variable $X = \inf(T_1, \ldots, T_N)$ follows the EL distribution with parameters p and β. As an example, such a random variable can model the lifetime of a system that failed when one of its components failed, assuming that it is dependent on a random number of independent components represented by N and that the lifetime of the i-th component is represented by T_i.

We leverage these characteristics of the EL distribution to create a new odd focds based on it. We present three special four-parameter distributions of the family that have very desirable statistical properties, such as versatile hazard rate shapes; increasing, decreasing, J, reversed-J, and bathtub shapes. Then, a complete mathematical treatment of the focds is derived, with several results on the pdf, moments, entropy (Rényi and Shannon entropy), order statistics, and stochastic ordering. By turning out some special distributions as models, we prove that they are more adequate to fit some data sets than notable competitors, with the same or more numbers of parameters, and the same baseline distribution as well. We explain this success by the original exponential-logarithmic definitions of the corresponding functions, offering some ability in the modeling that can be reached by other families.

The paper is composed of the following sections. In Section 2, we introduce the odd exponential-logarithmic focds. We present some special distributions in Section 3. The mathematical properties of the focds are derived in Section 4. For the inferential aspect, the maximum likelihood method is discussed in Section 5. The analysis of two real data sets is presented to illustrate the modeling potential of the focds in Section 6. Finally, the conclusion of the paper appears in Section 7.

2. The New Family

The proposed focds, called the odd EL generated (OEL-G) focds, is characterized by the cdf given by

$$F(x; p, \beta, \Im) = 1 - \frac{1}{\log(p)} \log\left[1 - (1-p)e^{-\beta \frac{G(x;\Im)}{1-G(x;\Im)}}\right], \quad x \in \mathbb{R}, \tag{2}$$

where $G(x; \Im)$ denotes the cdf of an absolutely continuous distribution based on a parameter vector denoted by $\Im$. We recall that $p \in (0,1)$. Its definition is based on the T-X transformation introduced by [1], the EL distribution previously presented and the odd transformation, i.e., we can write $F(x; p, \beta, \Im)$ as $F(x; p, \beta, \Im) = F_*(W[G(x; \Im)]; p, \beta)$, where $F_*(y; p, \beta)$ is the cdf of the EL distribution given by (1) and $W(y)$ is the following odd transformation: $W(y) = y/(1-y)$. One can also notice some compounding relations between the OEL-G and the OW-G and Pappas and Loukas generated (PAL-G) families by [6,28], respectively. Indeed, we can write $F(x; p, \beta, \Im)$ as

$$F(x; p, \beta, \Im) = 1 - \frac{1}{\log(p)} \log[1 - (1-p)S_o(x; \beta, \Im)],$$

where $S_o(x; \beta, \Im)$ denotes the survival function (sf) of the OW-G focds with parameters β and $\Im$, which also corresponds to the cdf of the PAL-G focds, with parameter p and the cdf of the OW-G focds as a baseline. However, to the best of our knowledge, the OEL-G focds as defined by (2) is new in the literature.

The sf of the OEL-G focds is given by $S(x; p, \beta, \Im) = 1 - F(x; p, \beta, \Im)$, hence

$$S(x; p, \beta, \Im) = \frac{1}{\log(p)} \log\left[1 - (1-p)e^{-\beta \frac{G(x;\Im)}{1-G(x;\Im)}}\right], \quad x \in \mathbb{R},$$

The appropriate pdf is given by deriving $F(x; p, \beta, \Im)$ from x; we get

$$f(x; p, \beta, \Im) = \left(\frac{1}{-\log(p)}\right) \frac{g(x; \Im)}{[1-G(x;\Im)]^2} \frac{(1-p)\beta e^{-\beta \frac{G(x;\Im)}{1-G(x;\Im)}}}{1 - (1-p)e^{-\beta \frac{G(x;\Im)}{1-G(x;\Im)}}}, \quad x \in \mathbb{R}, \tag{3}$$

where $g(x; \Im)$ refers to the pdf related to $G(x; \Im)$.

Also, the hrf of the OEL-G focds is specified by $h(x; p, \beta, \Im) = f(x; p, \beta, \Im)/S(x; p, \beta, \Im)$, hence

$$h(x; p, \beta, \Im) = \frac{g(x; \Im)}{[1-G(x;\Im)]^2} \frac{-(1-p)\beta e^{-\beta \frac{G(x;\Im)}{1-G(x;\Im)}}}{\left[1 - (1-p)e^{-\beta \frac{G(x;\Im)}{1-G(x;\Im)}}\right] \log\left[1 - (1-p)e^{-\beta \frac{G(x;\Im)}{1-G(x;\Im)}}\right]}, \quad x \in \mathbb{R}.$$

These two last functions are crucial to handling some statistical features of the OEL-G focds, such as the possible adequateness of the related models to various kinds of data.

3. Special Distributions

Three special four-parameter distributions of the OEL-G focds are described in this section, all defined with well-established baseline distributions, namely: the Weibull, gamma, and Fréchet distributions.

3.1. The OELW Distribution

The OEL Weibull (OELW) distribution is now introduced. It is defined by the cdf given by (2) with the Weibull distribution as baseline distribution, i.e., with the cdf given by $G(x; a, b) = 1 - e^{-(x/b)^a}$ and the pdf given by $g(x; a, b) = (a/b)(x/b)^{a-1}e^{-(x/b)^a}$, $a, b, x > 0$. When $x \leq 0$, the cdf and pdf are equal to 0. Thus, the cdf of the OELW distribution is given by

$$F(x; p, \beta, a, b) = 1 - \frac{1}{\log(p)} \log\left[1 - (1-p)e^{-\beta\left\{e^{(\frac{x}{b})^a} - 1\right\}}\right], \quad x > 0.$$

In this setting, the pdf is expressed as

$$f(x; p, \beta, a, b) = \left(\frac{1}{-\log(p)}\right) \frac{a}{b}\left(\frac{x}{b}\right)^{a-1} e^{(\frac{x}{b})^a} \frac{(1-p)\beta e^{-\beta\left\{e^{(\frac{x}{b})^a} - 1\right\}}}{1 - (1-p)e^{-\beta\left\{e^{(\frac{x}{b})^a} - 1\right\}}}, \quad x > 0.$$

The hrf is obtained as

$$h(x; p, \beta, a, b) = \frac{a}{b}\left(\frac{x}{b}\right)^{a-1} e^{(\frac{x}{b})^a} \frac{-(1-p)\beta e^{-\beta\left\{e^{(\frac{x}{b})^a} - 1\right\}}}{\left[1 - (1-p)e^{-\beta\left\{e^{(\frac{x}{b})^a} - 1\right\}}\right]\log\left[1 - (1-p)e^{-\beta\left\{e^{(\frac{x}{b})^a} - 1\right\}}\right]}, \quad x > 0.$$

The functions above are equal to 0 when $x \leq 0$. As graphical illustrations, Figure 1 shows some plots of $f(x; p, \beta, a, b)$ and $h(x; p, \beta, a, b)$, for selected values of the parameters p, β, a and b. These plots show that the pdf of the OELW distribution has a great shape flexibility. It can be left skewed, right skewed, J-shaped, reversed J-shape, and symmetric. Furthermore, the corresponding hrf can be increasing, decreasing, J, or bathtub in shape.

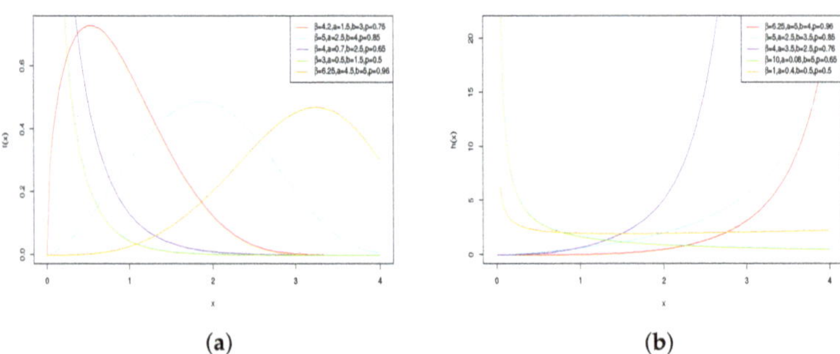

(a)   (b)

Figure 1. Examples of plots of the (**a**) pdf and (**b**) hrf of the OELW distribution for various values of p, β, a and b.

The applicability of the OLEW distribution will be highlighted in the application part of the paper (see Section 6).

3.2. Two Other Examples

To show other perspectives of lifetime modeling, two other special cases are briefly described below.

3.2.1. The OELGa Distribution

We now introduce the OEL gamma (OELGa) distribution. It is defined by the cdf given by (2) with the gamma distribution as baseline distribution, i.e., with the cdf given by $G(x; a, b) = (1/\Gamma(a))\gamma(a, bx)$ and the pdf given by $g(x; a, b) = (b^a/\Gamma(a))x^{a-1}e^{-bx}$, $a, b, x > 0$, where $\Gamma(a) = \int_0^{+\infty} t^{a-1}e^{-t}dt$ and $\gamma(a, bx) = \int_0^{bx} t^{a-1}e^{-t}dt$. When $x \leq 0$, the cdf and pdf are equal to 0. So, the cdf of the OELGa distribution is given by

$$F(x; p, \beta, a, b) = 1 - \frac{1}{\log(p)} \log\left[1 - (1-p)e^{-\beta \frac{\gamma(a,bx)}{\Gamma(a)-\gamma(a,bx)}}\right], \quad x > 0.$$

The related pdf is given as

$$f(x; p, \beta, a, b) = \left(\frac{1}{-\log(p)}\right) b^a x^{a-1} e^{-bx} \frac{\gamma(a, bx)}{[\Gamma(a) - \gamma(a, bx)]^2} \frac{(1-p)\beta e^{-\beta \frac{\gamma(a,bx)}{\Gamma(a)-\gamma(a,bx)}}}{1 - (1-p)e^{-\beta \frac{\gamma(a,bx)}{\Gamma(a)-\gamma(a,bx)}}}, \quad x > 0.$$

The hrf is given by

$$h(x; p, \beta, a, b) = b^a x^{a-1} e^{-bx} \frac{\gamma(a, bx)}{[\Gamma(a) - \gamma(a, bx)]^2} \times$$

$$\frac{-(1-p)\beta e^{-\beta \frac{\gamma(a,bx)}{\Gamma(a)-\gamma(a,bx)}}}{\left[1 - (1-p)e^{-\beta \frac{\gamma(a,bx)}{\Gamma(a)-\gamma(a,bx)}}\right] \log\left[1 - (1-p)e^{-\beta \frac{\gamma(a,bx)}{\Gamma(a)-\gamma(a,bx)}}\right]}, \quad x > 0.$$

The functions above are equal to 0 when $x \leq 0$. Figure 2 shows some plots of $f(x; p, \beta, a, b)$ and $h(x; p, \beta, a, b)$, for selected values of the parameters p, β, a and b. The plots indicate that the pdf of the OELGa distribution can be reverse J-shaped, symmetric, right skewed, left-skewed, and unimodal, whereas the hrf of the OELGa has J and increasing shapes.

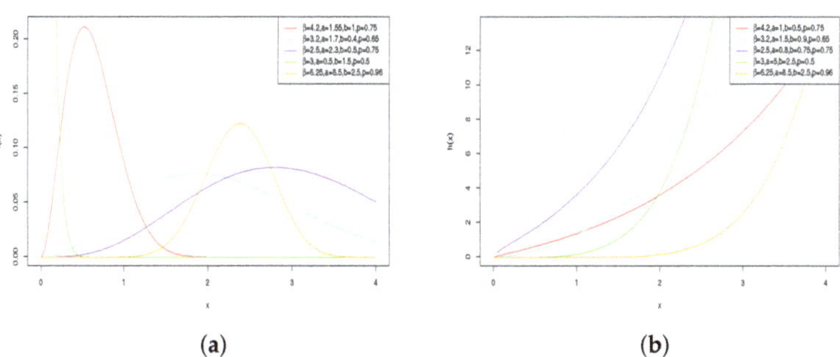

(a)   (b)

Figure 2. Examples of plots of the (**a**) pdf and (**b**) hrf of the OELGa distribution for various values of p, β, a and b.

3.2.2. The OELF Distribution

We now introduce the OEL Fréchet (OELF) distribution. It is defined by the cdf given by (2) with the Fréchet distribution as baseline distribution, i.e., with the cdf given by $G(x; a, b) = e^{-(x/b)^{-a}}$ and the pdf given by $g(x; a, b) = (a/b)(x/b)^{-a-1}e^{-(x/b)^{-a}}$, $a, b, x > 0$. When $x \leq 0$, the cdf and pdf are equal to 0. Hence, the cdf of the OELF distribution is given by

$$F(x; p, \beta, a, b) = 1 - \frac{1}{\log(p)} \log\left[1 - (1-p)e^{-\beta \frac{e^{-(\frac{x}{b})^{-a}}}{1-e^{-(\frac{x}{b})^{-a}}}}\right], \quad x > 0.$$

The pdf can be deduced as

$$f(x; p, \beta, a, b) = \left(\frac{1}{-\log(p)}\right)\frac{a}{b}\left(\frac{x}{b}\right)^{-a-1}\frac{e^{-(\frac{x}{b})^{-a}}}{[1 - e^{-(\frac{x}{b})^{-a}}]^2}\frac{(1-p)\beta e^{-\beta\frac{e^{-(\frac{x}{b})^{-a}}}{1-e^{-(\frac{x}{b})^{-a}}}}}{1 - (1-p)e^{-\beta\frac{e^{-(\frac{x}{b})^{-a}}}{1-e^{-(\frac{x}{b})^{-a}}}}}, \quad x > 0.$$

The hrf is expressed as

$$h(x; p, \beta, a, b) = \frac{a}{b}\left(\frac{x}{b}\right)^{-a-1}\frac{e^{-(\frac{x}{b})^{-a}}}{[1 - e^{-(\frac{x}{b})^{-a}}]^2} \times$$

$$\frac{-(1-p)\beta e^{-\beta\frac{e^{-(\frac{x}{b})^{-a}}}{1-e^{-(\frac{x}{b})^{-a}}}}}{\left[1 - (1-p)e^{-\beta\frac{e^{-(\frac{x}{b})^{-a}}}{1-e^{-(\frac{x}{b})^{-a}}}}\right]\log\left[1 - (1-p)e^{-\beta\frac{e^{-(\frac{x}{b})^{-a}}}{1-e^{-(\frac{x}{b})^{-a}}}}\right]}, \quad x > 0.$$

The functions above are equal to 0 when $x \leq 0$. Figure 3 shows some plots of $f(x; p, \beta, a, b)$ and $h(x; p, \beta, a, b)$, for selected values of the parameters p, β, a and b. The plots indicate that the OELF distribution can be reverse J-shaped, right skewed, left-skewed, and unimodal. On the other hand, the corresponding hrf has decreasing, increasing, J, reverse J- shapes.

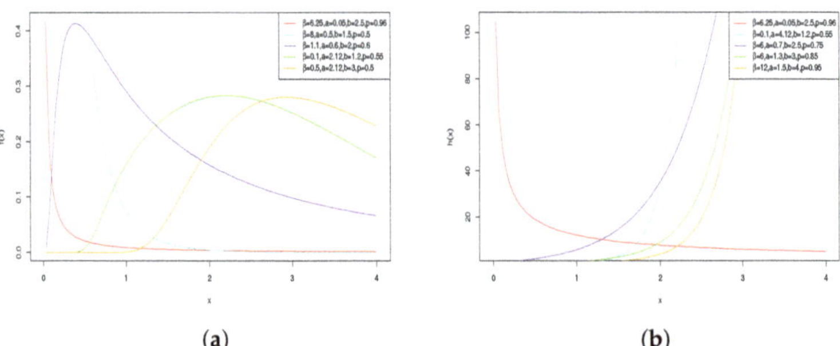

(a)   (b)

Figure 3. Examples of plots of the (a) pdf and (b) hrf of the OELF distribution for various values of p, β, a and b.

4. Mathematical Features

This section is devoted to some mathematical properties of the OEL-G focds. In the following, it is assumed that the criterion for interchanging summation and integration and the criterion for interchanging summation and differentiation are satisfied. Also, let us mention that most of the presented formulas can be handled in standard mathematical software (Mathematica, Maple, …).

4.1. Asymptotic Results

Here, we investigate some asymptotic results of the pdf and hrf of the OEL-G focds. First of all, as $x \to -\infty$ (or $G(x; \Im) \to 0$), we have

$$f(x; p, \beta, \Im) \sim \left(\frac{1-p}{-p\log(p)}\right)\beta g(x; \Im), \quad h(x; p, \beta, \Im) \sim \left(\frac{1-p}{-p\log(p)}\right)\beta g(x; \Im).$$

When $x \to +\infty$ (or $G(x; \mathfrak{I}) \to 1$), we have

$$f(x; p, \beta, \mathfrak{I}) \sim \left(\frac{1-p}{-\log(p)} \right) \beta \frac{g(x; \mathfrak{I})}{[1 - G(x; \mathfrak{I})]^2} e^{-\beta \frac{G(x; \mathfrak{I})}{1-G(x; \mathfrak{I})}}, \quad h(x; p, \beta, \mathfrak{I}) \sim \beta \frac{g(x; \mathfrak{I})}{[1 - G(x; \mathfrak{I})]^2}.$$

We thus see the role of the parameters β and p in the possible asymptotes for these functions. In particular, when $x \to +\infty$, we see that β has large impact on the convergence of $f(x; p, \beta, \mathfrak{I})$ due to the exponential term, whereas p has no effect on the limit of $h(x; p, \beta, \mathfrak{I})$. Also, the function $u(p) = (1-p)/[-p \log(p)]$ appearing multiple times is decreasing in p and convex, with $\lim_{p \to 0} u(p) = +\infty$ and $\lim_{p \to 1} u(p) = 0$.

4.2. Shapes of the pdf and hrf

The shapes of the pdf and hrf of the OEL-G focds can be described analytically. The critical point(s) of the pdf (also called mode(s)) of the OEL-G focds is(are) the root(s) of the following equation: $d[\log(f(x; p, \beta, \mathfrak{I}))]/dx = 0$, i.e.,

$$\frac{dg(x; \mathfrak{I})/dx}{g(x; \mathfrak{I})} + 2\frac{g(x; \mathfrak{I})}{1 - G(x; \mathfrak{I})} - \beta \frac{g(x; \mathfrak{I})}{[1 - G(x; \mathfrak{I})]^2} \frac{1}{\left[1 - (1-p)e^{-\beta \frac{G(x; \mathfrak{I})}{1-G(x; \mathfrak{I})}} \right]} = 0.$$

Similarly, the critical point(s) of the hrf of the OEL-G focds is(are) the root(s) of the following equation: $d[\log(h(x; p, \beta, \mathfrak{I}))]/dx = 0$, i.e.,

$$\frac{dg(x; \mathfrak{I})/dx}{g(x; \mathfrak{I})} + 2\frac{g(x; \mathfrak{I})}{1 - G(x; \mathfrak{I})} - \beta \frac{g(x; \mathfrak{I})}{[1 - G(x; \mathfrak{I})]^2} \frac{1}{\left[1 - (1-p)e^{-\beta \frac{G(x; \mathfrak{I})}{1-G(x; \mathfrak{I})}} \right]} \times$$

$$\left(1 + (1-p) \frac{e^{-\beta \frac{G(x; \mathfrak{I})}{1-G(x; \mathfrak{I})}}}{\log\left[1 - (1-p)e^{-\beta \frac{G(x; \mathfrak{I})}{1-G(x; \mathfrak{I})}} \right]} \right) = 0.$$

Mathematical software (R, Python, Mathematica, …) can be used to solve these two equations and determine whether the critical points are local maximums, minimums, or inflexion points for a given cdf $G(x; \mathfrak{I})$. It is the case for the proposed OELW, OELGa, and OELF distributions, where the equations above have no analytical solutions. For them, Figures 1–3, are informative on their global mode properties; these special distributions can be unimodal, with various hrf shapes.

4.3. Quantile Function

The qf of the OEL-G focds, say $Q(u; p, \beta, \mathfrak{I})$, satisfies the following functional equation: $F(Q(u; p, \beta, \mathfrak{I}); p, \beta, \mathfrak{I})) = Q(F(u; p, \beta, \mathfrak{I}); p, \beta, \mathfrak{I})) = u, u \in (0, 1)$. After some algebraic manipulations, we get

$$Q(u; p, \beta, \mathfrak{I}) = Q_G \left(\frac{\log\left(\frac{1-p}{1-p^{1-u}} \right)}{\beta + \log\left(\frac{1-p}{1-p^{1-u}} \right)}; \mathfrak{I} \right) \quad u \in (0, 1), \tag{4}$$

where $Q_G(u; \mathfrak{I})$ denotes the qf related to $G(x; \mathfrak{I})$. As a result, with appropriate values of u, quantiles of interest can be obtained. In particular, the median is reduced to

$$M = Q(0.5; p, \beta, \mathfrak{I}) = Q_G \left(\frac{\log(1 + \sqrt{p})}{\beta + \log(1 + \sqrt{p})}; \mathfrak{I} \right).$$

One can also use the quantile function for simulating values for a special OEL-G distribution. For any random variable U with the standard uniform distribution, $X = Q(U; p, \beta, \Im)$ has the cdf given by (2).

4.4. Expansions of the cdf and pdf

The cdf and pdf of the OEL-G focds are expressed here using exp-G cdfs and pdfs as defined by [29]. Then, the structural properties of the exp-G focds can be used to derive those of the OEL-G focds.

The following result is about the series expansion of the cdf.

Proposition 1. *Let $F(x; p, \beta, \Im)$ be the cdf given by (2). Then, assuming that $G(x; \Im) \in (0, 1)$, the following series expansion is valid:*

$$F(x; p, \beta, \Im) = \sum_{k,\ell=1}^{+\infty} \sum_{m=0}^{+\infty} a_{k,\ell,m} G(x; \Im)^{\ell+m},$$

where

$$a_{k,\ell,m} = \frac{1}{\log(p)} (-1)^{\ell+m} \binom{-\ell}{m} \frac{1}{k} \frac{1}{\ell!} (1-p)^k \beta^\ell k^\ell. \tag{5}$$

Proof. It follows from the Taylor theorem applied to the logarithmic function that $\log(1-x) = -\sum_{k=1}^{+\infty} \frac{1}{k} x^k$, $x \in (-1, 1)$, and some sum manipulations, that

$$F(x; p, \beta, \Im) = 1 + \frac{1}{\log(p)} \sum_{k=1}^{+\infty} \frac{1}{k} (1-p)^k e^{-\beta k \frac{G(x;\Im)}{1-G(x;\Im)}}$$

$$= \frac{1}{\log(p)} \sum_{k=1}^{+\infty} \frac{1}{k} (1-p)^k \left[e^{-\beta k \frac{G(x;\Im)}{1-G(x;\Im)}} - 1 \right].$$

For the term in brackets, the Taylor theorem applied to the exponential function, i.e., $e^x = \sum_{k=0}^{+\infty} \frac{1}{k!} x^k$, $x \in \mathbb{R}$, gives

$$e^{-\beta k \frac{G(x;\Im)}{1-G(x;\Im)}} - 1 = \sum_{\ell=1}^{+\infty} \frac{1}{\ell!} (-1)^\ell \beta^\ell k^\ell G(x; \Im)^\ell [1 - G(x; \Im)]^{-\ell}.$$

Now, the generalized binomial theorem, i.e., $(1-x)^v = \sum_{k=0}^{+\infty} \binom{v}{k} (-1)^k x^k$, $x \in (-1, 1)$, $v \in \mathbb{R}$, gives

$$[1 - G(x; \Im)]^{-\ell} = \sum_{m=0}^{+\infty} \binom{-\ell}{m} (-1)^m G(x; \Im)^m.$$

By combining all of the foregoing equalities, we get the desired result. The proof of Proposition 1 is now complete. $\square$

Corollary 1. *Owing to Proposition 1, upon differentiation of the involved functions, a series expansion for $f(x; p, \beta, \Im)$ is given by*

$$f(x; p, \beta, \Im) = \sum_{k,\ell=1}^{+\infty} \sum_{m=0}^{+\infty} b_{k,\ell,m} g(x; \Im) G(x; \Im)^{\ell+m-1},$$

where $b_{k,\ell,m} = (\ell + m) a_{k,\ell,m}$.

In comparison to the former analytical definition, for practical purposes (integration...), the expression of $f(x; p, \beta, \Im)$ in Corollary 1 can be more easy to handle through the following approximation:

$$f(x; p, \beta, \Im) \approx \sum_{k,\ell=1}^{M} \sum_{m=0}^{M} b_{k,\ell,m} g(x; \Im) G(x; \Im)^{\ell+m-1},$$

where M is a carefully chosen number.

4.5. Moments

Hereafter, we denote by X a random variable having the cdf of the OEL-G focds given by (2). Corollary 1 can be used to have a tractable expression for the moments of X, among other things. Indeed, for any integer r, the rth moment of X is given by

$$\mu_r' = E(X^r) = \int_{-\infty}^{+\infty} x^r f(x; p, \beta, \Im) dx = \sum_{k,\ell=1}^{+\infty} \sum_{m=0}^{+\infty} b_{k,\ell,m} \tau_{\ell,m,r},$$

where $\tau_{\ell,m,r} = \int_{-\infty}^{+\infty} x^r g(x; \Im) G(x; \Im)^{\ell+m-1} dx = \int_0^1 u^{\ell+m-1} [Q_G(u; \Im)]^r du$. For a given $G(x; \Im)$, this integral can be calculated or computed numerically. We refer to [30], where $\tau_{\ell,m,r}$ has been determined for some standard distributions (normal, beta, Weibull ...). For practical purposes, another remark concerns the infinity limit in the sums; as mentioned before, it can be substituted by a large positive integer.

As usual, the mean of X is obtained directly by $\mu = \mu_1'$. Also, the variance of X can be calculated using the following formula: $\sigma^2 = \mu_2' - \mu^2$.

In a similar vein, for $y \in \mathbb{R}$, the rth incomplete moment of X is given by

$$\mu_r'(y) = E(X^r 1_{\{X \leq y\}}) = \int_{-\infty}^{y} x^r f(x; p, \beta, \Im) dx = \sum_{k,\ell=1}^{+\infty} \sum_{m=0}^{+\infty} b_{k,\ell,m} \tau_{\ell,m,r}(y),$$

where $\tau_{\ell,m,r}(y) = \int_{-\infty}^{y} x^r g(x; \Im) G(x; \Im)^{\ell+m-1} dx = \int_0^{G(y;\Im)} u^{\ell+m-1} [Q_G(u; \Im)]^r du$. Then, one can express the mean deviations about the mean and about the median, as well as Bonferroni and Lorenz curves, which play a central role in life testing, reliability, and renewal theory.

Similarly, the moment generating function of X is given by

$$M(t) = E(e^{tX}) = \int_{-\infty}^{+\infty} e^{tx} f(x; p, \beta, \Im) dx = \sum_{k,\ell=1}^{+\infty} \sum_{m=0}^{+\infty} b_{k,\ell,m} \upsilon_{\ell,m}(t),$$

where $\upsilon_{\ell,m}(t) = \int_{-\infty}^{+\infty} e^{tx} g(x; \Im) G(x; \Im)^{\ell+m-1} dx = \int_0^1 u^{\ell+m-1} e^{tQ_G(u;\Im)} du$.

4.6. Skewness and Kurtosis

The skewness and kurtosis properties of the OEL-G focds can be explored via the four first moments or the use of the qf given by (4). The main measures defined by moments are the skewness and kurtosis parameters defined by

$$S = E\left[\left(\frac{X - \mu}{\sigma} \right)^3 \right] = \frac{\mu_3' - 3\mu_2'\mu + 2\mu^3}{\sigma^3}$$

and

$$K = E\left[\left(\frac{X - \mu}{\sigma} \right)^4 \right] = \frac{\mu_4' - 4\mu_3'\mu + 6\mu_2'\mu^2 - 3\mu^4}{\sigma^4}.$$

They can be expressed for a given baseline cdf $G(x; \Im)$.

Alternatively, if the moments do not exist (or in full generality), one can consider the measures defined with the qf. Examples are the Bowley skewness and the Moors kurtosis defined by, respectively,

$$S_* = \frac{Q(\frac{1}{4};p,\beta,\Im) + Q(\frac{3}{4};p,\beta,\Im) - 2Q(\frac{1}{2};p,\beta,\Im)}{Q(\frac{3}{4};p,\beta,\Im) - Q(\frac{1}{4};p,\beta,\Im)},$$

and

$$K_* = \frac{Q(\frac{7}{8};p,\beta,\Im) - Q(\frac{5}{8};p,\beta,\Im) + Q(\frac{3}{8};p,\beta,\Im) - Q(\frac{1}{8};p,\beta,\Im)}{Q(\frac{6}{8};p,\beta,\Im) - Q(\frac{2}{8};p,\beta,\Im)}.$$

We refer to [31,32] for more information on these quantile measures.

Table 1 provides the mean, variance, skewness S and kurtosis K (defined with the moments) of one of the members of the OEL-G focds, the OELW distribution, for different choices of parameter values.

Table 1. Moment measures of the OELW distribution for various choices of parameters.

Parameter	Mean	Variance	Skewness	Kurtosis
$a = 0.5$ $b = 1.5$ $\beta = 3$ $p = 0.5$	0.1375667	0.05076667	12.77926	21.5475
$a = 0.5$ $b = 1.5$ $\beta = 3$ $p = 0.05$	0.06894464	0.03067229	27.02119	42.1311
$a = 0.5$ $b = 1.5$ $\beta = 3$ $p = 0.012$	0.04940314	0.02260463	38.03328	57.77719
$a = 0.5$ $b = 2$ $\beta = 0.3$ $p = 0.2$	1.324836	6.267231	8.523013	13.13688
$a = 0.5$ $b = 0.02$ $\beta = 30$ $p = 0.012$	7.938928×10^{-8}	1.69951×10^{-10}	26,964.99	26,960.99
$a = 0.5$ $b = 5$ $\beta = 6.25$ $p = 0.96$	0.1737497	0.09645452	15.83522	27.64176

Table 1 indicates that, for fixed a, b and β, the mean and variance of the OELW distribution are decreasing functions with respect to p. Also, the OELW distribution tends to be skewed more to the right as p decreases.

4.7. Entropy

Entropy is a measure of the variation of uncertainty that finds numerous applications in various areas such as engineering, mathematical physics, and probability. One of the most famous useful entropy measures is the Rényi entropy, introduced by [33] and the

Shannon entropy by [34]. In the context of the OEL-G focds, the Rényi entropy of X is defined by

$$I_\delta(X) = \frac{1}{1-\delta} \log\left[\int_{-\infty}^{+\infty} f(x; p, \beta, \Im)^\delta dx\right],$$

where $\delta > 0$ and $\delta \neq 1$. As an alternative to direct computation, we now present an expression that depends on a tractable series expansion. In this regard, let us present and prove the following proposition, which can be viewed as an extension of Corollary 1.

Proposition 2. *Let $\delta \in \mathbb{R}$ and $f(x; p, \beta, \Im)$ be the pdf given by (3). Then, the following series expansion is valid:*

$$f(x; p, \beta, \Im)^\delta = \sum_{k,\ell,m=0}^{+\infty} c_{k,\ell,m}(\delta) g(x; \Im)^\delta G(x; \Im)^{\ell+m},$$

where

$$c_{k,\ell,m}(\delta) = \left(\frac{1}{-\log(p)}\right)^\delta \binom{-\delta}{k}\binom{-(2\delta+\ell)}{m}(-1)^{k+\ell+m}\frac{1}{\ell!}(1-p)^{\delta+k}\beta^{\delta+\ell}(\delta+k)^\ell.$$

Proof. We have

$$f(x; p, \beta, \Im)^\delta = \left(\frac{1}{-\log(p)}\right)^\delta \frac{g(x; \Im)^\delta}{[1-G(x; \Im)]^{2\delta}} \frac{(1-p)^\delta \beta^\delta e^{-\delta\beta\frac{G(x;\Im)}{1-G(x;\Im)}}}{\left[1-(1-p)e^{-\beta\frac{G(x;\Im)}{1-G(x;\Im)}}\right]^\delta}.$$

The generalized binomial formula demonstrates that

$$\left[1-(1-p)e^{-\beta\frac{G(x;\Im)}{1-G(x;\Im)}}\right]^{-\delta} = \sum_{k=0}^{+\infty}\binom{-\delta}{k}(-1)^k(1-p)^k e^{-k\beta\frac{G(x;\Im)}{1-G(x;\Im)}}.$$

By the Taylor series of the exponential function, we get

$$e^{-(\delta+k)\beta\frac{G(x;\Im)}{1-G(x;\Im)}} = \sum_{\ell=0}^{+\infty}\frac{1}{\ell!}(-1)^\ell(\delta+k)^\ell\beta^\ell G(x;\Im)^\ell[1-G(x;\Im)]^{-\ell}.$$

Furthermore, the generalized binomial formula gives

$$[1-G(x;\Im)]^{-(2\delta+\ell)} = \sum_{m=0}^{+\infty}\binom{-(2\delta+\ell)}{m}(-1)^m G(x;\Im)^m.$$

By combining all of the aforementioned equality, we get the desired result. $\square$

As a direct application of Proposition 2, the Rényi entropy is given by

$$I_\delta(X) = \frac{1}{1-\delta}\log\left[\sum_{k,\ell,m=0}^{+\infty} c_{k,\ell,m}(\delta)\int_{-\infty}^{+\infty} g(x;\Im)^\delta G(x;\Im)^{\ell+m}dx\right].$$

On the other side, the Shannon entropy of X is defined by

$$\eta(X) = -E\{\log[f(X; p, \beta, \Im)]\}.$$

It can be determined via the limit result: $\eta(X) = \lim_{\delta\to 1} I_\delta(X)$. However, this limit is not easy to handle. Some sum expressions can also be proved as an alternative. Indeed, we have

$$\eta(X) = -\log\left[\frac{1}{-\log(p)}\right] - \log(1-p) - \log(\beta) - E\{\log[g(X;\Im)]\} + 2E\{\log[1 - G(X;\Im)]\}$$
$$+ \beta E\left[\frac{G(X;\Im)}{1 - G(X;\Im)}\right] - \log(p)\{E[F(X;p,\beta,\Im)] - 1\}.$$

Now, by using Corollary 1, we have

$$E\{\log[g(X;\Im)]\} = \sum_{k,\ell=1}^{+\infty} \sum_{m=0}^{+\infty} b_{k,\ell,m}\kappa_{\ell,m},$$

where $\kappa_{\ell,m} = \int_{-\infty}^{+\infty} \log[g(x;\Im)]g(x;\Im)G(x;\Im)^{\ell+m-1}dx = \int_0^1 \log[g(Q_G(u;\Im);\Im)]u^{\ell+m-1}dx$. By using the Taylor series of the logarithmic function, we have

$$E\{\log[1 - G(X;\Im)]\} = -\sum_{i=1}^{+\infty} \frac{1}{i}E\left[G(X;\Im)^i\right].$$

By using the geometric series, it comes

$$E\left[\frac{G(X;\Im)}{1 - G(X;\Im)}\right] = \sum_{i=0}^{+\infty} E\left[G(X;\Im)^{i+1}\right].$$

By using Proposition 1, we have

$$E[F(X;p,\beta,\Im)] = \sum_{k,\ell=1}^{+\infty} \sum_{m=0}^{+\infty} a_{k,\ell,m}E\left[G(X;\Im)^{\ell+m}\right].$$

All the terms involving the expectation of exponentiated $G(X;\Im)$ are expressible by using the following results. For any $\zeta \geq 0$, by Corollary 1, we have

$$E\left[G(X;\Im)^\zeta\right] = \sum_{k,\ell=1}^{+\infty} \sum_{m=0}^{+\infty} b_{k,\ell,m} \int_{-\infty}^{+\infty} g(x;\Im)G(x;\Im)^{\zeta+\ell+m-1}dx$$
$$= \sum_{k,\ell=1}^{+\infty} \sum_{m=0}^{+\infty} b_{k,\ell,m} \frac{1}{\zeta+\ell+m}.$$

By putting all the above equalities together, we get a tractable expression for the Shannon entropy, and possible approximations can be derived for practical purposes.

4.8. Order Statistics

The following result concerns a distributional property of a mth order statistic related to the OEL-G focds.

Proposition 3. *Let $X_1, \ldots, X_n$ be a random sample of size n from X and $X_{m:n}$ be the corresponding mth order statistic, i.e., the mth random variable satisfying the inequalities $X_{1:n} \leq X_{2:n} \leq \ldots X_{m:n} \leq \ldots \leq X_{n:n}$, almost surely. The pdf of $X_{m:n}$ is then linearly represented in terms of pdfs of the exp-G focds.*

Proof. By definition, the pdf of $X_{m:n}$ is given by

$$f_{m:n}(x;p,\beta,\Im) = \frac{n!}{(m-1)!(n-m)!}F(x;p,\beta,\Im)^{m-1}[1 - F(x;p,\beta,\Im)]^{n-m}f(x;p,\beta,\Im).$$

Owing to the binomial formula, we can write

$$f_{m:n}(x; p, \beta, \Im) = \frac{n!}{(m-1)!(n-m)!} \sum_{d=0}^{m-1} \binom{m-1}{d} (-1)^d [1 - F(x; p, \beta, \Im)]^{d+n-m} f(x; p, \beta, \Im).$$

It follows from Corollary 1 that $f(x; p, \beta, \Im)$ can be expressed as a sum of pdfs of the exp-G focds. As a result, the proof concludes by demonstrating that $[1 - F(x; p, \beta, \Im)]^{d+n-m}$ can be expressed as a sum of cdfs of the exp-G focds, by exploiting the fact that the multiplication of a pdf and a cdf of the exp-G focds is a pdf of the exp-G focds, up to a constant factor. We have

$$[1 - F(x; p, \beta, \Im)]^{d+n-m} = \left(\frac{1}{-\log(p)} \right)^{d+n-m} \left\{ -\log \left[1 - (1-p) e^{-\beta \frac{G(x;\Im)}{1-G(x;\Im)}} \right] \right\}^{d+n-m}.$$

By the Taylor series of the integer power of the logarithmic function (see, for instance, http://functions.wolfram.com/ElementaryFunctions/Log/06/01/04/03/, accessed on 4 July 2022), we have

$$\left\{ -\log \left[1 - (1-p) e^{-\beta \frac{G(x;\Im)}{1-G(x;\Im)}} \right] \right\}^{d+n-m}$$
$$= (d+n-m) \sum_{k=0}^{+\infty} \sum_{j=0}^{k} \binom{k - (d+n-m)}{k} \binom{k}{j} (-1)^{j+k} \frac{1}{d+n-m-j} \times$$
$$u_{j,k} (1-p)^{d+n-m+k} e^{-(d+n-m+k)\beta \frac{G(x;\Im)}{1-G(x;\Im)}},$$

where $u_{j,k}$ can be determined recursively by $u_{j,0} = 1$ and, for $k \in \mathbb{N}^*$,

$$u_{j,k} = \frac{1}{k} \sum_{s=1}^{k} [k - s(j+1)] \frac{(-1)^{s+1}}{s+1} u_{j,k-s}.$$

We will now proceed in the same manner that we did in the proof of Proposition 1. The Taylor theorem applied to the exponential function gives

$$e^{-(d+n-m+k)\beta \frac{G(x;\Im)}{1-G(x;\Im)}} = \sum_{\ell=0}^{+\infty} \frac{1}{\ell!} (-1)^\ell \beta^\ell (d+n-m+k)^\ell G(x;\Im)^\ell [1 - G(x;\Im)]^{-\ell}.$$

It follows from the general binomial theorem that

$$[1 - G(x;\Im)]^{-\ell} = \sum_{m=0}^{+\infty} \binom{-\ell}{m} (-1)^m G(x;\Im)^m.$$

By combining the aforementioned equalities, we arrive at

$$[1 - F(x; p, \beta, \Im)]^{d+n-m} = \sum_{k,\ell,m=0}^{+\infty} w_{k,\ell,m} G(x;\Im)^{\ell+m},$$

where

$$w_{k,\ell,m} = \left(\frac{1}{-\log(p)} \right)^{d+n-m} (d+n-m) \sum_{j=0}^{k} \binom{k - (d+n-m)}{k} \binom{k}{j} \binom{-\ell}{m} (-1)^{j+k+\ell+m} \times$$
$$\frac{1}{d+n-m-j} u_{j,k} (1-p)^{d+n-m+k} \frac{1}{\ell!} \beta^\ell (d+n-m+k)^\ell.$$

We thus have a linear representation of $[1 - F(x; p, \beta, \Im)]^{d+n-m}$ in terms of cdfs of the exp-G focds, ending the proof of Proposition 3. $\square$

Thanks to Proposition 3, one can determine various mathematical properties for the mth order statistic, such as moments, incomplete moments, entropy, and so on.

4.9. Stochastic Ordering

Here, a stochastic ordering result involving the OEL-G focds is investigated. First of all, some elementary relations are presented below. The complete theory can be found in [35]. Let X_1 and X_2 be two random variables having the sfs and pdfs given by $S_1(x)$ and $S_2(x)$, and $f_1(x)$ and $f_2(x)$, respectively. Then, X_1 is said to be "smaller than X_2" in the following senses:

1. stochastic order, denoted by $X_1 \leq_{st} X_2$, if $S_1(x) \leq S_2(x)$ for all x,
2. hazard rate order, denoted by $X_1 \leq_{hr} X_2$, if $S_1(x)/S_2(x)$ is decreasing in x,
3. likelihood ratio order, denoted by $X_1 \leq_{lr} X_2$, if $f_1(x)/f_2(x)$ is decreasing in x.

Then, we have the following implications:

$$(X_1 \leq_{lr} X_2) \Rightarrow (X_1 \leq_{hr} X_2) \Rightarrow (X_1 \leq_{st} X_2).$$

A stochastic ordering result on the OEL-G focds is presented below.

Proposition 4. *Let X_1 having the cdf given by (2) with $p = p_1$ and X_2 having the cdf given by (2) with $p = p_2$. Then, if $p_1 \leq p_2$, we have $X_1 \leq_{lr} X_2$ (implying $X_1 \leq_{hr} X_2$ and $X_1 \leq_{st} X_2$). The equality in the likelihood ratio order is satisfied if and only if $p_1 = p_2$.*

Proof. Let $f(x; p_1, \beta, \Im)$ and $f(x; p_2, \beta, \Im)$ be the pdfs of X_1 and X_2, respectively. Then, by using (3), we have

$$\frac{f(x; p_1, \beta, \Im)}{f(x; p_2, \beta, \Im)} = \left(\frac{\log(p_2)}{\log(p_1)}\right)\left(\frac{1 - p_1}{1 - p_2}\right)\frac{1 - (1 - p_2)e^{-\beta\frac{G(x;\Im)}{1 - G(x;\Im)}}}{1 - (1 - p_1)e^{-\beta\frac{G(x;\Im)}{1 - G(x;\Im)}}}.$$

Hence, by differentiation, we obtain

$$\frac{d}{dx}\frac{f(x; p_1, \beta, \Im)}{f(x; p_2, \beta, \Im)} = \left(\frac{\log(p_2)}{\log(p_1)}\right)\left(\frac{1 - p_1}{1 - p_2}\right)\beta(p_1 - p_2)\frac{g(x; \Im)}{[1 - G(x; \Im)]^2}\frac{e^{-\beta\frac{G(x;\Im)}{1 - G(x;\Im)}}}{\left[1 - (1 - p_1)e^{-\beta\frac{G(x;\Im)}{1 - G(x;\Im)}}\right]^2}.$$

Now, observe that the sign of $d[f(x; p_1, \beta, \Im)/f(x; p_2, \beta, \Im)]/dx$ is the same to the sign of $p_1 - p_2$. So, if $p_1 \leq p_2$, $f(x; p_1, \beta, \Im)/f(x; p_2, \beta, \Im)$ is decreasing with respect to x, implying the desired result. The proof of Proposition 4 is now complete. $\square$

5. Maximum Likelihood Estimation

In this section, we examine the statistical practice of the OEL-G model.

5.1. Method

To begin, we determine the (standard) maximum likelihood estimates (MLEs) of the parameters p, β, and $\Im$.

Let $x_1, \ldots, x_n$ be observed values from X. Then, the log-likelihood function for $\Theta = (p, \beta, \Im)$ is given by

$$\ell_n = n \log\left[\frac{1}{-\log(p)}\right] + n \log(1 - p) + n \log(\beta) + \sum_{i=1}^{n} \log[g(x_i; \Im)] - 2\sum_{i=1}^{n} \log[1 - G(x_i; \Im)]$$

$$- \beta \sum_{i=1}^{n} \frac{G(x_i; \Im)}{1 - G(x_i; \Im)} - \sum_{i=1}^{n} \log\left[1 - (1 - p)e^{-\beta\frac{G(x_i;\Im)}{1 - G(x_i;\Im)}}\right].$$

The first derivatives of ℓ_n with respect to p, β and $\Im$ are

$$\frac{\partial \ell_n}{\partial p} = -\frac{n}{p \log(p)} - \frac{n}{1-p} - \sum_{i=1}^{n} \frac{e^{-\beta \frac{G(x_i;\Im)}{1-G(x_i;\Im)}}}{1-(1-p)e^{-\beta \frac{G(x_i;\Im)}{1-G(x_i;\Im)}}},$$

$$\frac{\partial \ell_n}{\partial \beta} = \frac{n}{\beta} - \sum_{i=1}^{n} \frac{G(x_i;\Im)}{1-G(x_i;\Im)} \frac{1}{1-(1-p)e^{-\beta \frac{G(x_i;\Im)}{1-G(x_i;\Im)}}}$$

and

$$\frac{\partial \ell_n}{\partial \Im} = \sum_{i=1}^{n} \frac{g^{(\Im)}(x_i;\Im)}{g(x_i;\Im)} + 2\sum_{i=1}^{n} \frac{G^{(\Im)}(x_i;\Im)}{1-G(x_i;\Im)} - \beta \sum_{i=1}^{n} \frac{G^{(\Im)}(x_i;\Im)}{[1-G(x_i;\Im)]^2} \frac{1}{1-(1-p)e^{-\beta \frac{G(x_i;\Im)}{1-G(x_i;\Im)}}},$$

where $g^{(\Im)}(x_i;\Im) = \partial g(x_i;\Im)/\partial \Im$ and $G^{(\Im)}(x_i;\Im) = \partial G(x_i;\Im)/\partial \Im$.

The MLEs of Θ, say $\hat{\Theta} = (\hat{p}, \hat{\beta}, \hat{\Im})$, are the simultaneous solutions of the following equations: $\partial \ell_n/\partial p = 0$, $\partial \ell_n/\partial \beta = 0$ and $\partial \ell_n/\partial \Im = 0$. These MLEs do not have analytical expressions, but they can be computed numerically using well-established algorithms available in mathematical software. Moreover, assuming that there are r components in $\Im$, with $\Im = (\Im_1, \ldots, \Im_r)$, the corresponding observed information matrix at $\Theta = \Theta_*$ is given by

$$J(\Theta_*) = -\begin{pmatrix} \frac{\partial^2 \ell_n}{\partial p^2} & \frac{\partial^2 \ell_n}{\partial p \partial \beta} & \frac{\partial^2 \ell_n}{\partial p \partial \Im_1} & \cdots & \frac{\partial^2 \ell_n}{\partial p \partial \Im_r} \\ \cdot & \frac{\partial^2 \ell_n}{\partial \beta^2} & \frac{\partial^2 \ell_n}{\partial \beta \partial \Im_1} & \cdots & \frac{\partial^2 \ell_n}{\partial \beta \partial \Im_r} \\ \cdot & \cdot & \frac{\partial^2 \ell_n}{\partial \Im_1^2} & \cdots & \frac{\partial^2 \ell_n}{\partial \Im_1 \partial \Im_r} \\ \cdot & \cdot & \cdot & \cdots & \cdot \\ \cdot & \cdot & \cdot & \cdots & \cdot \\ \cdot & \cdot & \cdot & \cdots & \frac{\partial^2 \ell_n}{\partial \Im_r^2} \end{pmatrix}_{\Theta=\Theta_*}.$$

Under some standard conditions of regularity, when n is large, the sub-jacent distribution of $\hat{\Theta}$ can be assimilated to the following Gaussian distribution: $\mathcal{N}_{r+2}(\Theta, J(\hat{\Theta})^{-1})$, where $J(\hat{\Theta})$ is the observed information matrix at $\Theta = \hat{\Theta}$. Confidence intervals and statistical tests for the model parameters can be constructed from this result. Further details on the maximum likelihood estimation in the setting of odd focds can be found in [16,36].

5.2. Simulation

Here, we perform a simulation study evaluating the performance of the MLEs presented above for the OELW distribution for selected values of the parameters a, b, p, and β. The simulation experiment was repeated 1000 times each with sample sizes of 30, 60, and 100, and parameter combinations are

(I) $a = 0.5, b = 0.2, p = 0.05$ and $\beta = 0.01$
(II) $a = 0.55, b = 0.3, p = 0.04$ and $\beta = 0.01$
(III) $a = 0.5, b = 0.4, p = 0.8$ and $\beta = 0.1$

Table 2 presents the average estimates (AEs), average bias (Bias), and mean square error (MSE) values of parameters for different sample sizes.

Table 2. AEs, Bias, and MSE of parameters based on 1000 simulations of the OELW distribution.

	n	Parameter	AEs	Bias	MSE
I	30	a	0.4917	−0.0083	0.0138
		b	0.2122	0.0122	0.0032
		p	0.1716	0.1216	0.4697
		β	0.0174	0.0074	0.0315
	60	a	0.4983	−0.0017	0.0003
		b	0.2015	0.0015	0.0002
		p	0.0614	0.0114	0.0108
		β	0.0121	0.0021	0.0004
	100	a	0.5002	0.0002	3.9531×10^{-5}
		b	0.1998	−0.0002	3.9531×10^{-5}
		p	0.0487	−0.0013	0.0017
		β	0.0094	−0.0006	0.0004
II	30	a	0.0373	−0.5127	0.2823
		b	0.0211	−0.2789	0.0840
		p	0.0075	−0.0324	0.0174
		β	0.0007	−0.0092	0.0012
	60	a	0.5481	−0.0019	0.0020
		b	0.3002	0.0002	0.0009
		p	0.0460	0.0060	0.0751
		β	0.0106	0.0006	0.0028
	100	a	0.5498	−0.0002	3.3018×10^{-5}
		b	0.3001	8.4933×10^{-6}	3.9531×10^{-5}
		p	0.0410	0.0010	0.0009
		β	0.0102	0.0002	4.8256×10^{-5}
III	30	a	0.0327	−0.4673	0.2377
		b	0.0230	−0.3770	0.1507
		p	0.0467	−0.7532	0.3422
		β	0.0044	−0.0956	0.0105
	60	a	0.3342	−0.1658	0.0930
		b	0.2598	−0.1402	0.0559
		p	0.5333	−0.2666	0.2138
		β	0.0621	−0.0379	0.0058
	100	a	0.5013	0.0013	0.0017
		b	0.3998	−0.0002	4.2791×10^{-5}
		p	0.8053	0.0053	0.0703
		β	0.0995	−0.0005	0.0003

It can be noted that as the sample size increases, the bias decays towards zero and the MSE decreases. That is, the random versions of the MLEs are asymptotically unbiased and consistent. Therefore, the maximum likelihood method works quite well to estimate the parameters of the OELW distribution.

6. Applications

In this section, we show how the OELW model can be used in real-world data analysis applications. We fit the OELW distribution to two data sets and compare the results with those of the fitted four or five-parameter distributions also based on the Weibull distribution, namely the log-logistic Weibull (LLoGW) distribution by [37] and exponentiated generalized modified Weibull (EGMW) distribution by [38]. The Akaike information criterion (AIC), Bayesian information criterion (BIC), Anderson-Darling (A^*), Cramér-von Mises (W^*) and the values of the Kolmogorov-Smirnov (K-S) statistic and the corresponding p-values (p-Vs) are used to compare the three models after we estimate the unknown parameters of each model using the maximum likelihood method of estimation. In addition, for three data sets, the observed Fisher information matrix for the OELW distribution is provided.

6.1. The Survival Data Sets

The first real data set is a subset of the findings of [39]. It is based on the survival periods (in years) of 46 patients who received just chemotherapy. The data are provided below.

{0.047; 0.115; 0.121; 0.132; 0.164; 0.197; 0.203; 0.260; 0.282; 0.296; 0.334; 0.395; 0.458; 0.466; 0.501; 0.507; 0.529; 0.534; 0.540; 0.641; 0.644; 0.696; 0.841; 0.863; 1.099; 1.219; 1.271; 1.326; 1.447; 1.485; 1.553; 1.581; 1.589; 2.178; 2.343; 2.416; 2.444; 2.825; 2.830; 3.578; 3.658; 3.743; 3.978; 4.003; 4.033}

A summary of measures of descriptive statistics is provided in Table 3.

Table 3. Descriptive statistics of the survival data set.

Minimum	Mean	Median	Variance	Skewness	Kurtosis	Maximum
0.047	1.341	0.841	1.5540	0.9721	2.6638	4.033

Table 4 gives the relevant numerical summaries for the three fits based on the survival data set.

Table 4. Estimated values, log-likelihood, AIC, and BIC for the survival data set.

Distribution	Estimates	$-\log(L)$	AIC	BIC	A^*	W^*	K-S	p-V
OELW	$\hat{a} = 1.2911$ $\hat{b} = 1.8757$ $\hat{p} = 0.0086$ $\hat{\beta} = 0.1485$	55.6795	119.3589	126.5856	0.5029	0.0794	0.1000	0.7213
LLoGW	$\hat{s} = 5.4074$ $\hat{c} = 0.9984$ $\hat{\beta} = 1.0963$ $\hat{\alpha} = 0.5519$	58.1248	124.2497	131.4763	0.5297	0.0802	0.1086	0.6245
EGMW	$\hat{\alpha} = 0.0262$ $\hat{\theta} = 23.6778$ $\hat{\beta} = 1.1346$ $\hat{\mu} = 7.4771$ $\hat{\lambda} = 0.9329$	58.0787	126.1574	135.1907	0.5339	0.0813	0.109	0.6197

Figure 4 gives the graphs of the estimated pdfs and cdfs of the considered distributions. Figure 5 gives the probability-probability (PP) plot of the OELW distribution.

The observed Fisher information matrix for the OELW distribution is given by

$$J(\Theta_*) = \begin{pmatrix} 46614.05 & -4027.669 & 1022.565 & 460.6903 \\ . & 715.1325 & 49.82654 & -149.2407 \\ . & . & 80.76499 & -136.9867 \\ . & . & . & 33.75663 \end{pmatrix}.$$

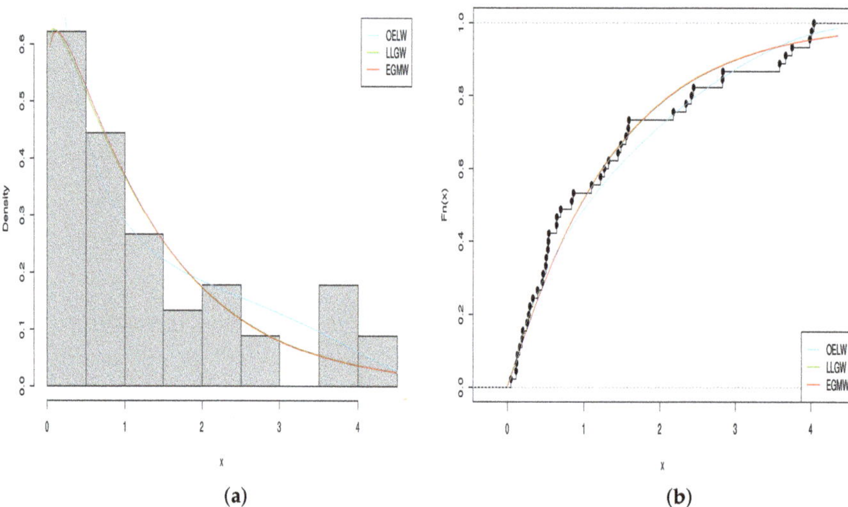

(a)   (b)

Figure 4. Obtained plots of the (**a**) estimated pdfs and (**b**) estimated cdfs of the considered distributions for the survival data set.

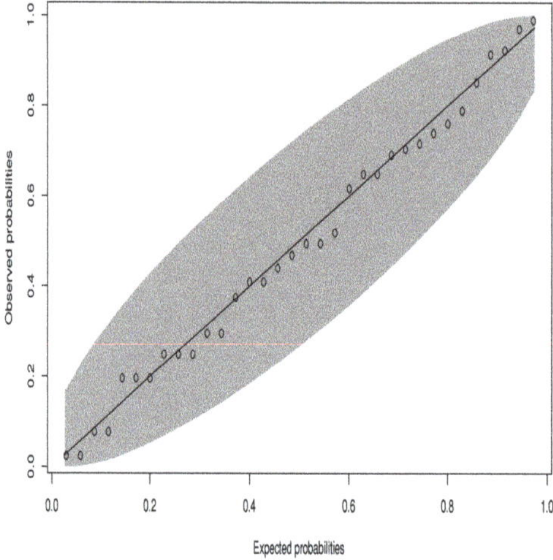

Figure 5. PP plot of the survival data set.

6.2. The Vinyl Chloride Data Set

The second data set represents 34 observations of the vinyl chloride data (in mg/L) that was obtained from clean-up gradient groundwater monitoring wells. The data are obtained from by [40] and are given below.

{5.1; 1.2; 1.3; 0.6; 0.5; 2.4; 0.5; 1.1; 8; 0.8; 0.4; 0.6; 0.9; 0.4; 2; 0.5; 5.3; 3.2; 2.7; 2.9; 2.5; 2.3; 1; 0.2; 0.1; 0.1; 1.8; 0.9; 2; 4; 6.8; 1.2; 0.4; 0.2}

A summary of measures of descriptive statistics is provided in Table 5.

Table 5. Descriptive statistics of the vinyl chloride data set.

Minimum	Mean	Median	Variance	Skewness	Kurtosis	Maximum
0.100	1.879	1.150	3.8126	1.6037	5.0054	8.000

Table 6 gives the relevant numerical summaries for the three fits based on the vinyl chloride data set.

Table 6. Estimated values, log-likelihood, AIC, and BIC for the vinyl chloride data set.

Distribution	Estimates	$-\log(L)$	AIC	BIC	A^*	W^*	K-S	p-V
OELW	$\hat{a} = 1.9409$ $\hat{b} = 8.7259$ $\hat{p} = 0.0023$ $\hat{\beta} = 2.0977$	54.2109	116.4218	122.5273	0.2002	0.0289	0.0785	0.9849
LLoGW	$\hat{s} = 9.7787$ $\hat{c} = 5.0155$ $\hat{\beta} = 0.9910$ $\hat{\alpha} = 0.5270$	55.354	118.708	124.8134	0.2881	0.0438	0.0931	0.9301
EGMW	$\hat{\alpha} = 0.0158$ $\hat{\theta} = 33.6386$ $\hat{\beta} = 1.0908$ $\hat{\mu} = 2.2873$ $\hat{\lambda} = 0.8806$	55.3950	120.79	128.4218	0.3159	0.0517	0.0973	0.9043

Figure 6 gives the graph of the estimated pdfs and cdfs of the considered distributions. Figure 7 gives the PP plot of the OELW distribution.

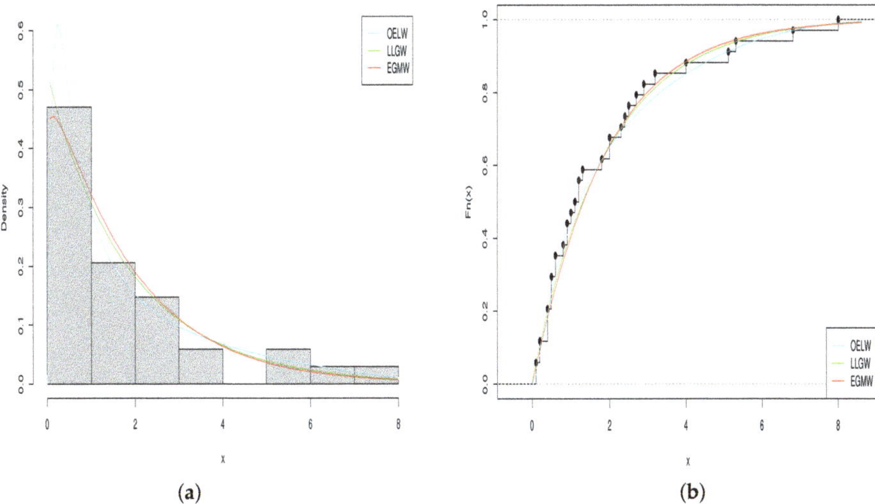

Figure 6. Obtained plots of the (a) estimated pdfs and (b) estimated cdfs of the considered distributions for the vinyl chloride data set.

The observed Fisher information matrix for the OELW distribution is

$$J(\Theta_*) = \begin{pmatrix} 357010.6 & -570.5293 & 3586.215 & 267.4771 \\ . & 2.083823 & -5.523455 & -1.17793 \\ . & . & 37.51406 & 432.3013 \\ . & . & . & 0.3955599 \end{pmatrix}.$$

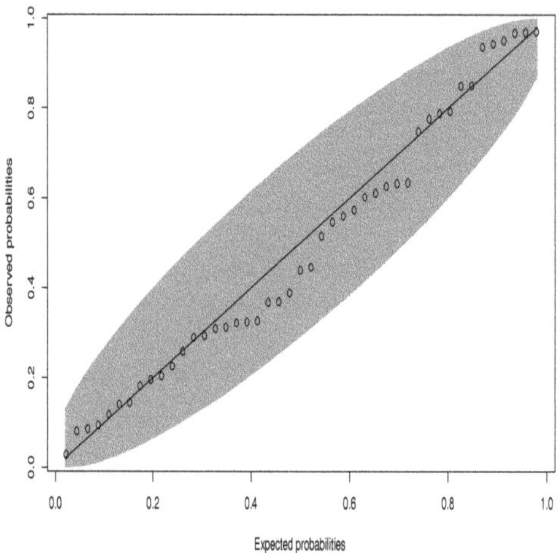

Figure 7. PP plot of the vinyl chloride data set.

6.3. Carbon Dioxide Data Sets

The third data set contains the annual mean growth rate of carbon dioxide during the period of 1959 to 2016 in Mauna Loa, Hawaii. The measurements are given in parts per million by year (ppm/yr). These data are taken from the given website https://www.esrl.noaa.gov/gmd/ccgg/trends/gr.html/, accessed on 4 July 2022. They are given below.

{0.94; 0.50; 0.96; 0.64; 0.71; 0.32; 1.06; 1.28; 0.70; 1.06; 1.35; 1.00; 0.81; 1.74; 1.18; 0.95; 1.06; 0.83; 2.15; 1.31; 1.82; 1.68; 1.43; 0.86; 2.36; 1.51; 1.21; 1.47; 2.06; 2.24; 1.24; 1.20; 1.05; 0.49; 1.36; 1.95; 2.01; 1.24; 1.91; 2.97; 0.92; 1.62; 1.62; 2.51; 2.27; 1.59; 2.57; 1.69; 2.31; 1.54; 2.00; 2.30; 1.92; 2.65; 1.99; 2.17; 2.95; 3.03}

A summary of measures of descriptive statistics is provided in Table 7.

Table 7. Descriptive statistics of the carbon dioxide data set.

Minimum	Mean	Median	Variance	Skewness	Kurtosis	Maximum
0.320	1.556	1.490	0.4457	0.3413	2.3618	3.030

Table 8 gives the relevant numerical summaries for the three fits based on the carbon dioxide data set.

Figure 8 gives the graph of the estimated pdfs and cdfs of the considered distributions. Figure 9 gives the PP plot of the OELW distribution.

Table 8. Estimated values, log-likelihood, AIC, BIC, *A**, *W**, K-S and p-V for the carbon dioxide data set.

Distribution	Estimates	$-\log(L)$	AIC	BIC	A^*	W^*	K-S	p-V
OELW	$\hat{a} = 3.7419$ $\hat{b} = 2.8634$ $\hat{p} = 0.0143$ $\hat{\beta} = 1.2442$	55.6190	119.2381	127.4798	0.1204	0.0132	0.0461	0.9997
LLoGW	$\hat{s} = 5.7793$ $\hat{c} = 2.5283$ $\hat{\beta} = 2.5456$ $\hat{\alpha} = 0.2272$	56.68579	121.3716	129.6134	0.2053	0.0339	0.0682	0.9499
EGMW	$\hat{\alpha} = 0.0601$ $\hat{\theta} = 25.4632$ $\hat{\beta} = 6.3678$ $\hat{\mu} = 0.0151$ $\hat{\lambda} = 6.6891$	56.16804	122.3361	132.6383	0.2144	0.0271	0.0650	0.967

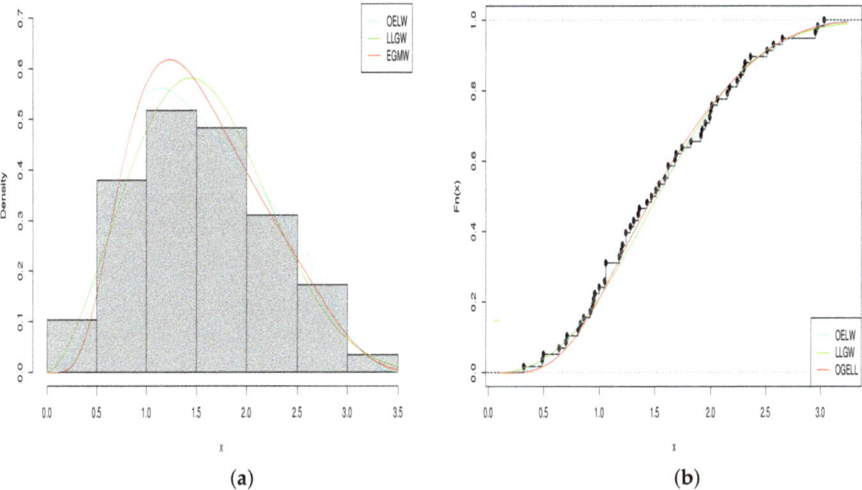

Figure 8. Obtained plots of the (**a**) estimated pdfs and (**b**) estimated cdfs of the considered distributions for the carbon dioxide data set.

The observed Fisher information matrix for the OELW distribution is

$$
J(\Theta_*) = \begin{pmatrix}
18811.95 & -362.08 & 444.2855 & 606.4949 \\
\cdot & 13.93288 & -6.794833 & -30.04717 \\
\cdot & \cdot & 11.62585 & 626.3218 \\
\cdot & \cdot & \cdot & 25.90807
\end{pmatrix}.
$$

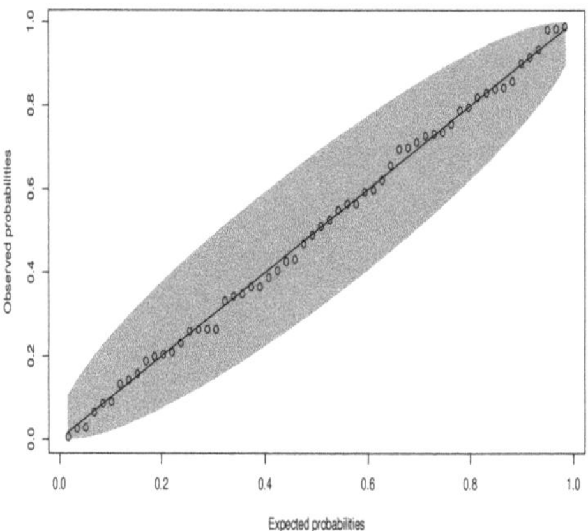

Figure 9. PP plot of the carbon dioxide data set.

In Tables 4, 6 and 8, the MLEs of the parameters for the fitted distributions, along with log-likelihood, AIC, BIC, A^*, (W^*) and K-S with p-V values are presented for three different data sets, respectively. From these tables, it is quite obvious that for the three data sets, the OELW distribution is the best model with the lowest values of AIC, BIC, A^*, W^*, K-S, and highest p-V of the K-S statistics. Hence, the OELW distribution turns out to be a better model than the LLoGW and EGMW models. A visual comparison of the closeness of the fitted pdfs with the observed histogram of the data, fitted cdfs with empirical cdfs, and PP plots for different data is presented in Figures 4–9, respectively. These plots indicate that the proposed distributions provide a closer fit to these data.

7. Conclusions

The OEL-G family of continuous distributions is a new family that we introduced and analyzed in this research. It has the feature of combining the functionalities of the logarithmic and odd transformations, of the EL distributions, and odd transformations, respectively. We gave explicit formulations for the moments, generating function, skewness, kurtosis, entropies, and order statistics, as well as a convenient linear representation for the probability density function. The OELW distribution, which is a subset of the OEL-G family, has been given special attention. Then there are statistical applications. The OELW model parameters are estimated using the maximum likelihood method, and the observed Fisher information matrix is explained. When compared to the famous LLoGW and EGMW distributions, three examples of real-life data fitting demonstrate good results in favor of the suggested distribution. Based on the findings, the proposed family might be regarded as a valuable addition to the field's existing knowledge.

Author Contributions: Conceptualization, C.C., L.T., M.J. and K.V.J.; methodology, C.C., L.T., M.J. and K.V.J.; software, C.C., L.T., M.J. and K.V.J.; validation, C.C., L.T., M.J. and K.V.J.; formal analysis, C.C., L.T., M.J. and K.V.J.; investigation, C.C., L.T., M.J. and K.V.J.; resources, C.C., L.T., M.J. and K.V.J.; data curation, C.C., L.T., M.J. and K.V.J.; writing—original draft preparation, C.C., L.T., M.J. and K.V.J.; writing—review and editing, C.C., L.T., M.J. and K.V.J.; visualization, C.C., L.T., M.J. and K.V.J. All authors have read and agreed to the published version of the manuscript.

Funding: This research received no external funding.

Acknowledgments: The authors would like to thank the three reviewers for the constructive comments on the first version of the paper.

Conflicts of Interest: The authors declare no conflict of interest.

References

1. Alzaatreh, A.; Famoye, F.; Lee, C. A new method for generating families of continuous distributions. *Metron* **2013**, *71*, 63–79. [CrossRef]
2. Tomy, L.; Jose, M.; Jose, M. The T-X family of distributions: A retrospect. *Think India J.* **2019**, *22*, 9407–9420.
3. Gleaton, J.U.; Lynch, J.D. Properties of generalized log-logistic families of lifetime distributions. *J. Probab. Stat. Sci.* **2006**, *4*, 51–64.
4. Torabi, H.; Montazari, N.H. The gamma-uniform distribution and its application. *Kybernetika* **2012**, *48*, 16–30.
5. Cordeiro, G.M.; Ortega, E.M.M.; da Cunha, D.C.C. The exponentiated generalized class of distributions. *J. Data Sci.* **2013**, *11*, 1–27. [CrossRef]
6. Bourguignon, M.; Silva, R.B.; Cordeiro, G.M. The Weibull-G family of probability distributions. *J. Data Sci.* **2014**, *12*, 53–68. [CrossRef]
7. Tahir, M.H.; Cordeiro, G.M.; Alizadeh, M.; Mansoor, M.; Zubair, M.; Hamedani, G.G. The odd generalized exponential family of distributions with applications. *J. Stat. Distrib. Appl.* **2015**, *2*, 1–28. [CrossRef]
8. Rosaiah, K.; Rao, G.S.; Sivakumar, D.C.U.; Kalyani, K. The Odd Generalized Exponential Log Logistic Distribution. *Int. J. Math. Stat. Invent.* **2016**, *4*, 21–29.
9. Braga, A.S.; Cordeiro, G.M.; Ortega, E.M.M.; da-Cruz, J.N. The odd log-logistic normal distribution: Theory and applications. *J. Stat. Pract.* **2016**, *10*, 311–335. [CrossRef]
10. Haghbin, H.; Ozel, G.; Alizadeh, M.; Hamedani, G.G. A new generalized odd log-logistic family of distributions. *Commun. Stat. Theory Methods* **2017**, *46*, 9897–9920. [CrossRef]
11. Haq, M.A.; Elgarhy, M. The odd Fréchet-G family of probability distributions. *J. Stat. Appl. Probab.* **2018**, *7*, 189–203. [CrossRef]
12. Hosseini, B.; Afshari, M.; Alizadeh, M. The generalized odd gamma-G family of distributions: Properties and applications. *Austrian J. Stat.* **2018**, *47*, 69–89. [CrossRef]
13. Afify, A.Z.; Cordeiro, G.M.; Maed, M.E.; Alizadeh, M.; Al-Mofleh, H.; Nofal, Z.M. The generalized odd Lindley-G family: Properties and applications. *An. Acad. Bras. Cienc.* **2019**, *91*, 1–22. [CrossRef]
14. Jamal, F.; Reyad, H.; Chesneau, C.; Nasir, A.; Othman, S. The Marshall-Olkin odd Lindley-G family of distributions: Theory and applications. *Punjab Univ. J. Math.* **2019**, *51*, 111–125.
15. Bakouch, H.S.; Chesneau, C.; Khan, M.N. The extended odd family of probability distributions with practice to a submodel. *Filomat* **2019**, *33*, 3855–3867. [CrossRef]
16. Chesneau, C.; Djibrila, S. The generalized odd inverted exponential-G family of distributions: Properties and applications. *Eurasian Bull. Math.* **2019**, *2*, 86–110.
17. El-Morshedy, M.; Eliwa, M.S. The odd flexible Weibull-H family of distributions: Properties and estimation with applications to complete and upper record data. *Filomat* **2019**, *33*, 2635–2652. [CrossRef]
18. Badr, M.A.; Elbatal, I.; Jamal, F.; Chesneau, C.; Elgarhy, M. The transmuted odd Fréchet-G family of distributions: Theory and applications. *Mathematics* **2020**, *8*, 958. [CrossRef]
19. Nasir, M.A.; Tahir, M.H.; Chesneau, C.; Jamal, F.; Shah, M.A.A. The odd generalized gamma-G family of distributions: Properties, regressions and applications. *Statistica* **2020**, *80*, 3–38.
20. Chesneau, C.; El Achi, T. Modified odd Weibull family of distributions: Properties and applications. *J. Indian Soc. Probab. Stat.* **2020**, *21*, 259–286. [CrossRef]
21. Al-Marzouki, S.; Jamal, F.; Chesneau, C.; Elgarhy, M. Topp-Leone odd Fréchet generated family of distributions with applications to Covid-19 data sets. *Comput. Model. Eng. Sci.* **2020**, *125*, 437–458.
22. Mi, Z.; Hussain, S.; Chesneau, C. On a special weighted version of the odd Weibull-generated class of distributions. *Math. Comput. Appl.* **2021**, *26*, 62. [CrossRef]
23. Altun, E.; Korkmaz, M.C.; El-Morshedy, M.; Eliwa, M.S. A new flexible family of continuous distributions: The additive odd-G family. *Mathematics* **2021**, *9*, 1837. [CrossRef]
24. Eliwa, M.S.; El-Morshedy, M.; Ali, S. Exponentiated odd Chen-G family of distributions: Statistical properties, Bayesian and non-Bayesian estimation with applications. *J. Appl. Stat.* **2021**, *48*, 1948–1974. [CrossRef]
25. Jamal, F.; Handique, L.; Ahmed, A.H.N.; Khan, S.; Shafiq, S.; Marzouk, W. The generalized odd linear exponential family of distributions with applications to reliability theory. *Math. Comput. Appl.* **2022**, *27*, 55. [CrossRef]
26. Jamal, F.; Chesneau, C.; Aidi, K. The sine extended odd Fréchet-G family of distribution with applications to complete and censored data. *Math. Slovaca* **2022**, *71*, 961–982. [CrossRef]
27. Tahmasbi, R.; Rezaei, S. A two-parameter lifetime distribution with decreasing failure rate. *Comput. Stat. Data Anal.* **2008**, *52*, 3889–3901. [CrossRef]
28. Pappas, V.; Adamidis, K.; Loukas, S. A family of lifetime distributions. *Int. J. Qual. Stat. Reliab.* **2012**, *2012*, 760687. [CrossRef]
29. Gupta, R.D.; Kundu, D. Generalized exponential distributions. *Aust. N. Z. J. Stat.* **1999**, *41*, 173–188. [CrossRef]
30. Cordeiro, G.M.; Nadarajah, S. Closed-form expressions for moments of a class of beta generalized distributions. *Braz. J. Probab. Stat.* **2011**, *25*, 14–33. [CrossRef]
31. Moors, J.J.A. A quantile alternative for Kurtosis. *Statistician* **1988**, *37*, 25–32. [CrossRef]
32. Galton, F. *Inquiries into Human Faculty and Its Development*; Macmillan Company: London, UK, 1883.

33. Rényi, A. On measures of entropy and information. In Proceedings of the 4th Berkeley Symposium on Mathematical Statistics and Probability, Berkeley, CA, USA, 20 June–30 July 1960; University of California Press: Berkeley, CA, USA, 1961; Volume 1, pp. 547–561.
34. Shannon, C.E. Prediction and entropy of printed English. *Bell Syst. Tech. J.* **1951**, *30*, 50–64. [CrossRef]
35. Shaked, M.; Shanthikumar, J.G. *Stochastic Orders*; Wiley: New York, NY, USA, 2007.
36. El-Morshedy, M.; Alshammari, F.S.; Tyagi, A.; Elbatal, I.; Hamed, Y.S.; Eliwa, M.S. Bayesian and frequentist inferences on a type I half-logistic odd Weibull generator with applications in engineering. *Entropy* **2021**, *23*, 446. [CrossRef]
37. Oluyede, B.O.; Foya, S.; Warahena-Liyanage, G.; Huang, S. The Log-logistic Weibull Distribution with Applications to Lifetime Data. *Austrian J. Stat.* **2016**, *45*, 43–69. [CrossRef]
38. Aryal, G.; Elbatal, I. On the exponentiated generalized modified Weibull distribution. *Commun. Stat. Appl. Methods* **2015**, *22*, 333–348. [CrossRef]
39. Bekker, A.; Roux, J.; Mostert, P. A generalization of the compound Rayleigh distribution: Using a Bayesian methods on cancer survival times. *Commun. Stat. Theory Methods* **2000**, *29*, 1419–1433. [CrossRef]
40. Bhaumik, D.K.; Kapur, K.; Gibbons, R.D. Testing Parameters of a Gamma Distribution for Small Samples. *Technometrics* **2009**, *51*, 326–334. [CrossRef]

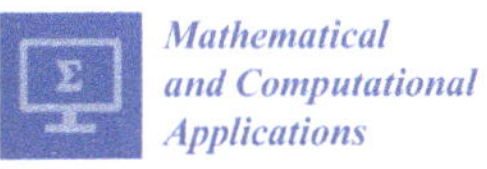

Mathematical and Computational Applications

Article

Bivariate Generalized Half-Logistic Distribution: Properties and Its Application in Household Financial Affordability in KSA

Marwa K. H. Hassan [1] and **Christophe Chesneau** [2,*]

1   Department of Mathematics, Faculty of Education, Ain Shams University, Cairo 11566, Egypt
2   Department of Mathematics, LMNO, CNRS-Université de Caen, Campus II, Science 3, CEDEX, 14032 Caen, France
*   Correspondence: christophe.chesneau@unicaen.fr

Abstract: The generalized half-logistic distribution is ideal to fit the lifetime of some products, such as ball bearings and electrical insulation. In this paper, we aim to extend this scope by creating a motivated bivariate version. We thus introduce the bivariate generalized half-logistic distribution using the Farlie Gumbel Morgenstern (FGM) copula, which is called the FGM bivariate generalized half-logistic distribution (FGMBGHLD for short). In particular, the FGMBGHLD finds application in describing bivariate lifetime datasets that have weak correlations between variables. Some statistical properties and functions of our new distribution, such as the product moments, moment generating function, reliability function, and hazard rate function, are derived. We discuss the maximum likelihood estimation method of the FGMBGHLD parameters. As an application of the FGMBGHLD in reliability, we consider the stress–strength model when the stress and strength random variables are dependent. We also derive the point and interval estimates of the stress–strength coefficient. Finally, we use the data from the household income and expenditure survey of KSA 2018 for Saudi households by administrative region to demonstrate the practicability of the proposed model. A comparison with a modern bivariate Weibull distribution is performed.

Keywords: Farlie Gumbel Morgenstern (FGM) copula; generalized half-logistic distribution (GHLD); reliability parameter; Monte Carlo simulation; statistical properties; household financial affordability

MSC: 62N01; 62N02; 62E10

Citation: Hassan, M.K.H.; Chesneau, C. Bivariate Generalized Half-Logistic Distribution: Properties and Its Application in Household Financial Affordability in KSA. *Math. Comput. Appl.* **2022**, *27*, 72. https://doi.org/10.3390/mca27040072

Academic Editor: Sandra Ferreira

Received: 19 July 2022
Accepted: 17 August 2022
Published: 19 August 2022

Publisher's Note: MDPI stays neutral with regard to jurisdictional claims in published maps and institutional affiliations.

1. Introduction

In various fields, such as life testing, reliability, and biological and engineering sciences, there is a need for flexible lifetime distributions with various probability density and hazard rate properties. To this end, Mudholkar et al. (1995) [1] introduced the exponentiated Weibull family of distributions, which includes unimodal distributions with bathtub hazard rates as well as a broader class of monotone hazard rates. Alternative distributions have been examined since, presenting slightly different features. Gupta and Kundu (1999) [2] proposed a generalized exponential distribution. Olopade (2008) [3] considered two distributions, named type-I and type-III generalized half-logistic distributions. Kantam et al. (2014) [4] proposed a type-II generalized half-logistic distribution (GHLD-II for short). For the purpose of this paper, a brief presentation of the GHLD-II is necessary. On the mathematical plan, the probability density function (PDF), cumulative distribution function (CDF), and reliability function of the GHLD-II with scale parameter σ and power parameter μ are given by

$$f(x) = f(x; \sigma, \mu) = \frac{\mu \left(2\, e^{-\frac{x}{\sigma}}\right)^{\mu}}{\sigma \left(1 + e^{-\frac{x}{\sigma}}\right)^{\mu+1}}, \quad 0 < x < \infty, \quad \sigma > 0, \; \mu > 0 \tag{1}$$

$$F(x) = F(x; \sigma, \mu) = 1 - \left(\frac{2\, e^{-\frac{x}{\sigma}}}{1 + e^{-\frac{x}{\sigma}}} \right)^{\mu} \tag{2}$$

and

$$R(x) = R(x; \sigma, \mu) = \left(\frac{2\, e^{-\frac{x}{\sigma}}}{1 + e^{-\frac{x}{\sigma}}} \right)^{\mu}. \tag{3}$$

Thus, the GHLD-II is developed through the exponentiation of the reliability function of the half-logistic distribution (see Balakrishnan (1985) [5]).

The flexibility of the GHLD-II is mainly in the mode and tail of the distribution, making it an interesting distribution for the modeling of lifetime phenomena. It is proven to define a better model than the exponential, Weibull, and half-logistic models (see Kantam et al. (2014) [4]).

The first objective of this paper is to derive a comprehensive bivariate generalized half-logistic distribution (BGHLD for short) using the copula approach and study its statistical properties, such as PDF, CDF, product moments, moment generating function, and hazard rate function. Many authors discuss the same idea but other distributions; see Almetwally et al. (2020) [6], Almetwally and Muhammed (2020) [7], and Muhammed et al. (2021) [8]. In view of the impact of the GHLD-II in the recent literature, we derive that bivariate versions have a promising future in terms of modeling and data analysis. Now, in order to detail and motivate the construction of our BGHLD, let us present some basics of the notion of the copula. As a first approach, we can say that a copula is a multivariate CDF for which the marginal distribution of each variable is uniform on the interval $(0, 1)$. It describes the dependence between random variables. The definitions below provide more technical details.

Definition 1. *Let us consider a random vector $(X_1, \ldots, X_d)$ and the marginal CDFs denoted by $F_i(x) = P(X_i < x)$, for $i = 1, \ldots, d$. Then, using probability integral transform (PIT) for each component, the distribution of the random vector $(U_1, \ldots, U_d) = (F_1(X_1), \ldots, F_d(X_d))$ belongs to the $(unif(0, 1))^d$ family of distributions, and the copula related to $(X_1, \ldots, X_d)$ is defined as the joint CDF of $(U_1, \ldots, U_d)$, i.e.,*

$$C(u_1, \ldots, u_d) = P(U_1 \le u_1, \ldots, U_d \le u_d), \tag{4}$$

with $(u_1, \ldots, u_d) \in [0, 1]^d$.

Definition 2. *$C: [0, 1]^d \to [0, 1]$ is a d-dimensional copula if it is a CDF with*

$$C(u_1, \ldots, u_{i-1}, 0, u_{i+1}, \ldots, u_d) = 0, \quad C(1, \ldots, u, 1, \ldots, 1) = u, \tag{5}$$

with $(u_1, \ldots, u_d) \in [0, 1]^d$ and $u \in [0, 1]$. In the bivariate case, $C : [0, 1]^2 \to [0, 1]$ is a bivariate copula if $C(0, u) = C(u, 0) = 0$, $C(1, u) = C(u, 1) = u$ and $C(u_2, v_2) - C(u_2, v_1) - C(u_1, v_2) + C(u_1, v_1) \ge 0$ for all $0 \le u_1 \le u_2 \le 1$ and $0 \le v_1 \le v_2 \le 1$.

The Sklar theorem, established by Sklar (1959) [9], is pivotal in copula theory. It states that, for two random variables X_1 and X_2 with marginal CDFs $F_1(x_1)$ and $F_2(x_2)$ and marginal PDFS $f_1(x_1)$ and $f_2(x_2)$, respectively, the CDF and PDF of (X_1, X_2) are given by

$$F(x_1, x_2) = C(F_1(x_1), F_2(x_2)) \tag{6}$$

and

$$f(x_1, x_2) = f(x_1)f(x_2)c(F_1(x_1), F_2(x_2)), \tag{7}$$

respectively, where $c(u_1, u_2)$ denotes the copula density related to $C(u_1, u_2)$, i.e., $c(u_1, u_2) = \partial^2 C(u_1, u_2)/(\partial u_1 \partial u_2)$.

Gumbel (1960) [10] discussed one of the most popular parametric families of copulas, called the Farlie Gumbel Morgenstern (FGM) copula. The FGM copula and its density are specified by

$$C(u_1, u_2) = u_1 u_2 (1 + \theta(1 - u_1)(1 - u_2)), \quad -1 \leq \theta \leq 1, \tag{8}$$

and

$$c(u_1, u_2) = 1 + \theta(1 - 2u_1)(1 - 2u_2), \tag{9}$$

respectively. The parameter θ can be thought of as a dependence parameter that is dependent on the underlying random variables, with the independent case being $\theta = 0$. The FGM copula is thus simple, flexible, and can be adapted when dealing with the construction of bivariate distributions with complicated marginal distributions in terms of functions. It is used in our study to create our BGHLD, which we naturally call the FGMBGHLD.

The second objective is to develop the maximum likelihood (ML) estimation method of the FGMBGHLD parameters. Finally, the third goal is to derive the corresponding stress–strength model, but when and how this makes sense: in the dependent case, which can occur in engineering, operations research, quality control, education, economics, and insurance. Domma and Giordano (2013) [11] provided an example. In this paper, we are interested in economics, where X and Y are household income and consumption, and $R = P(Y < X)$ is a measure of household financial affordability.

This paper is organized as follows. In Section 2, the FGMBGHLD is described. In Section 3, we derive some statistical properties of the FGMBGHLD. In Section 4, we exploit the copula approach to take into account the dependence of stress and strength variables in evaluating R. In Section 5, the ML estimation method for the FGMBGHLD parameters is discussed. In Section 6, point and interval estimations for R are elaborated. In Section 7, a Monte Carlo simulation study is performed to study the behavior of different estimates. In Section 8, the estimation of R is applied to KSA data (year 2018) to measure the household financial affordability for Saudi households by administrative region, with comparison to a modern bivariate Weibull distribution. The conclusion of this paper appears in Section 9.

2. FGM Bivariate Generalized Half-Logistic Distribution (FGMBGHLD)

Applying the Sklar theorem as stated in Equations (6) and (7) with Equations (1) and (2), and the FGM copula in Equations (8) and (9), we obtain the CDF and PDF of a random vector (Y_1, Y_2) following the FGMBGHLD. They are given by

$$F(y_1, y_2) = \left(1 - \left(\frac{2\,e^{-\frac{y_1}{\sigma_1}}}{1 + e^{-\frac{y_1}{\sigma_1}}}\right)^{\mu_1}\right)\left(1 - \left(\frac{2\,e^{-\frac{y_2}{\sigma_2}}}{1 + e^{-\frac{y_2}{\sigma_2}}}\right)^{\mu_2}\right)\left(1 + \theta\left(\frac{2\,e^{-\frac{y_1}{\sigma_1}}}{1 + e^{-\frac{y_1}{\sigma_1}}}\right)^{\mu_1}\left(\frac{2\,e^{-\frac{y_2}{\sigma_2}}}{1 + e^{-\frac{y_2}{\sigma_2}}}\right)^{\mu_2}\right) \tag{10}$$

and

$$f(y_1, y_2) = \frac{\mu_1\left(2\,e^{-\frac{y_1}{\sigma_1}}\right)^{\mu_1}}{\sigma_1\left(1 + e^{-\frac{y_1}{\sigma_1}}\right)^{\mu_1+1}}\frac{\mu_2\left(2\,e^{-\frac{y_2}{\sigma_2}}\right)^{\mu_2}}{\sigma_2\left(1 + e^{-\frac{y_2}{\sigma_2}}\right)^{\mu_2+1}}\left[1 + \theta\left(2\left(\frac{2\,e^{-\frac{y_1}{\sigma_1}}}{1 + e^{-\frac{y_1}{\sigma_1}}}\right)^{\mu_1} - 1\right)\left(2\left(\frac{2\,e^{-\frac{y_2}{\sigma_2}}}{1 + e^{-\frac{y_2}{\sigma_2}}}\right)^{\mu_2} - 1\right)\right], \tag{11}$$

respectively, with the restrictions of the variables and parameters already mentioned.

In order to illustrate the effect of the dependence parameter θ on the shape of these functions, Figure 1 shows the three-dimensional plots of the PDF and CDF with different values of θ (positive and negative).

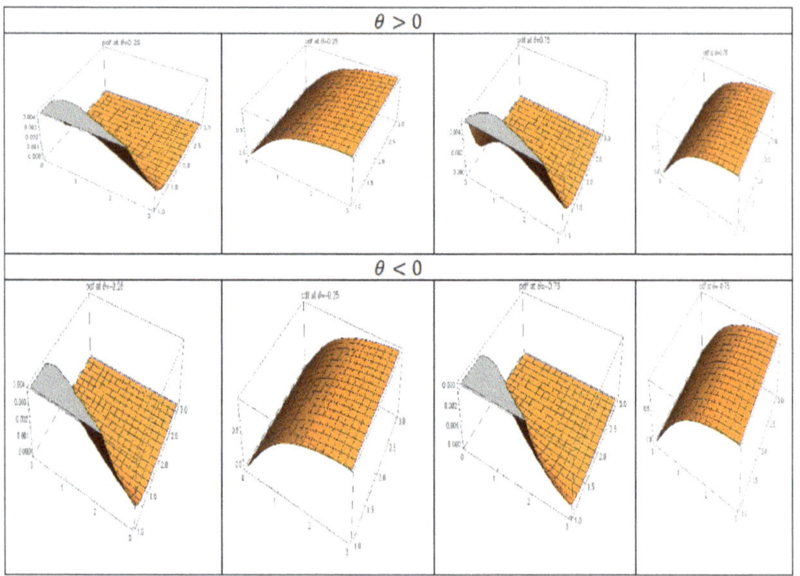

Figure 1. Three-dimensional plots for the PDF and CDF of the FGMBGHLD with different values of θ (for $\mu_1 = \mu_2 = 0.5$, $\sigma_1 = 0.2$ and $\sigma_2 = 0.1$).

From Figure 1, we see that the variable variations of θ play a significant role; the PDF can take different forms in the space, with various skewness and kurtosis.

3. Statistical Properties of the FGMBGHLD

Here, we discuss some statistical properties of the FGMBHLD as defined by Equations (10) and (11). The marginal distributions, product moments, moment generating function, conditional distribution, generating random variables, and reliability function are derived.

3.1. Marginal PDFs

From a random vector (Y_1, Y_2) following the FGMBHLD, for $i = 1, 2$, the distribution of Y_i has the following PDF:

$$f_i(y_i) = \int_0^\infty f(y_i, y_j) dy_j, \quad i \neq j, \ j = 1, 2. \tag{12}$$

Thus, more concretely, Y_1 has the following PDF:

$$f_1(y_1) = \frac{\mu_1 \left(2 e^{-\frac{y_1}{\sigma_1}} \right)^{\mu_1}}{\sigma_1 \left(1 + e^{-\frac{y_1}{\sigma_1}} \right)^{\mu_1 + 1}}, \quad y_1 > 0, \ \mu_1, \ \sigma_1 > 0 \tag{13}$$

and Y_2 has the following PDF:

$$f_2(y_2) = \frac{\mu_2 \left(2 e^{-\frac{y_2}{\sigma_2}} \right)^{\mu_2}}{\sigma_2 \left(1 + e^{-\frac{y_2}{\sigma_2}} \right)^{\mu_2 + 1}}, \quad y_2 > 0, \ \mu_2, \ \sigma_2 > 0. \tag{14}$$

On the other hand, for $i \neq j$ with $i, j = 1, 2$, the general formula for the conditional PDF of Y_i given $Y_j = y_j$ is

$$f(y_i|y_j) = f_i(y_i)[1 + \theta(1 - 2\,F_i(y_i))(1 - 2\,F_j(y_j))], \quad -1 \leq \theta \leq 1, \tag{15}$$

where $F_i(y_i)$ and $F_j(y_j)$ denote the CDFs of Y_i and Y_j, respectively.

Similarly, the conditional CDF of Y_i given $Y_j = y_j$ is

$$F(y_i|y_j) = F_i(y_i)[1 + \theta(1 - F_i(y_i)) - 2\,F_j(y_j) + 2\,F_i(y_i)F_j(y_j)]. \tag{16}$$

We omit their analytical expressions for the FGMBHLD for the sake of brevity.

3.2. Moment Generating Function

The moment generating function of (Y_1, Y_2) following the FGMBHLD is obtained as

$$M_{(Y_1,Y_2)}(t_1, t_2) = E\left(e^{t_1 Y_1} e^{t_2 Y_2}\right) = \int_0^\infty \int_0^\infty e^{t_1 y_1} e^{t_2 y_2} f(y_1, y_2) dy_1 dy_2$$

$$= \frac{\mu_1 \mu_2 \, 2^{\mu_1 + \mu_2}}{(1 + t_1 \sigma_1)(1 + t_2 \sigma_2)}\left[2^{1+\mu_1}\theta\, W_1\left(2^{1+\mu_2}\, W_2 - W_3\right) - \left(W_4\left(2^{1+\mu_2}W_2 - (1+\theta)W_3\right)\right)\right], \tag{17}$$

where

$$W_1 =\, _2F_1(-1 - 2\mu_1, -1 - t_1\sigma_1; -t_1\sigma_1; -1), \tag{18}$$

$$W_2 =\, _2F_1(-1 - 2\mu_2, -1 - t_2\sigma_2; -t_2\sigma_2; -1), \tag{19}$$

$$W_3 =\, _2F_1(-1 - \mu_2, -1 - t_2\sigma_2; -t_2\sigma_2; -1) \tag{20}$$

and

$$W_4 =\, _2F_1(-1 - \mu_1, -1 - t_1\sigma_1; -t_1\sigma_1; -1), \tag{21}$$

where $_2F_1(a, b; c; z)$ refers to the (generalized) hypergeometric function.

The parameters t_1 and t_2 must be selected such that the above quantities exist in the mathematical sense, which is the case for $t_1 \leq 0$ and $t_2 \leq 0$ among other more technical cases.

3.3. Product Moments

To obtain the product moments about the origin of (Y_1, Y_2) following the FGMBHLD, for any positive real numbers r_1 and r_2, we calculate

$$\mu_{r_1 r_2} = E(Y_1^{r_1} Y_2^{r_2}) = \int_0^\infty \int_0^\infty y_1^{r_1} y_2^{r_2} f(y_1, y_2) dy_1 dy_2 = \frac{\mu_1 \mu_2 \, 2^{\mu_1 + \mu_2}}{\sigma_1 \sigma_2}(A + B + C + D), \tag{22}$$

where

$$A = (1 + \theta)\left[\sum_{s_1=0}^\infty \binom{1 + \mu_1}{s_1}\left(\frac{-1 + s_1}{\sigma_1}\right)^{-1-r_1}\Gamma(1 + r_1)\right]\left[\sum_{s_2=0}^\infty \binom{1 + \mu_2}{s_2}\left(\frac{-1 + s_2}{\sigma_2}\right)^{-1-r_2}\Gamma(1 + r_2)\right], \tag{23}$$

$$B = \theta 2^{2+(\mu_1+\mu_2)}\left[\sum_{s_3=0}^\infty \binom{1 + 2\mu_1}{s_3}\left(\frac{-1 + s_3}{\sigma_1}\right)^{-1-r_1}\Gamma(1 + r_1)\right]\left[\sum_{s_4=0}^\infty \binom{1 + 2\mu_2}{s_4}\left(\frac{-1 + s_4}{\sigma_2}\right)^{-1-r_2}\Gamma(1 + r_2)\right], \tag{24}$$

$$C = -\theta 2^{(\mu_1+1)}\left[\sum_{s_3=0}^\infty \binom{1 + 2\mu_1}{s_3}\left(\frac{-1 + s_3}{\sigma_1}\right)^{-1-r_1}\Gamma(1 + r_1)\right]\left[\sum_{s_2=0}^\infty \binom{1 + \mu_2}{s_2}\left(\frac{-1 + s_2}{\sigma_2}\right)^{-1-r_2}\Gamma(1 + r_2)\right] \tag{25}$$

and

$$D = -\theta 2^{(\mu_2+1)}\left[\sum_{s_1=0}^{\infty}\binom{1+\mu_1}{s_1}\left(\frac{-1+s_1}{\sigma_1}\right)^{-1-r_1}\Gamma(1+r_1)\right]\left[\sum_{s_4=0}^{\infty}\binom{1+2\mu_2}{s_4}\left(\frac{-1+s_4}{\sigma_2}\right)^{-1-r_2}\Gamma(1+r_2)\right]. \tag{26}$$

It is understood that $\Gamma(x)$ refers to the standard gamma function, with $\Gamma(m+1) = m!$ for any integer m. From the product moments, various measures of moment skewness and kurtosis can be presented. On this topic, see, for instance, Almetwally et al. (2020) [6], Almetwally and Muhammed (2020) [7], and Muhammed et al. (2021) [8].

3.4. Reliability and Hazard Rate Functions

The reliability function of a bivariate distribution with an associated copula is defined by the copula composed with its marginal reliability functions. See Osmetti and Chiodini (2011) [12]. Hence, based on (Y_1, Y_2) following the FGMBHLD, it is expressed as

$$R(y_1, y_2) = C(R_1(y_1),\, R_2(y_2)), \tag{27}$$

where $R_1(y_1)$ and $R_2(y_2)$ denote the reliability functions of Y_1 and Y_2, respectively. According to the FGM copula, we obtain

$$R(y_1, y_2) = R_1(y_1)R_2(y_2)[1 + \theta(1 - R_1(y_1))(1 - R_2(y_2))], \quad -1 \leq \theta \leq 1. \tag{28}$$

For the FGMBHLD, the reliability function is

$$R(y_1, y_2) = \left(\frac{2\,e^{-\frac{y_1}{\sigma_1}}}{1+e^{-\frac{y_1}{\sigma_1}}}\right)^{\mu_1}\left(\frac{2\,e^{-\frac{y_2}{\sigma_2}}}{1+e^{-\frac{y_2}{\sigma_2}}}\right)^{\mu_2}\left[1 + \theta\left(1 - \left(\frac{2\,e^{-\frac{y_1}{\sigma_1}}}{1+e^{-\frac{y_1}{\sigma_1}}}\right)^{\mu_1}\right)\left(1 - \left(\frac{2\,e^{-\frac{y_2}{\sigma_2}}}{1+e^{-\frac{y_2}{\sigma_2}}}\right)^{\mu_2}\right)\right]. \tag{29}$$

Moreover, Basu (1971) [13] defined the bivariate hazard rate function as

$$h(y_1, y_2) = \frac{f(y_1, y_2)}{R(y_1, y_2)}. \tag{30}$$

For the FGMBHLD, the hazard rate function is indicated as

$$h(y_1, y_2) = \frac{\mu_1\,\mu_2}{\sigma_1\,\sigma_2\left(1 + e^{-\frac{y_1}{\sigma_1}}\right)\left(1 + e^{-\frac{y_2}{\sigma_2}}\right)}\,\frac{1 + \theta\left(2\left(\frac{2\,e^{-\frac{y_1}{\sigma_1}}}{1+e^{-\frac{y_1}{\sigma_1}}}\right)^{\mu_1} - 1\right)\left(2\left(\frac{2\,e^{-\frac{y_2}{\sigma_2}}}{1+e^{-\frac{y_2}{\sigma_2}}}\right)^{\mu_2} - 1\right)}{1 + \theta\left(1 - \left(\frac{2\,e^{-\frac{y_1}{\sigma_1}}}{1+e^{-\frac{y_1}{\sigma_1}}}\right)^{\mu_1}\right)\left(1 - \left(\frac{2\,e^{-\frac{y_2}{\sigma_2}}}{1+e^{-\frac{y_2}{\sigma_2}}}\right)^{\mu_2}\right)}. \tag{31}$$

4. Reliability for Dependence Stress–Strength Model

Domma and Giordano (2013) [11] introduced the concept of dependence via the stress–strength model. They calculated the reliability measure under the hypothesis that the bivariate distribution of the stress and strength variables, modeled by the random variables X and Y, is defined by joining their respective marginal CDFs $F(x)$ and $G(y)$ for any copula. In this setting, the measure R for dependent X and Y can be defined as

$$R = P(Y < X) = \int_0^{\infty}\int_0^{x} c(F(x), G(y))f(x)g(y)dy\,dx, \tag{32}$$

where $f(x)$ and $g(y)$ denote the PDFs of X and Y, respectively, and $c(u_1, u_2)$ the copula density. Using the FGM copula, we have the following relationship: $R = R_1 + \theta\,D$, where

$$R_1 = \int_0^{\infty}\int_0^{x} f(x)g(y)dy\,dx = \int_0^{\infty} G(x)f(x)dx = E[G(X)] \tag{33}$$

and

$$D = \int_0^\infty \int_0^x (1 - 2F(x))(1 - 2\,G(y))dF(x)dG(y) = E[G(X)(1 - G(X))(1 - 2F(X))]. \tag{34}$$

Now, we calculate R when X and Y have possibly non-identical GHLD with the CDFs $F(x) = 1 - \left(\frac{2\,e^{-\frac{x}{\sigma}}}{1+e^{-\frac{x}{\sigma}}} \right)^{\mu_1}$ and $G(y) = 1 - \left(\frac{2\,e^{-\frac{y}{\sigma}}}{1+e^{-\frac{y}{\sigma}}} \right)^{\mu_2}$, respectively. Hence, σ is common to the two marginal distributions. In this case, after some integral developments, we obtain

$$R = \frac{\mu_2}{\mu_1 + \mu_2} + \theta\,\mu_1 \left(\frac{1}{2\,\mu_1 + \mu_2} - \frac{2}{\mu_1 + \mu_2} + \frac{1}{\mu_1 + 2\,\mu_2} \right). \tag{35}$$

5. Estimation Method for the Distribution Parameters

In this section, we present the ML method for estimating the FGMBHLD parameters.

Let $(x_1, y_1) \ldots (x_n, y_n)$ be a random sample from a random vector (X, Y) following the FGMBHLD with the parameters μ_1, μ_2, σ_1, σ_2, and θ. Hence, in particular, X follows the GBHLD(μ_1, σ_1) and Y follows the GBHLD(μ_2, σ_2). Elaal and Jarwan (2017) [14] introduced the ML estimation method for bivariate distributions based on copula. The basis consists of constructing the log-likelihood function as

$$Ln\,L = \sum_{i=1}^{n} \ln[f(x_i)\,g(y_i)c(F(x_i), G(y_i))], \tag{36}$$

where $F(x)$ and $G(y)$ are the CDFs of X and Y, and $f(x)$ and $g(y)$ are their respective PDFs, and $c(u_1, u_2)$ refers to the copula density. The ML estimates (MLEs) of the involved parameters are obtained by maximizing this function with respect to these parameters.

Under the setting of the FGMBHLD, we have

$$Ln\,L = n\,Ln\,(\mu_1) + n\,Ln\,(\mu_2) + n\,(\mu_1 + \mu_2)\ln(2) - \frac{\mu_1}{\sigma_1} \sum_{i=1}^{n} x_i - \frac{\mu_2}{\sigma_2} \sum_{i=1}^{n} y_i - n\ln(\sigma_1)$$

$$- n\,\ln(\sigma_2) - (\mu_1 + 1) \sum_{i=1}^{n} \ln\left(1 + e^{-\frac{x_i}{\sigma_1}} \right) - (\mu_2 + 1) \sum_{i=1}^{n} \ln\left(1 + e^{-\frac{y_i}{\sigma_2}} \right)$$

$$+ \sum_{i=1}^{n} \ln(1 + \theta\phi(x_i, \mu_1, \sigma_1)\,\eta(y_i, \mu_2, \sigma_2)), \tag{37}$$

where $\phi(x_i, \mu_1, \sigma_1) = 1 - 2F(x_i)$ and $\eta(y_i, \mu_2, \sigma_2) = 1 - 2G(y_i)$.

The MLEs of the parameters μ_1, μ_2, σ_1, σ_2 and θ, say $\widehat{\mu}_1$, $\widehat{\mu}_2$, $\widehat{\sigma}_1$, $\widehat{\sigma}_2$, and $\widehat{\theta}$, are those maximizing this function. They can be obtained by differentiation. To be more precise, by differentiating the log-likelihood with respect to the distribution parameters, we obtain

$$\frac{\partial Ln\,L}{\partial \mu_1} = \frac{n}{\mu_1} + n\,\ln(2) - \frac{1}{\sigma_1} \sum_{i=1}^{n} x_i - \sum_{i=1}^{n} \ln\left(1 + e^{-\frac{x_i}{\sigma_1}} \right) + \sum_{i=1}^{n} \frac{\theta\,\eta(y_i, \mu_2, \sigma_2)\,\phi_{\mu_1}(x_i, \mu_1, \sigma_1)}{(1 + \theta\phi(x_i, \mu_1, \sigma_1)\,\eta(y_i, \mu_2, \sigma_2))}, \tag{38}$$

$$\frac{\partial Ln\,L}{\partial \mu_2} = \frac{n}{\mu_2} + n\,\ln(2) - \frac{1}{\sigma_2} \sum_{i=1}^{n} y_i - \sum_{i=1}^{n} \ln\left(1 + e^{-\frac{y_i}{\sigma_2}} \right) + \sum_{i=1}^{n} \frac{\theta\,\phi(x_i, \mu_1, \sigma_1)\eta_{\mu_2}(y_i, \mu_2, \sigma_2)}{(1 + \theta\phi(x_i, \mu_1, \sigma_1)\,\eta(y_i, \mu_2, \sigma_2))}, \tag{39}$$

$$\frac{\partial Ln\,L}{\partial \sigma_1} = \frac{\mu_1}{\sigma_1^2} \sum_{i=1}^{n} x_i - \frac{n}{\sigma_1} + \frac{(\mu_1 + 1)}{\sigma_1^2} \sum_{i=1}^{n} \frac{x_i\,e^{-\frac{x_i}{\sigma_1}}}{1 + e^{-\frac{x_i}{\sigma_1}}} + \sum_{i=1}^{n} \frac{\theta\,\eta(y_i, \mu_2, \sigma_2)\,\phi_{\sigma_1}(x_i, \mu_1, \sigma_1)}{(1 + \theta\phi(x_i, \mu_1, \sigma_1)\,\eta(y_i, \mu_2, \sigma_2))}, \tag{40}$$

$$\frac{\partial Ln\,L}{\partial \sigma_2} = \frac{\mu_2}{\sigma_2^2} \sum_{i=1}^{n} y_i - \frac{n}{\sigma_2} + \frac{(\mu_2 + 1)}{\sigma_2^2} \sum_{i=1}^{n} \frac{y_i\,e^{-\frac{y_i}{\sigma_2}}}{1 + e^{-\frac{y_i}{\sigma_2}}} + \sum_{i=1}^{n} \frac{\theta\,\phi(x_i, \mu_1, \sigma_1)\eta_{\sigma_2}(y_i, \mu_2, \sigma_2)}{(1 + \theta\phi(x_i, \mu_1, \sigma_1)\,\eta(y_i, \mu_2, \sigma_2))} \tag{41}$$

and

$$\frac{\partial Ln\,L}{\partial \theta} = \sum_{i=1}^{n} \frac{\phi(x_i, \mu_1, \sigma_1)\eta(y_i, \mu_2, \sigma_2)}{1 + \theta\phi(x_i, \mu_1, \sigma_1)\,\eta(y_i, \mu_2, \sigma_2)}, \tag{42}$$

where

$$\phi_{\mu_1}(x_i, \mu_1, \sigma_1) = 2^{\mu_1+1}\left(\frac{1}{1+e^{\frac{x_i}{\sigma_1}}}\right)^{\mu_1}\left(-\frac{x_i}{\sigma_1} + \ln(2) - \ln\left(1 + e^{-\frac{x_i}{\sigma_1}}\right)\right), \tag{43}$$

$$\eta_{\mu_2}(y_i, \mu_2, \sigma_2) = 2^{\mu_2+1}\left(\frac{1}{1+e^{\frac{y_i}{\sigma_2}}}\right)^{\mu_2}\left(-\frac{y_i}{\sigma_2} + \ln(2) - \ln\left(1 + e^{-\frac{y_i}{\sigma_2}}\right)\right), \tag{44}$$

$$\phi_{\sigma_1}(x_i, \mu_1, \sigma_1) = \frac{\mu_1\, x_i 2^{\mu_1+1}\left(e^{-\frac{x_i}{\sigma_1}}\right)^{\mu_1-1}\left(1 + e^{-\frac{x_i}{\sigma_1}}\right)^{-\mu_1}}{\sigma_1^2\left(1 + e^{\frac{x_i}{\sigma_1}}\right)} \tag{45}$$

and

$$\eta_{\sigma_2}(y_i, \mu_2, \sigma_2) = \frac{\mu_2\, y_i 2^{\mu_2+1}\left(e^{-\frac{y_i}{\sigma_2}}\right)^{\mu_2-1}\left(1 + e^{-\frac{y_i}{\sigma_2}}\right)^{-\mu_2}}{\sigma_2^2\left(1 + e^{\frac{y_i}{\sigma_1}}\right)}. \tag{46}$$

By setting the above first partial derivatives of $Ln\, L$ to zero, we obtain $\widehat{\mu}_1$, $\widehat{\mu}_2$, $\widehat{\sigma}_1$, $\widehat{\sigma}_2$ and $\widehat{\theta}$. Since we cannot obtain a closed form for these estimates, a numerical method must be used.

6. Estimation of the Stress–Strength Distribution Parameter

In this section, we introduce the MLE for $R = P(Y < X)$. Moreover, we derive a motivated asymptotic confidence interval and a bootstrap confidence interval for it.

6.1. Maximum Likelihood Estimate of R

From observed data $(x_1, y_1) \ldots (x_n, y_n)$, which are taken from a random vector (X, Y) following the FGMBHLD with the parameters μ_1, μ_2, σ_1, σ_2, and θ, with $\sigma = \sigma_1 = \sigma_2$, we consider the MLEs $\widehat{\mu}_1$, $\widehat{\mu}_2$, $\widehat{\sigma}$ and $\widehat{\theta}$ of these parameters, respectively. Then, based on Equation (35) and the invariance property, the MLE of R is obtained by substitution as

$$R_{MLE} = \frac{\widehat{\mu}_2}{\widehat{\mu}_1 + \widehat{\mu}_2} + \widehat{\theta}\,\widehat{\mu}_1\left(\frac{1}{2\,\widehat{\mu}_1 + \widehat{\mu}_2} - \frac{2}{\widehat{\mu}_1 + \widehat{\mu}_2} + \frac{1}{\widehat{\mu}_1 + 2\,\widehat{\mu}_2}\right). \tag{47}$$

6.2. Asymptotic Confidence Interval (ACI)

We now aim to compute the ACI for R with a large sample. Let $\Theta = (\mu_1, \mu_2, \sigma, \theta)$, and Θ_i be the i-th component of this vector. First, we construct the Fisher information matrix as follows:

$$I(\Theta) = I(\mu_1, \mu_2, \sigma, \theta) = \begin{bmatrix} I_{11} & \cdots & I_{14} \\ \vdots & \ddots & \vdots \\ I_{41} & \cdots & I_{44} \end{bmatrix}, \tag{48}$$

where $I_{ij} = -E\left[\frac{\partial^2 Ln\, L(X,Y)}{\partial\Theta_i\partial\Theta_j}\right]$, $i, j = 1, \ldots, 4$, Θ_i refereing to the ith component of Θ.

Second, we construct the variance–covariance matrix by replacing the distribution parameters by their MLEs, and we obtain

$$\widehat{V} = \begin{bmatrix} \widehat{V}_{11} & \cdots & \widehat{V}_{14} \\ \vdots & \ddots & \vdots \\ \widehat{V}_{41} & \cdots & \widehat{V}_{44} \end{bmatrix}, \tag{49}$$

where $\widehat{V}_{ij} = -\frac{\partial^2 Ln\, L}{\partial\Theta_i\partial\Theta_j}\Big|_{\Theta=\widehat{\Theta}}$, $i, j = 1, \ldots, 4$.

To obtain the ACI of R, the following two theorems are useful.

Theorem 1. *As $n \to \infty$, we have*

$$\left(\sqrt{n}(\widehat{\mu}_1 - \mu_1), \sqrt{n}(\widehat{\mu}_2 - \mu_2), \sqrt{n}(\widehat{\sigma} - \sigma), \sqrt{n}\left(\widehat{\theta} - \theta\right)\right) \to N_4(0, A^{-1}), \tag{50}$$

where $A = \dfrac{\widehat{V}}{n}$.

Proof. The theorem can be demonstrated using the asymptotic properties of MLEs of the distribution parameters under regularity conditions and the multivariate central limit theorem. $\square$

Theorem 2. *As $n \to \infty$, we have $\sqrt{n}\left(\widehat{R} - R\right) \to N(0, B)$, where $B = b^T A^{-1} b$,*

$$b = \left(\frac{\partial R}{\partial \mu_1}, \frac{\partial R}{\partial \mu_2}, \frac{\partial R}{\partial \sigma}, \frac{\partial R}{\partial \theta}\right). \tag{51}$$

Proof. The proof is based on Theorem 1 and the application of the delta method. $\square$

According to Xu and Long (2007) [15], a $100(1 - \alpha)\%$ ACI of R is

$$\left(\widehat{R} - Z_{\alpha/2}\sqrt{\frac{\widehat{B}}{n}}, \widehat{R} + Z_{\alpha/2}\sqrt{\frac{\widehat{B}}{n}}\right), \tag{52}$$

where $Z_{\alpha/2}$ denotes the value providing an area of $\frac{\alpha}{2}$ in the upper tail of the standard normal distribution, and $\widehat{B} = \widehat{b}^T A^{-1} \widehat{b}$, where $\widehat{b}$ is defined as Equation (51) with substitution of the unknown parameters by the corresponding MLEs.

7. Simulation

In this section, a Monte Carlo simulation study is introduced to describe the point and interval estimation of R.

7.1. Random Variate Generation

Nelsen (2006) [16] discussed the generation of a sample from a specified joint distribution using the conditional distribution method. In the setting of the FGMBHLD, it consists of the following steps:

(i)   Generate u and v independently from a uniform $(0, 1)$ distribution.
(ii)   Put $y_1 = \sigma_1 \ln(1 - 2(1 - u)^{\mu_1})$.
(iii)   Put $F(y_2|y_1) = v$ to find y_2 using numerical simulation.
(iv)   Repeat (i) to (iii) n-times to obtain (y_{1j}, y_{2j}), $j = 1, \ldots, n$.

The obtained n pair of values are thus generated values from (Y_1, Y_2) following the FGMBHLD.

7.2. Bootstrap Confidence Interval (BCI)

Efron (1982) [17] proposed the bootstrap percentile method (Boot-P) as follows:

(i)   Select the simple random sample (x_i, y_i), $i = 1, \ldots, n$.
(ii)   Re-sample the simple random sample (x_i, y_i) with replacement.
(iii)   Obtain the new simple random sample (x_i^*, y_i^*).
(iv)   Compute R^*.
(v)   Repeat step (i)–(iv) B-times and compute $R_1^*, \ldots, R_n^*$.
(vi)   Arrange $R_1^*, \ldots, R_n^*$, from the smallest to the largest $R_{(1)}^*, \ldots, R_{(n)}^*$.
(vii)   Construct a $100(1 - \alpha)\%$ ACI of R as

$$\left(R_{\frac{\alpha}{2}, B}^*, R_{\left(1 - \frac{\alpha}{2}\right), B}^*\right). \tag{53}$$

7.3. Experiment

1. Assume some true values of the parameters μ_1, μ_2, σ, θ and compute the corresponding true values of R.
   Case 1: If $\mu_1 = 0.5$, $\mu_2 = 1.5$, $\sigma = 1$, $\theta = -0.75$, then $R = 0.8678$.
   Case 2: If $\mu_1 = 0.5$, $\mu_2 = 1.5$, $\sigma = 1$, $\theta = -0.25$, then $R = 0.7892$.
   Case 3: If $\mu_1 = 0.5$, $\mu_2 = 1.5$, $\sigma = 1$, $\theta = 0.25$, then $R = 0.7107$.
   Case 4: If $\mu_1 = 0.5$, $\mu_2 = 1.5$, $\sigma = 1$, $\theta = 0.75$, then $R = 0.6321$.
2. Use the algorithm in Section 7.2 to generate different sample sizes with $n = 30, 50, 70$ and 100, with 10,000 replications. All computations are obtained using Mathematica 11.1.
3. Calculate R_{MLE} according to the methodology in Section 6.1 and the "average R_{MLE}", say R^*_{MLE}, based on all the samples at a fixed size.
4. Evaluate the ACI and BCI according to the methodology in Sections 6.2 and 7.2.
5. Study the behavior of R_{MLE} by evaluating the bias defined by the "average of $(R_{MLE} - R)$" and the mean square error (MSE) indicated as the "average of $(R_{MLE} - R)^2$".
6. In the context of interval estimation, we compare the ACI and BCI using the asymptotic confidence length (ACL) and converge probability (CP).

The results of the simulation study are presented in Table 1.

Table 1. Results of the Monte Carlo simulation study.

Sample Size	R_{true}	R^*_{MLE}	MLE		ACI		BCI	
			Bias	MSE	ACL	CP	ACL	CP
$\mu_1 = 0.5$, $\mu_2 = 1.5$, $\sigma = 1$, $\theta = -0.75$								
$n = 30$	0.8678	0.3960	−0.0228	0.0157	0.205	0.935	0.598	0.780
$n = 50$		0.6463	−0.0064	0.0020	0.474	0.831	0.838	0.690
$n = 70$		0.7841	−0.0084	0.0049	0.397	0.846	0.858	0.684
$n = 100$		0.3392	−0.0037	0.0014	0.223	0.980	0.503	0.819
$\mu_1 = 0.5$, $\mu_2 = 1.5$, $\sigma = 1$, $\theta = -0.25$								
$n = 30$	0.7892	0.1027	−0.0157	0.0074	0.792	0.856	0.511	0.818
$n = 50$		0.4670	−0.0044	0.0009	0.418	0.864	0.463	0.842
$n = 70$		0.1989	−0.0011	0.0001	0.251	0.926	0.524	0.810
$n = 100$		0.4110	−0.0052	0.0027	0.226	0.932	0.282	0.903
$\mu_1 = 0.5$, $\mu_2 = 1.5$, $\sigma = 1$, $\theta = 0.25$								
$n = 30$	0.7107	0.0858	−0.0208	0.0130	0.171	0.940	0.730	0.731
$n = 50$		0.3216	−0.0077	0.0030	0.367	0.861	0.806	0.693
$n = 70$		0.7090	−0.0001	0.0001	0.537	0.980	0.095	0.970
$n = 100$		0.6757	−0.0003	0.0001	0.099	0.967	0.126	0.960
$\mu_1 = 0.5$, $\mu_2 = 1.5$, $\sigma = 1$, $\theta = 0.75$								
$n = 30$	0.6321	0.6630	0.0010	0.0000	0.272	0.912	0.189	0.941
$n = 50$		0.5554	−0.0015	0.0001	0.308	0.894	0.194	0.922
$n = 70$		0.2641	−0.0052	0.0019	0.291	0.887	0.136	0.946
$n = 100$		0.6775	0.0004	0.0000	0.108	0.965	0.0931	0.970

From Table 1, we can conclude that:

1. At $n = 100$, the value of the MSE becomes very small.
2. In general, the length of the ACI becomes smaller than the length of the BCI.

3. When the ACL decreases, the CP increases.
4. The CP in almost all cases of the ACI is more than the CP in the BCI.

Hence, from the above results, the behavior of the MLEs is good for large samples. Moreover, the ACI is more suitable than the BCI for the stress–strength model.

8. Application: Household Financial Affordability in KSA 2018

In this section, we introduce a real application of the stress–strength model in an economic data setting, where X and Y represent household income and consumption, respectively. Here, $R = P(Y < X)$ is a household's financial affordability. We use the data from the household income and expenditure survey of KSA 2018. The survey period was from 28 February 2017 to 31 March 2018 in each month. In this study, we are interested in studying the behavior of R when X represents the average household monthly income by administrative region for Saudi households and Y represents the average household monthly consumption expenditure by administrative region for Saudi households, in order to measure the financial affordability for Saudi households by administrative region in 2018. The data are shown in Table 2. Table 3 presents the descriptive statistics for the data.

Table 2. Average household monthly income (X) and consumption expenditure (Y) by administrative region for Saudi households in 2018.

Administrative Region	Income	Consumption
Riyadh	16,011	15,917
Makkah	14,648	14,256
Madinah	12,016	118,322
Al-Qassim	15,322	14,371
Eastern Region	17,872	17,665
Asir	11,817	11,666
Tabuk	11,024	10,890
Hail	11,571	11,461
North Board	12,051	11,876
Jazan	15,199	15,071
Najran	11,388	11,376
Al-Baha	13,728	13,605
Aljouf	14,193	14,101
Total	14,823	14,584

Table 3. Descriptive statistics for the income and consumption.

Measure	Income	Consumption
Min	11,024	10,890
Max	17,872	17,665
Median	13,728	13,605
SE	592.605	574.401
Skewness	0.529	0.637
Kurtosis	−0.686	−0.378
Mean	13,603.076	13,391.307

To achieve our aim, we demonstrate the practicability of our proposed model. The Anderson–Darling (AD) goodness of fit statistic value is used to confirm that the GHLD is suitable for the income and consumption data; the corresponding p-values are almost equal to 1. Moreover, the quantile–quantile (Q–Q) plot is used to confirm this statement, as shown in Figure 2.

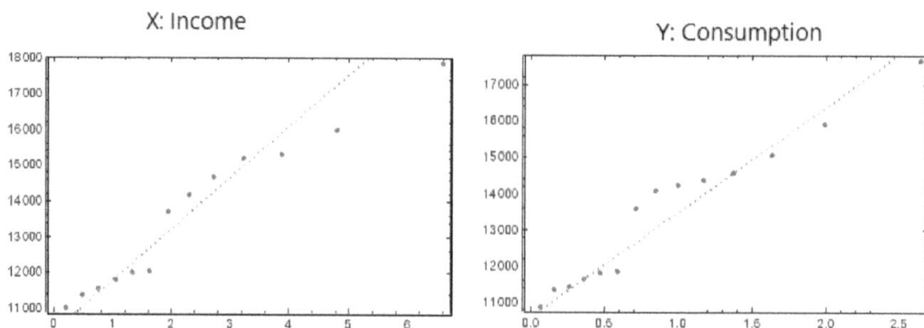

Figure 2. Q–Q plot for the income and consumption.

Now, we evaluate $R = P(Y < X)$ in the following two cases:

Case 1: If X and Y are independent with X following the GHLD(μ_1, σ) and Y following the GHLD(μ_2, σ), and the dependent parameter θ is set as 0;

Case 2: If X and Y are dependent with (X,Y) following the FGMBGHLD.

We calculate, in both cases, the MLEs of the distribution parameters and R, as well as the ACI and ACL. The results are shown in Table 4.

Table 4. The MLEs, ACIs, and ACLs of the distribution parameters for the income and consumption.

Case	MLE	MLE for R	ACI	ACL
Case 1	$\widehat{\mu}_1 = 0.0143$ $\widehat{\mu}_2 = 0.0270$ $\widehat{\sigma} = 0.3529$	0.3462	$(0.2979, 0.3945)$	0.0965
Case 2	$\widehat{\mu}_1 = 0.0135$ $\widehat{\mu}_2 = 0.0201$ $\widehat{\sigma} = 0.1403$ $\widehat{\theta} = 0.4713$	0.2149	$(0.1248, 0.3050)$	0.1802

From Table 4, we can conclude that:

1. Since θ is estimated as 0.4713, and is therefore positive, then the relation between X and Y is positive, as we see in Figure 3.
2. The measure of affordability when X and Y are dependent is less than when X and Y are independent, so the case of dependent variables is more realistic.

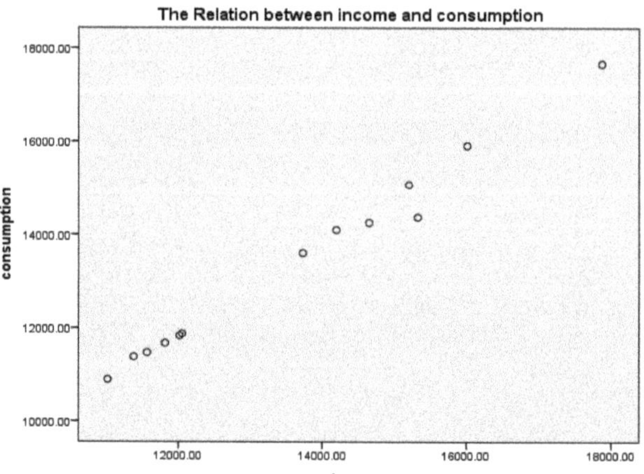

Figure 3. The scatter plot for the income and consumption of KSA, year 2018.

Finally, Figure 4 shows the (estimated) PDF and CDF of the FGMGBHLD with the estimated parameters from the considered data.

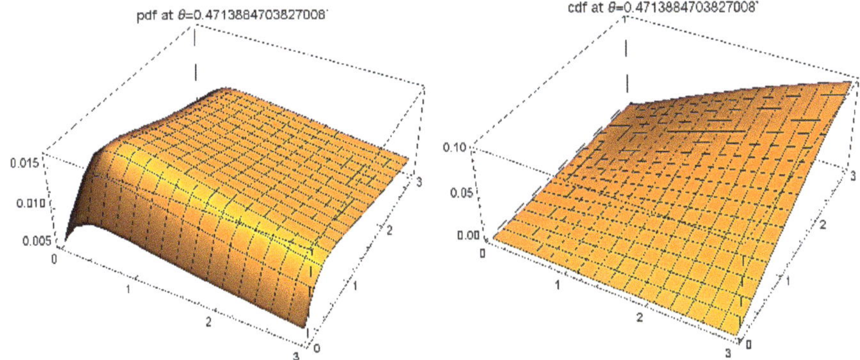

Figure 4. The estimated PDF and CDF of the FGMBGHLD for the income and consumption of KSA, year 2018.

It can be noted that the PDF seems unimodal (bump effect) with a long two-dimensional tail. With the FGMBGHLD, the equation behind the calculated PDF and CDF can be employed for further modeling.

To conclude this section, in order to show the performance of our new distribution on KSA data, we compare it with the bivariate Weibull distribution (BWD) as presented in Almetwally et al. (2020) [6]. First, we use the goodness of fit test and Q–Q plot to show that the BWD is a good fit to the KSA data. From the AD goodness of fit test, we find that the p-value equals 0.082 and 0.125 for the two considered data sets, respectively. As a result, the BWD fits the KSA data well. Figure 5 illustrates this conclusion.

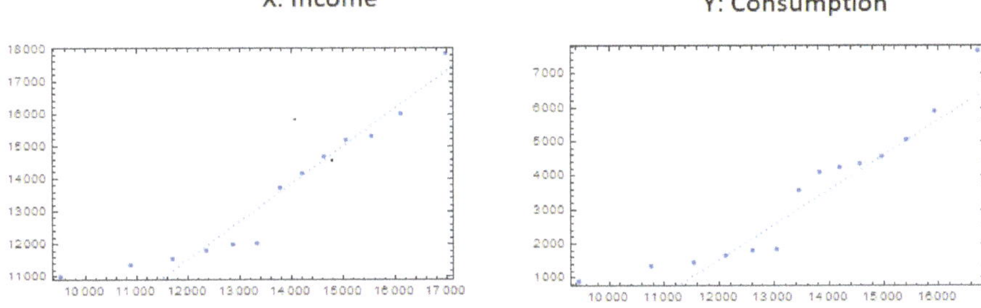

Figure 5. Q–Q plot for the BWD for the income and consumption of KSA, year 2018.

Now, we repeat our application but replace our proposed distribution by the BWD. Table 5 shows the result of the MLEs, R, ACIs, and ACLs of the distribution parameters and stress–strength model in the following two cases:

Case 1: If X and Y are independent with X following the $Weibull(\alpha_1, \beta)$ and Y following the $Weibull(\alpha_2, \beta)$;

Case 2: If X and Y are dependent with (X, Y) following the BWD.

Table 5. The MLEs, ACIs, and ACLs of the BWD parameters for the income and consumption of KSA, year 2018.

Case	MLE	MLE for R	ACI	ACL
Case 1	$\widehat{\alpha}_1 = 6.5$ $\widehat{\alpha}_2 = 7.5$ $\widehat{\beta} = 1.45$	0.4642	(0.4631, 0.6461)	0.1820
Case 2	$\widehat{\alpha}_1 = 6.5$ $\widehat{\alpha}_2 = 7.5$ $\widehat{\beta} = 1.45$ $\widehat{\theta} = 0.1082$	0.4275	(0.3456, 0.6058)	0.2602

From the ACL viewpoint, we can compare the performance of our distribution and the BWD on the KSA data. Thus, from Tables 4 and 5, we observe that the ACLs for our proposed distribution are lower than those of the BWD for both cases.

We complete this result by using the AD test for copula-based distributions as described in Genest et al. (2013) [18]. Table 6 shows the p-values of this AD test for our distribution and the BWD (dependent case for both).

Table 6. AD test for the proposed distribution and the BWD.

Distribution	p-Value
FGMBGHLD	0.4999
BWD	0.2067

The lower p-value is obtained for the FGMBGHLD distribution. Based on the results above, we can confirm that the proposed distribution is more suitable than the BWD for the considered KSA data.

9. Conclusions

In this paper, we introduced the bivariate distribution using the FGM copula approach, abbreviated as FGMBGHLD. We studied some of its statistical properties, such as the PDF, CDF, product moments, moment generating function, reliability function, and hazard rate function. In a multivariate statistical setting (and bivariate in particular), it is well known that the maximum likelihood estimation method gives unique estimates (under some regularity conditions) and guarantees their asymptotic performance from the unbiased and normality viewpoints. For these reasons, we developed it for the FGMBGHLD. We also applied the FGMBGHLD in a real-life data analysis scenario. We investigated the stress–strength model represented by R when the stress and strength variables are dependent and have the FGMBGHLD as a joint distribution. A simulated study was performed to study the behavior of the maximum likelihood estimate of R. Confidence intervals were constructed using two different techniques. Finally, we provided a real application of the considered (dependent) stress–strength model when X and Y measure the household financial affordability in KSA 2018 for Saudi households by administrative region. The obtained results are quite good and competitive with those of a valuable competitor (the bivariate Weibull distribution as introduced by [6]). Research perspectives include the application of the FGMBGHLD to more different bivariate data types, its multivariate version, and the development of regression model types.

Author Contributions: Conceptualization, M.K.H.H. and C.C.; methodology, M.K.H.H. and C.C.; software, M.K.H.H. and C.C.; validation, M.K.H.H. and C.C.; formal analysis, M.K.H.H. and C.C.; investigation, M.K.H.H. and C.C.; resources, M.K.H.H. and C.C.; data curation, M.K.H.H. and C.C.; writing—original draft preparation, M.K.H.H. and C.C.; writing—review and editing, M.K.H.H. and C.C.; visualization, M.K.H.H. and C.C. All authors have read and agreed to the published version of the manuscript.

Funding: This research received no external funding.

Data Availability Statement: Not applicable.

Acknowledgments: Not applicable.

Conflicts of Interest: The authors declare no conflict of interest.

References

1. Mudholkar, G.S.; Srivastava, D.; Freimer, M. Exponentiated Weibull family: A reanalysis of the bus-motor failure data. *Technometrics* **1995**, *37*, 436–445. [CrossRef]
2. Gupta, R.D.; Kundu, D. Generalized exponential distributions. *Aust. N. Z. J. Stat.* **1999**, *41*, 173–188. [CrossRef]
3. Olapade, A.K. On Type III Generalized Half Logistic Distribution. *arXiv* **2008**, arXiv:0806.1580v1.
4. Kantam, R.R.L.; Ramakrishna, V.; Ravikumar, M.S. Estimation and Testing in Type-II Generalized Half Logistic Distribution. *J. Mod. Appl. Stat. Methods* **2014**, *13*, 267–277. [CrossRef]
5. Balakrishnan, N. Order statistics from the half logistic distribution. *J. Stat. Comput. Simul.* **1985**, *20*, 287–309. [CrossRef]
6. Almetwally, E.M.; Muhammed, H.Z.; El-Sherpieny, E.S. Bivariate Weibull Distribution: Properties and Different Methods of Estimation. *Ann. Data Sci.* **2020**, *7*, 163–193. [CrossRef]
7. Almetwally, E.M.; Muhammed, H.Z. On a bivariate Fréchet distribution. *J. Stat. Appl. Probab. Lett.* **2020**, *9*, 71–91.
8. Muhammed, H.Z.; El-Sherpieny, E.S.; Almetwally, E.M. Dependency Measures For New Bivariate Models Based on Copula Function. *Inf. Sci. Lett.* **2021**, *10*, 511–526.
9. Sklar, A. Fonctions de répartition à *n* dimensions et leurs marges. *Publ. Inst. Stat. Univ. Paris* **1959**, *8*, 229–231.
10. Gumbel, E.J. Bivariate exponential distributions. *J. Am. Stat. Assoc.* **1960**, *55*, 698–707. [CrossRef]
11. Domma, F.; Giordano, S. A copula-based approach to account for dependence in stress-strength models. *Stat. Pap.* **2013**, *54*, 807–826. [CrossRef]
12. Osmetti, S.A.; Chiodini, P.M. A method of moments to estimate bivariate survival functions: The copula approach. *Statistica* **2011**, *71*, 469–488.
13. Basu, A.P. Bivariate failure rate. *J. Am. Stat. Assoc.* **1971**, *66*, 103–104. [CrossRef]
14. Al Turk, L.I.; Elaal, M.K.A.; Jarwan, R.S. Inference of bivariate generalized exponential distribution based on copula functions. *Appl. Math. Sci.* **2017**, *11*, 1155–1186. [CrossRef]
15. Xu, J.; Long, J.S. Using the Delta Method to Construct Confidence Intervals for Predicted Probabilities, Rates, and Discrete Changes. Available online: https://jslsoc.sitehost.iu.edu/stata/ci_computations/spost_deltaci.pdf (accessed on 18 July 2022).
16. Nelsen, R.B. *An Introduction to Copulas*; Springer: New York, NY, USA, 2006.
17. Efron, B. The Jackknife, the Bootstrap and Other Re-Sampling Plans. In *CBMS-NSF Reginal Conference Series in Applied Mathematics*; SIAM: Philadelphia, PA, USA, 1982; Volume 38.
18. Genest, C.; Huang, W.; Dufour, J.M. A regularized goodness-of-fit test for copulas. *J. Soc. Fr. Stat.* **2013**, *154*, 64–77.

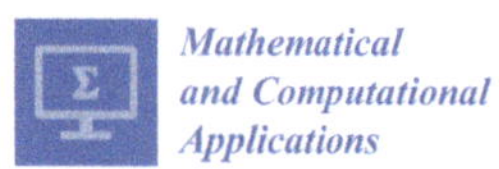

Mathematical and Computational Applications

Article

Flexible Parametric Accelerated Hazard Model: Simulation and Application to Censored Lifetime Data with Crossing Survival Curves

Abdisalam Hassan Muse [1,2,*], **Christophe Chesneau** [3,*], **Oscar Ngesa** [1,4] and **Samuel Mwalili** [1,5]

[1] Institute for Basic Sciences, Technology and Innovation (PAUSTI), Pan African University, Nairobi 62000-00200, Kenya

[2] Faculty of Science and Humanities, School of Postgraduate Studies and Research, Amoud University, Borama 25263, Somalia

[3] Department of Mathematics, LMNO, CNRS-Université de Caen, Campus II, Science 3, 14032 Caen, France

[4] Department of Mathematics and Physical Sciences, Taita Taveta University, Voi 635-80300, Kenya

[5] Department of Statistics and Actuarial Science, Jomo Kenyatta University of Agriculture and Technology (JKUAT), Nairobi 62000-00200, Kenya

* Correspondence: abdisalam.hassan@amoud.edu.so (A.H.M.); christophe.chesneau@unicaen.fr (C.C.)

Abstract: This study aims to propose a flexible, fully parametric hazard-based regression model for censored time-to-event data with crossing survival curves. We call it the accelerated hazard (AH) model. The AH model can be written with or without a baseline distribution for lifetimes. The former assumption results in parametric regression models, whereas the latter results in semi-parametric regression models, which are by far the most commonly used in time-to-event analysis. However, under certain conditions, a parametric hazard-based regression model may produce more efficient estimates than a semi-parametric model. The parametric AH model, on the other hand, is inappropriate when the baseline distribution is exponential because it is constant over time; similarly, when the baseline distribution is the Weibull distribution, the AH model coincides with the accelerated failure time (AFT) and proportional hazard (PH) models. The use of a versatile parametric baseline distribution (generalized log-logistic distribution) for modeling the baseline hazard rate function is investigated. For the parameters of the proposed AH model, the classical (via maximum likelihood estimation) and Bayesian approaches using noninformative priors are discussed. A comprehensive simulation study was conducted to assess the performance of the proposed model's estimators. A real-life right-censored gastric cancer dataset with crossover survival curves is used to demonstrate the tractability and utility of the proposed fully parametric AH model. The study concluded that the parametric AH model is effective and could be useful for assessing a variety of survival data types with crossover survival curves.

Keywords: Bayesian inference; hazard-based regression model; survival analysis; accelerated hazard model; generalized log-logistic distribution; crossover survival curves; censored data; maximum likelihood estimation.

Citation: Muse, A.H.; Chesneau, C.; Ngesa, O.; Mwalili, S. Flexible Parametric Accelerated Hazard Model: Simulation and Application to Censored Lifetime Data with Crossing Survival Curves. *Math. Comput. Appl.* **2022**, *27*, 104. https://doi.org/10.3390/mca27060104

Academic Editor: Sandra Ferreira

Received: 2 November 2022
Accepted: 29 November 2022
Published: 30 November 2022

Publisher's Note: MDPI stays neutral with regard to jurisdictional claims in published maps and institutional affiliations.

1. Introduction

In the analysis of lifetime data, hazard-based regression models have played a pivotal role. Such models produce a much more versatile framework for modeling survival data. They also make it conceivable to easily interpret the parameters from a practical perspective. When using regression models to analyze lifetime data, the Cox proportional hazard (PH) [1,2] model is the most widely assumed semi-parametric framework. The PH model's main assumption is that the hazard ratios are proportional over time. When such assumptions are not validated by data, alternative survival regression models, such as the accelerated failure time (AFT) [3,4], and proportional odds (PO) [5] models might be applied in the analysis. However, none of them are appropriate for capturing lifetime data with crossing survival and hazard curves [6].

This kind of issue is frequently associated with clinical trials, including control and treatment groups. The survival function (SF) of one group may degrade swiftly while the SF of the other group decays slowly. The curves tend to meet at some point, resulting in an inversion in terms of who is on the bottom/top. The study of this change is essential in many clinical studies because determining the crossing time reveals when the target treatment for an illness can be judged beneficial [6].

In practice, time-to-event data with crossing survival curves can occur for a variety of reasons. Crossing survival curves, according to Breslow [7], may occur when a treatment has an early rapid benefit and then becomes equally or worse than placebo treatment after such a time period. Additionally, as described in Diao et al. [8], crossing survival curves may occur in clinical studies when a particular intensive treatment (i.e., surgery) may have negative consequences at first but show good results in the long term.

Several techniques have been presented in the literature to handle this crossover feature in time-to-event data. The most often used are based on regression coefficients that change over time; see, for instance, Egge and Zahl [9], Putter et al. [10], Shyur et al. [11], and Zhang et al. [12]. Two recent works considering the modeling and analysis of time-to-event data with crossing survival curves are [6,13]. For this type of problem, Chen and Wang [14] developed a semi-parametric two-sample framework. The two-sample feature refers to a scenario in which there is a control, and a treatment group, which can be readily represented by a binary variable. The AH model is an intriguing choice because it formulates similarly to the PH and AFT models. In their model, they leave the baseline hazard rate function (HRF) undefined. As an alternative to the PO or AFT models, their model relaxes the proportional hazard assumption while still allowing for the inclusion of both time-independent and time-dependent factors.

Although they offered an exploratory visual examination of the model's suitability, they did not completely cover statistical model checking of the proposed model. Chen and Jewell [15] presented the AH model and its applicability to censored survival data. They used the AH model to analyze real data from a randomized clinical study of biodegradable carmustine polymers for the treatment of brain cancer. This analysis illustrated the model's useful applications and the recommended test statistics.

The semi-parametric AH model estimators, on the other hand, include the unknown distribution in the asymptotic variance. Thus, numerically demanding approaches are required to make an inference about this parameter. As a result, Lee [16] suggested a straightforward estimation method for the semi-parametric AH model in which estimators are asymptotically normal with a distribution-free asymptotic variance. This also yields several lack-of-fit tests. These tests are similar to Gill–Schumacher tests in that the estimating functions are assessed at two separate weight functions, generating two estimators that are close to each other. They demonstrated that the estimators and tests perform well for some weight functions using numerical experiments. For more information about the estimators and tests for the semi-parametric AH model, we refer to [17].

Cox [1] pioneered the use of semi-parametric hazard-based regression models for univariate time-to-event data with the PH model. Rubio et al. [18] and Khan [19] presented two influential papers that propose the use of extended lifetime distributions to substitute the baseline hazard in a time-to-event analysis. The formulation of parametric hazard-based regression models is a central issue in Lawless [20]. The authors explored the benefits of using parametric hazard-based regression models. It is noticed that the baseline-modified distribution should be chosen based on its flexibility to incorporate varied failure rate shapes. A few examples include: Muse et al. [21], Muse et al. [22], Ashraful-Ul-Alam and Khan [23], Alvares and Rubio [24], Muse et al. [25], Al-aziz et al. [26], and Khan and Khosa [27].

Despite the numerous advantages of the semi-parametric AH framework, its implementation in applications appears to be restricted, owing to the technical difficulties in implementing theoretical breakthroughs. Estimation for the covariance matrices is challenging when the data are censored because the asymptotic covariance matrices for the regression estimators in this model involve the unknown baseline HRF and its derivative.

However, censored data present a new technological barrier. Numerically demanding approaches, such as resampling techniques, can be used to approximate the covariance matrices. However, they are inefficient in actual settings due to their high computing cost [28].

The current study presents a fully parametric hazard-based regression model to fit the AH model to address the aforementioned concerns. The fundamental idea is to represent the baseline hazard by using a generalized log-logistic (GLL) distribution that is closed under both the AFT [25] and PH [22] frameworks and may incorporate various hazard rate shapes data including monotone and non-monotone shapes. Another advantage of the baseline is that it encompasses some of the most parametric distributions used in reliability and survival studies, such as log-logistic (LL), Burr XII with both 2-parameter and 3-parameter cases, Weibull, and exponential distributions. The shared tractability of parametric regression models and the adaptability of semi-parametric regression models is another appealing aspect of the suggested parametric AH model.

Thus, the main contribution of this study is to introduce and study a novel, flexible, parametric AH model to incorporate right-censored lifetime data with crossing survival curves. This is done by assuming the GLL lifetime distribution to deal with the baseline hazard in the parametric AH model. To the best of the author's knowledge, we emphasize that using the parametric AH model with GLL baseline distribution hazard to extend the original AH semi-parametric model has never been considered in the literature. The methods are studied by using the classical and Bayesian frameworks for a more comprehensive presentation of models for all statistical audiences to consider. A detailed simulation study is also being developed. This entails introducing one binary and one continuous covariate into the baseline hazard. The reader should be aware that the majority of the single covariate scenarios have been researched in prominent references, such as [8].

Additionally, the following are some significant benefits of the methodology proposed here.

i.   It possesses the adaptability of parametric survival regression models.
ii. It offers a continuous SF that makes it simple to find where two survival curves overlap.
iii. It allows different shapes for the HRF and has the tractability of a parametric survival regression model.

The following is a brief description of the sections that compose the article. Section 2 discusses the formulation of the parametric AH model and associated probabilistic functions. Section 3 presents the baseline distribution under consideration, as well as alternative competing lifetime distributions, including some of its special cases. The proposed parametric AH model with GLL baseline distribution HRF and its submodels are presented in Section 4. Section 5 discusses the model inferential procedures. Section 6 performs the simulation studies. Section 7 demonstrates a real-life, right-censored cancer dataset with crossed survival curves. Section 8 concludes the study with some farewell remarks and suggests future research.

2. AH Model Formulation

In order to handle lifetime data with crossing of hazard and survival curves, Chen and Weng [14] proposed a hazard-based regression model known as the AH model that is expressed as follows:

$$h(t;x) = h_0\big(t\psi(x'\beta)\big) = h_0\Big(te^{x'\beta}\Big), \tag{1}$$

where $\psi(x'\beta) = e^{x'\beta}$ is the link function of the explanatory variables, $x = (x_1, x_2, \ldots, x_p)$ is a vector of covariates, $\beta' = (\beta_1, \beta_2, \ldots, \beta_p)$ is a vector of coefficients of regression, and $h_0(t)$ corresponds to the baseline hrf.

In this model, $e^{x'\beta}$ characterizes how the explanatory variables into x change the time scale of the underlying HRF. In this case, $\beta < 0$ or $\beta > 0$ imply deceleration or acceleration of the HRF's time scale, respectively. For example, if one explanatory variable has a value of 1 for a treatment group and a value of 0 for a control group, then $e^{\beta} = \frac{1}{2}$ indicates that the HRF of the treatment group advances in half the time as those in the control group. The same is true for $e^\beta = 2$, which indicates that the HRF of the treatment group advances twice as quickly as those in the control group. There are no differences between the groups, according to $e^\beta = 1$.

The AH model offers some appealing and intriguing characteristics. The AH model, unlike the AFT and PH models, can handle the crossing of survival and hazard curves [29]. Furthermore, the AH framework enables both the control and treatment groups' hazard curves to begin at the same time point. This is especially beneficial in randomized controlled trials, because it is more reasonable to hypothesize that the hazard or risk between groups is comparable at $t = 0$ [30].

The inability of the parametric AH model to incorporate situations where the HRF is constant over time is a limitation that is not shared by the AFT and PH models (e.g., exponential distribution) [28]. As a result, before implementing this model, it is crucial to assess the non-constancy of the baseline function. The AH model, like the AFT and PH models, has coincidences when the baseline HRF is a Weibull distribution [31].

As an alternative, the cumulative hazard function (CHF) can be used to represent the parametric AH model as follows:

$$H(t;x) = H_0\left(te^{x'\beta}\right)e^{-x'\beta},\tag{2}$$

where $H_0(t)$ denotes the baseline CHF.

The other probabilistic functions for the parametric AH model, associated with Equation (2), can be expressed as follows.

The sf for the parametric AH model is

$$S(t;x) = \left[S_0\left(te^{x'\beta}\right)\right]^{e^{-x'\beta}},\tag{3}$$

where $S_0(t)$ denotes the baseline SF. The cumulative distribution function (CDF) for the parametric AH model is

$$F(t;x) = 1 - \left[S_0\left(te^{x'\beta}\right)\right]^{e^{-x'\beta}}.\tag{4}$$

The probability density function (PDF) for the parametric AH model is

$$f(t;x) = f_0\left(te^{x'\beta}\right)\left[S_0\left(te^{x'\beta}\right)\right]^{e^{-x'\beta}},\tag{5}$$

where $f_0(t)$ denotes the baseline PDF.

3. Baseline Hazard

Standard parametric models using several prominent survival distributions are commonly used in survival data analysis. The LL distribution is one of the most commonly utilized in oncology research, owing to the flexibility of its HRF and the ability to estimate its parameters. We frequently have datasets in medical research that demand more advanced parametric models. To do this, the literature has introduced new classes of parametric distributions based on the modification of the LL distribution. Specific situations include the GLL distribution [32], Kumaraswamy LL (KuLL) distribution [33], heavy-tailed LL (HTLL) distribution [34], tan LL (TLL) distribution [35], a novel LL (NLL) distribution [36], arctan LL distribution [37], and an extended LL (ELL) distribution [38], among others [39].

For fully parametric hazard-based regression models, we must assume a parametric form for the baseline, of which there are an infinite number of options, and which one is appropriate will generally depend on the situation. We analyze a general-purpose candidate, the chosen GLL distribution presented by Khan and Khosa [27], in this paper. The GLL distribution is constructed by using the AH framework, and it is then contrasted with various baseline hazards that can take into account different hazard rate shapes as well as some of its special case distributions.

The HRF and the CHF of the GLL distribution are expressed as follows:

$$h_{GLL}(t;\theta) = \frac{\alpha k(kt)^{\alpha-1}}{1+(\eta t)^{\alpha}}, \quad t \geq 0, \quad k,\alpha,\eta > 0, \tag{6}$$

$$H_{GLL}(t;\theta) = \frac{k^{\alpha}}{\eta^{\alpha}}\log[1+(\eta t)^{\alpha}], \quad t \geq 0, \quad k,\alpha,\eta > 0, \tag{7}$$

where θ represents the vector of the involved parameters.

The HRF in Equation (6) consists of different submodels of the GLL distribution [32]. These distributions are listed as follows:

Log-logistic (LL): when $k = \eta$, Equation (6) reduces to the hrf of the LL distribution, which is

$$h_{LL}(t;\theta) = \frac{\alpha k(kt)^{\alpha-1}}{1+(kt)^{\alpha}}, \quad t \geq 0, \quad k,\alpha > 0. \tag{8}$$

Burr-XII (BXII): when $\eta = 1$, equation (6) reduces to the hrf of the BXII-2 distribution, which is

$$h_{BXII}(t;\theta) = \frac{\alpha k(kt)^{\alpha-1}}{1+t^{\alpha}}, \quad t \geq 0, \quad k,\alpha > 0. \tag{9}$$

Weibull (W): when $\eta \to 0$, Equation (6) reduces to the hrf of the W distribution, which is

$$h_{W}(t;\theta) = \alpha k(kt)^{\alpha-1}, \quad t \geq 0, \quad k,\alpha > 0. \tag{10}$$

In this work, we compare the proposed baseline hazard to its submodels as well as three additional baseline hazard candidates that can be incorporated for both monotone and nonmonotone hazard rate shapes: the power generalized Weibull (PGW) model [40], exponentiated Weibull (EW) model [41], and the generalized gamma (GG) model [42]. The corresponding distributions have comparable levels of adaptability and tractability. The following are the HRF functions for the PGW, EW, and GG distributions, respectively:

$$h_{PGW}(t;\theta) = \frac{\alpha}{\eta k^{\alpha}}t^{\alpha-1}\left[1+\left(\frac{t}{k}\right)^{\alpha}\right]^{\left(\frac{1}{\eta}-1\right)}, \quad t \geq 0, \quad k,\alpha,\eta > 0, \tag{11}$$

$$h_{GG}(t;\theta) = \frac{\frac{\eta}{\Gamma\left(\frac{\alpha}{\eta}\right)k^{\alpha}}t^{\alpha-1}e^{-\left(\frac{t}{k}\right)^{\eta}}}{1-\frac{\gamma\left(\frac{\alpha}{\eta},\left(\frac{t}{k}\right)^{\eta}\right)}{\Gamma\left(\frac{\alpha}{\eta}\right)}}, \quad t \geq 0, \quad k,\alpha,\eta > 0, \tag{12}$$

where $\gamma(t,x)$ and $\Gamma(x)$ denote the incomplete and complete gamma functions, respectively, and

$$h_{EW}(t;\theta) = \frac{\alpha k\eta(kt)^{\alpha-1}\left[1-e^{-(kt)^{\alpha}}\right]^{\eta-1}e^{e^{-(kt)^{\alpha}}}}{1-\left[1-e^{-(kt)^{\alpha}}\right]^{\eta}}, \quad t \geq 0, \quad k,\alpha,\eta > 0. \tag{13}$$

We also used the gamma (G) and log-normal (LN) distributions, two additional popular classical distributions used in survival and reliability research.

4. The Proposed Model

There are several approaches to expressing parametric hazard-based regression models. The AH model formulation is one such strategy. The GLL hazard-based regression model can be written in the context of the AH framework by substituting the exponential function for the link function in Equation (1). We recall that the HRF under the AH framework is computed as follows:

$$h(t) = h_0\left(te^{x'\beta}\right).$$

We begin with the GLL baseline distribution HRF with parameters α, η, and k (with the AH model notations). The HRF with an explanatory variable vector x is as follows:

$$h(t; \theta, \beta, x) = h_0\left(te^{x'\beta}; \theta\right) = \frac{\alpha k(kt^*)^{\alpha-1}}{1+(\eta t^*)^{\alpha}}, \tag{14}$$

which is the GLL HRF with $t^* = te^{x'\beta}$ once more. In addition, the other survival probabilistic functions for the GLL–AH framework are expressed as follows.

The SF for the GLL–AH model is

$$S(t; \theta, \beta, x) = \left[S_0\left(te^{x'\beta}; \theta\right)\right]^{e^{-x'\beta}} = \left[1 + \left(\eta te^{x'\beta}\right)^{\alpha}\right]^{\frac{k^\alpha e^{-x'\beta}}{\eta^\alpha}}. \tag{15}$$

The CDF for the GLL–AH model is

$$F(t; \theta, \beta, x) = 1 - \left[S_0\left(te^{x'\beta}; \theta\right)\right]^{e^{-x'\beta}} = 1 - \left[1 + \left(\eta te^{x'\beta}\right)^{\alpha}\right]^{\frac{k^\alpha e^{-x'\beta}}{\eta^\alpha}}. \tag{16}$$

The CHF for the GLL–AH model is

$$H(t; \theta, \beta, x) = H_0\left(te^{x'\beta}; \theta\right)e^{-x'\beta} = \left(\frac{k^\alpha}{\eta^\alpha}\log\left[1 + \left(\eta te^{x'\beta}\right)^{\alpha}\right]\right)e^{-x'\beta}. \tag{17}$$

The PDF for the GLL–AH model is

$$f(t; \theta, \beta, x) = f_0\left(te^{x'\beta}; \theta\right)\left[S_0\left(te^{x'\beta}; \theta\right)\right]^{e^{-x'\beta}} = \frac{\alpha k\left(kte^{x'\beta}\right)^{\alpha-1}}{\left[1 + \left(\eta te^{x'\beta}\right)^{\alpha}\right]^{\frac{k^\alpha}{\eta^\alpha}+1}}\left[1 + \left(\eta te^{x'\beta}\right)^{\alpha}\right]^{\frac{k^\alpha e^{-x'\beta}}{\eta^\alpha}}. \tag{18}$$

4.1. Submodels

The proposed parametric hazard-based GLL–AH model framework has three submodels that are also closed under the AH framework.

4.1.1. Submodel I: $\eta = 1$

If we put $\eta = 1$ in Equation (14), we get the HRF of the BXII–AH model, which is expressed mathematically as

$$h(t; x) = \frac{\alpha k\left(kte^{x'\beta}\right)^{\alpha-1}}{1 + \left(te^{x'\beta}\right)^{\alpha}}. \tag{19}$$

4.1.2. Submodel II: $\eta = k$

If we put $\eta = k$ in Equation (14), we are referred to the HRF of the LL–AH model, which is stated mathematically as

$$h(t; x) = \frac{\alpha k\left(kte^{x'\beta}\right)^{\alpha-1}}{1 + \left(kte^{x'\beta}\right)^{\alpha}}. \tag{20}$$

4.1.3. Submodel III: $\eta^\alpha \to 0$.

If we put $\eta^\alpha \to 0$ in Equation (14), we are referred to the HRF of the W–AH model, which is stated mathematically as

$$h(t;x) = \alpha k \left(kte^{x'\beta} \right)^{\alpha-1}. \tag{21}$$

5. Inferential Procedures

In this section, the parameters of the proposed parametric AH model with GLL baseline distribution HRF are estimated by using a classical approach (via the maximum likelihood estimation (MLE) method) and Bayesian inference using noninformative priors.

5.1. Classical Approach

We are concerned in this subsection with a full likelihood function for the proposed parametric AH model. The likelihood function is an important component not only in the Bayesian approach but also in classical inference, in which the standard approach for estimating parameters involves maximizing it. Consider both noninformative and independent (right) censorship.

Suppose there are n individuals with survival times denoted by $T_1, T_2, \ldots, T_n$. Assuming that the data are subject to right censoring, we observe $t_i = \min(T_i, RC_i)$, where $RC_i > 0$ being the censoring time for individual i. Letting $\delta_i = I(T_i \le RC_i)$ that equals 1 if $T_i \le RC_i$ and 0 otherwise, the observed data for individual i consists of $\{t_i, \delta_i, x_i\}, i = 1, 2, \ldots, n$, where t_i is a survival time or censoring time according to whether $\delta_i = 1$ or 0, respectively, and $x_i = (x_{i1}, x_{i2}, \ldots, x_{ip})'$ is a $p \times 1$ column vector of external covariates.

When considering a parametric AH model, the censored likelihood function can be written as follows:

$$
\begin{aligned}
L(\theta, \beta; D) &= \prod_{i=1}^{n} [f(t_i; \theta, \beta, x_i)]^{\delta_i} [S(t_i; \theta, \beta, x_i)]^{1-\delta_i} \\
&= \prod_{i=1}^{n} \left[\frac{h(t_i; \theta, \beta, x_i)}{S(t_i; \theta, \beta, x_i)} \right]^{\delta_i} [S(t_i; \theta, \beta, x_i)]^{1-\delta_i} \\
&= \prod_{i=1}^{n} [h(t_i; \theta, \beta, x_i)]^{\delta_i} S(t_i; \theta, \beta, x_i) \\
&= \prod_{i=1}^{n} [h(t_i; \theta, \beta, x_i)]^{\delta_i} \exp[-H(t_i; \theta, \beta, x_i)] \\
&= \prod_{i=1}^{n} \left[h_0\left(t_i e^{x_i'\beta}; \theta \right) \right]^{\delta_i} \exp\left[-H_0\left(t_i e^{x_i'\beta}; \theta \right) e^{-x_i'\beta} \right],
\end{aligned} \tag{22}
$$

where $D = (t_i, \delta_i, x_i, i = 1, 2, \ldots, n)$ represents the observed data, which includes survival times, censoring time, and covariates. In our expression, we recall that θ is the vector of baseline distributional parameters, and β is the regression coefficients. An iterative optimization approach can be used to produce the MLE (e.g., the Newton–Raphson algorithm). Because the MLEs are approaching normalcy, various hypothesis tests and interval constructions of model parameters are conceivable.

The log-likelihood function is expressed as follows:

$$\ell(\theta, \beta; D) = \sum_{i=1}^{n} \delta_i \log\left[h_0\left(te^{x_i'\beta}; \theta \right) \right] - \sum_{i=1}^{n} H_0\left(t_i e^{x_i'\beta}; \theta \right) e^{-x_i'\beta}. \tag{23}$$

The GLL–AH model's full log-likelihood function is expressed as follows:

$$\ell(\theta, \beta; D) = \sum_{i=1}^{n} \delta_i \log(\alpha) + \sum_{i=1}^{n} \delta_i \alpha \log(k) + (\alpha - 1) \sum_{i=1}^{n} \delta_i \log\left(t_i e^{x_i'\beta}\right)$$
$$- \sum_{i=1}^{n} \delta_i \log\left[1 + \left(\eta t_i e^{x_i'\beta}\right)^{\alpha}\right] - \left(\frac{k}{\eta}\right)^{\alpha} \sum_{i=1}^{n} e^{-x_i'\beta} \log\left[1 + \left(\eta t_i e^{x_i'\beta}\right)^{\alpha}\right]. \tag{24}$$

To obtain the MLE of $\theta' = (k, \alpha, \eta)$, and β, we can directly maximize Equation (24) with respect to (k, α, η), and β. Alternatively, we can express the first derivative of the log-likelihood function in order to solve the nonlinear equations below for the log-likelihood function's first derivative.

With this aim, let us set $\varphi = (k, \alpha, \eta, \beta)$. Then the first derivatives of the log-likelihood functions are as follows:

$$\frac{\partial \ell(\varphi)}{\partial \alpha} = \frac{1}{\alpha} \sum_{i=1}^{n} \delta_i + \sum_{i=1}^{n} \delta_i \log(k) + \sum_{i=1}^{n} \delta_i \log\left(t_i e^{x_i'\beta}\right)$$
$$- \sum_{i=1}^{n} \delta_i \frac{\left(\eta t_i e^{x_i'\beta}\right)^{\alpha} \log\left(\eta t_i e^{x_i'\beta}\right)}{1 + \left(\eta t_i e^{x_i'\beta}\right)^{\alpha}}$$
$$- \left(\frac{k}{\eta}\right)^{\alpha} \log(k) \sum_{i=1}^{n} e^{-x_i'\beta} \log\left[1 + \left(\eta t_i e^{x_i'\beta}\right)^{\alpha}\right] \tag{25}$$
$$+ \left(\frac{k}{\eta}\right)^{\alpha} \log(\eta) \sum_{i=1}^{n} e^{-x_i'\beta} \log\left[1 + \left(\eta t_i e^{x_i'\beta}\right)^{\alpha}\right]$$
$$- \left(\frac{k}{\eta}\right)^{\alpha} \sum_{i=1}^{n} \frac{e^{-x_i'\beta} \left(\eta t_i e^{x_i'\beta}\right)^{\alpha} \log\left(\eta t_i e^{x_i'\beta}\right)}{1 + \left(\eta t_i e^{x_i'\beta}\right)^{\alpha}},$$

$$\frac{\partial \ell(\varphi)}{\partial \eta} = -\left(\frac{\alpha}{\eta}\right) \sum_{i=1}^{n} \delta_i \frac{\left(\eta t_i e^{x_i'\beta}\right)^{\alpha}}{1 + \left(\eta t_i e^{x_i'\beta}\right)^{\alpha}}$$
$$+ \left(\frac{\alpha}{\eta}\right) \left(\frac{k}{\eta}\right)^{\alpha} \sum_{i=1}^{n} e^{-x_i'\beta} \log\left[1 + \left(\eta t_i e^{x_i'\beta}\right)^{\alpha}\right] \tag{26}$$
$$- \left(\frac{\alpha}{\eta}\right) \left(\frac{k}{\eta}\right)^{\alpha} \sum_{i=1}^{n} \frac{e^{-x_i'\beta} \left(\eta t_i e^{x_i'\beta}\right)^{\alpha}}{1 + \left(\eta t_i e^{x_i'\beta}\right)^{\alpha}},$$

$$\frac{\partial \ell(\varphi)}{\partial k} = \left(\frac{\alpha}{k}\right) \sum_{i=1}^{n} \delta_i - \left(\frac{\alpha}{k}\right) \left(\frac{k}{\eta}\right)^{\alpha} \sum_{i=1}^{n} e^{-x_i'\beta} \log\left[1 + \left(\eta t_i e^{x_i'\beta}\right)^{\alpha}\right] \tag{27}$$

and

$$\frac{\partial \ell(\varphi)}{\partial \beta_j} = (\alpha - 1) \sum_{i=1}^{n} \delta_i x_{ij} - \alpha \sum_{i=1}^{n} \delta_i x_{ij} \frac{\left(\eta t_i e^{x_i'\beta}\right)^{\alpha}}{1 + \left(\eta t_i e^{x_i'\beta}\right)^{\alpha}}$$
$$+ \left(\frac{k}{\eta}\right)^{a} \sum_{i=1}^{n} x_{ij} \log\left[1 + \left(\eta t_i e^{x_i'\beta}\right)^{\alpha}\right]. \tag{28}$$

5.2. Bayesian Approach

In this subsection, the prior distributions for the parameters of the proposed model are first established, and these distributions are then multiplied by the likelihood function to create the Bayesian model.

5.2.1. Prior Distribution

The formulation of a prior distribution is a crucial step in every Bayesian approach. This is especially true for fully parametric survival regression models. Because we lack prior knowledge from historical data or from prior experiments, we set the prior scenario in this study using a noninformative independent gamma distribution, Gamma $(10, 10)$, as the baseline distribution parameters. Gamma distributions are flexible and include noninformative priors (uniform) and the marginal priors distribution for each regression coefficient is taken as a normal distribution centered at zero with a wide known variance $(0, 100)$. Numerous study articles in the literature, such as [19,22,24–26,43], take these priors into account. Here, we consider

$$\pi(\alpha) \sim G(a_1, b_1) = \frac{b_1^{a_1}}{\Gamma(a_1)} \alpha^{a_1 - 1} e^{-b_1 \alpha}; a_1, b_1, \alpha > 0, \tag{29}$$

$$\pi(\eta) \sim G(a_2, b_2) = \frac{b_2^{a_2}}{\Gamma(a_2)} \eta^{a_2 - 1} e^{-b_2 \eta}; a_2, b_2, \eta > 0, \tag{30}$$

$$\pi(k) \sim G(a_3, b_3) = \frac{b_3^{a_3}}{\Gamma(a_3)} k^{a_3 - 1} e^{-b_3 k}; a_3, b_3, k > 0. \tag{31}$$

From historical data of the baseline distribution, it is simple to determine the hyperparametric values of the prior distributions [32]. When the explanatory variables are assumed to have a prior normal distribution, we have the following regression coefficients:

$$\pi(\beta') \sim N(a_4, b_4). \tag{32}$$

The joint prior distribution of all unknown parameters has a PDF given by

$$\pi(\alpha, k, \eta, \beta') = \pi(\alpha)\pi(\eta)\pi(k)\pi(\beta'). \tag{33}$$

5.2.2. Likelihood Function

The likelihood function for the GLL general hazard model is computed as follows:

$$L_{GLL-AH}(\theta, \beta; D) = = \prod_{i=1}^{n} \left[h_0\left(t_i e^{x_i' \beta}; \theta\right) \right]^{\delta_i} \exp\left[-H_0\left(t_i e^{x_i' \beta}; \theta\right) e^{-x_i' \beta} \right]$$

$$= \prod_{i=1}^{n} \left[\frac{\alpha k \left(k t_i e^{x_i' \beta}\right)^{\alpha-1}}{1 + \left(\eta t_i e^{x_i' \beta}\right)^{\alpha}} \right]^{\delta_i} \tag{34}$$

$$\exp\left[-\left\{ \frac{k^{\alpha}}{\eta^{\alpha}} \log\left[1 + \left(\eta t_i e^{x_i' \beta}\right)^{\alpha} \right] \right\} e^{-x_i' \beta} \right].$$

5.2.3. Posterior Distribution

The joint posterior PDF is expressed as the multiplication of the likelihood function in Equation (34) and the prior distribution in Equation (33):

$$p(\alpha, k, \eta, \beta; t) \propto \prod_{i=1}^{n} \left[\frac{\alpha k \left(k t_i e^{x_i' \beta}\right)^{\alpha-1}}{1 + \left(\eta t_i e^{x_i' \beta}\right)^{\alpha}} \right]^{\delta_i}$$

$$\exp\left[-\left\{ \frac{k^{\alpha}}{\eta^{\alpha}} \log\left[1 + \left(\eta t_i e^{x_i' \beta}\right)^{\alpha} \right] \right\} e^{-x_i' \beta} \right] \tag{35}$$

$$\times \pi(\alpha, k, \eta, \beta'),$$

where the prior specification for the unknown parameters is represented by the first four terms on the right-hand side of the equation.

The joint posterior PDF is analytically intractable because of how challenging it is to integrate. Therefore, the inference can be supported by the Markov chain Monte Carlo (McMC) simulation methods, including the Gibbs sampler and Metropolis–Hastings algorithms, which can be used to generate samples from which features of the relevant marginal distributions can be inferred.

6. Simulation Study

In this section, we offer a thorough Monte Carlo (MC) simulation analysis to assess how well the suggested model performs in terms of estimating the parameters of the baseline distribution and the regression coefficients. There are two inferential techniques used in the analysis.

I.   Procedure I: An MLE estimate technique.
II. Procedure II: A Bayesian estimation technique with independent gamma priors for the baseline distribution parameters and a normal prior for the regression coefficients, as well as non-informative priors.

Two explanatory variables in an AH regression framework were considered in all simulations: one binary covariate (x_1) generated from Bernoulli (0.5) distribution and one continuous covariate (x_2) generated from the standard normal distribution. Regression parameter values were chosen to be $\beta = (0.75, 0.5)$ corresponding to the covariate vector $x = (x_1, x_2)'$.

The GLL baseline distribution hazard was used to generate the survival data, and the exponential distribution with a rate parameter equal to the censoring proportion of 10% was used to generate the censoring times.

We were particularly interested in the performance and accuracy of the proposed model's estimators in the simulation exercise, specifically the bias, standard error, and mean square error. The simulation's findings were derived from 500 replications with 50, 100, 300, and 500 samples for each parameter value. The results are shown in Table 1, which includes the mean estimate (est), standard error (SE), average bias (AB), mean square error (MSE), and coverage probability for the MLE estimates for both inferential techniques. The estimates' averages are extremely close, and generally, the AB and MSE are less as sample size rises. Additionally, as sample sizes are increased, estimates for all evaluated parameters perform better. We also note that, compared to MLE estimates, Bayesian estimates have a lower SE.

Similar results were obtained from a simulation analysis with around 20% censored observations for each dataset (data not shown). In conclusion, our simulation work has shown that the suggested parametric AH model may prove to be a highly helpful parametric hazard-based regression model to accurately represent survival data with or without crossover survival curves.

Table 1. Simulation study for GLL–AH regression model. True values (True), Estimates (Est.), standard error (SE), average bias (AB), mean square error (MSE), and coverage probability (CP 95%) are presented for the parameters.

	True	Est.	SE	AB	MSE	CP	Est.	SE	AB	MSE	$\hat{R}$
					Set I $n = 50$						
M_2				MLE Approach				Bayesian			
β_1	0.75	0.800	0.100	0.050	0.037	93.85	0.790	0.002	0.040	0.036	1.002
β_2	0.5	0.558	0.042	0.058	0.024	94.50	0.512	0.003	0.012	0.011	1.002
α	1.50	1.590	0.010	0.090	0.008	95.20	1.505	0.001	0.005	0.003	1.000
k	0.75	0.900	0.435	0.150	0.063	92.05	0.850	0.005	0.100	0.045	1.002
η	1.20	1.265	0.011	0.065	0.046	94.25	1.212	0.000	0.012	0.004	1.003

Table 1. *Cont.*

	True	Est.	SE	AB	MSE	CP	Est.	SE	AB	MSE	$\hat{R}$
					Set II $n = 100$						
M_2				MLE approach				Bayesian			
β_1	0.75	0.790	0.100	0.040	0.036	94.10	0.770	0.001	0.020	0.018	1.000
β_2	0.5	0.530	0.030	0.030	0.024	94.80	0.510	0.002	0.010	0.010	1.001
α	1.50	1.610	0.040	0.110	0.087	93.40	1.553	0.001	0.053	0.041	1.003
k	0.75	0.850	0.250	0.100	0.056	93.20	0.800	0.004	0.050	0.037	1.002
η	1.20	1.250	0.008	0.050	0.034	94.80	1.205	0.000	0.005	0.003	1.001
					Set III $n = 300$						
	True	Est.	SE	AB	MSE	CP	Est.	SE	AB	MSE	$\hat{R}$
M_2				MLE approach				Bayesian			
β_1	0.75	0.78	0.092	0.030	0.032	94.40	0.768	0.001	0.018	0.016	1.000
β_2	0.5	0.525	0.013	0.025	0.021	93.90	0.503	0.001	0.003	0.002	1.000
α	1.50	1.592	0.021	0.042	0.030	93.85	1.506	0.001	0.006	0.006	1.001
k	0.75	0.844	0.212	0.094	0.049	93.46	0.798	0.003	0.048	0.036	1.000
η	1.20	1.252	0.008	0.052	0.034	94.60	1.205	0.000	0.005	0.003	1.001
	True	Est.	SE	AB	MSE	CP	Est.	SE	AB	MSE	$\hat{R}$
					Set IV $n = 500$						
M_2				MLE approach				Bayesian			
β_1	0.75	0.775	0.065	0.025	0.017	95.10	0.752	0.000	0.002	0.002	1.000
β_2	0.5	0.526	0.013	0.026	0.021	94.00	0.503	0.001	0.003	0.002	1.000
α	1.50	1.550	0.040	0.050	0.037	94.70	1.503	0.001	0.003	0.001	1.000
k	0.75	0.825	0.110	0.075	0.048	94.07	0.780	0.003	0.030	0.027	1.001
η	1.20	1.205	0.005	0.005	0.003	95.04	1.203	0.000	0.003	0.001	1.001

7. Applications

This section examines a right-censored dataset from an oncology clinical trial with crossover survival curves to show how the proposed parametric AH model can be used to model lifetime data with crossing survival curves. First, the Rstan package's Bayesian analysis of the AH model and its competing models, such as the PH, PO, and AFT models, is provided. After performing a traditional analysis with the MLE technique, add model comparison. Next, by using a frequentist estimation approach, regression analyses were conducted by using the proposed baseline hazard (GLL), power generalized Weibull (PGW), generalized gamma (GG), exponentiated Weibull (EW), log-logistic (LL), and Weibull (W) distributions as a baseline to AH models, and the fits were compared by using information criteria (Akaike information criterion (AIC), Consistent AIC (CAIC), and Hannan–Quinn information criterion (QIC)). The GLL–AH and its submodels are then used to do a Bayesian analysis.

7.1. Gastric Cancer Dataset

We look at the Gastrointestinal Tumor Study Group's gastric cancer data collection (1982). This dataset has frequently been used in studies involving crossing survival curves, particularly in the field related to survival analysis. A few instances include Demarqui and Mayrink [6] and Diao et al. [8]. The dataset is freely accessible under the label "gastric" by using the R package AmoudSurv [44].

This oncology clinical trial includes 90 patients who have been diagnosed with locally advanced gastric cancer. The patients were randomly assigned to the following groups: (i) a control group, which included 45 patients who got chemotherapy; and (ii) a treatment group, which included 45 patients who received radiation therapy along with chemotherapy. In this study, these patients were followed for around 5 years. For each patient, three variables are reported in the datasets: the response time, which indicates failure (time to death) or right censoring (the censoring proportion in this data set is around 12.22%), a binary failure indicator, which identifies patients who experienced the event of interest, and a group binary indicator with 1, indicating the type of treatment.

Figure 1 shows the overall survival curve for the gastric cancer dataset as well as the survival curves for the two types of therapies (chemotherapy vs. chemotherapy mixed with radiotherapy) used to treat locally unresectable gastric cancer. Close inspection reveals crossovers and crossings between the curves, which supports the AH model's efficacy and suitability for this data analysis. The fundamental non-parametric plots for the survival time of the gastric cancer dataset are presented in Figure 2.

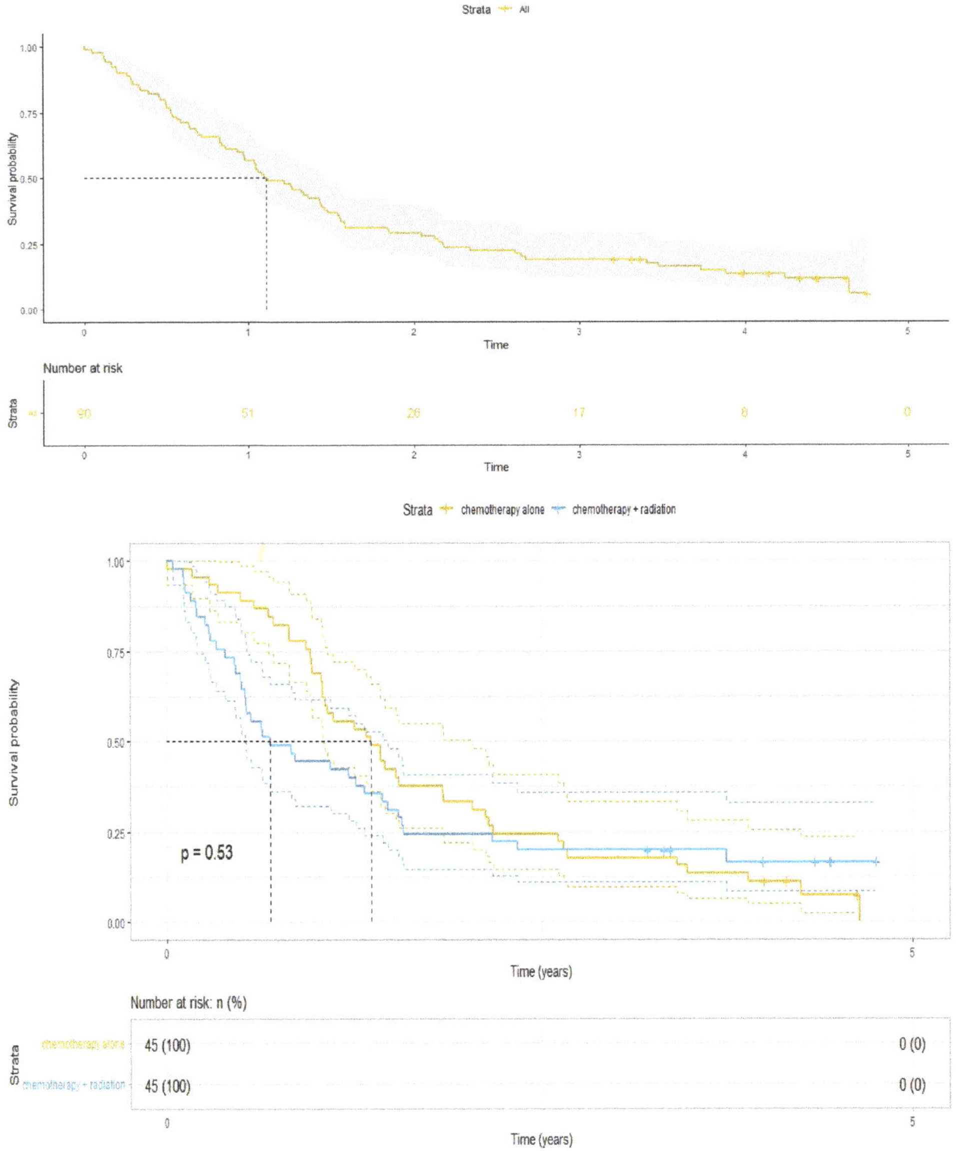

Figure 1. Illustrating the overall survival curve and the crossing survival curves for the two types of treatment.

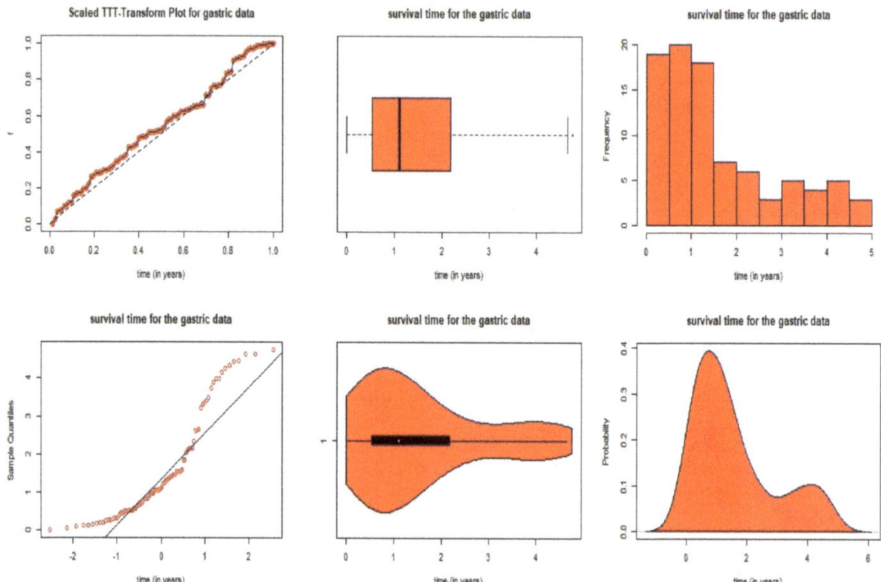

Figure 2. Fundamental plots for the survival time of the gastric cancer dataset.

7.2. Classical Analysis

The MLE estimates for baseline distribution parameters and coefficients of regression for the proposed AH model with different baseline distributions and other survival regression models with the GLL baseline distribution are provided in Tables 2 and 3.

Table 2 summarizes the statistics for the GLL–AH model and other survival regression models, including the PH, PO, and AFT models with all GLL baseline distributions. Based on the information criterion values, we conclude that the GLL–AH model has the lowest AIC, CAIC, and HQIC values compared to the other survival regression models, which indicates that the GLL–AH model outperforms its competing models.

Table 2. Results from the fitted proposed fully parametric AH regression model and other survival regression models with the GLL baseline distribution to gastric cancer dataset.

Models	Parameter(s)	Estimate	SE	AIC	CAIC	HQIC
GLL-AH	β	2.690	0.021	**244.318**	**242.845**	**248.351**
	α	1.505	0.040			
	k	0.542	0.036			
	η	0.133	0.022			
GLL-PO	β	0.750	0.101	251.816	250.522	255.848
	α	1.382	0.100			
	k	0.650	0.074			
	η	0.500	0.042			
GLL-PH	β	0.130	0.241	255.565	254.345	259.598
	α	1.302	0.140			
	k	0.759	0.136			
	η	0.580	0.222			
GLL-AFT	β	0.540	0.135	252.139	250.851	256.171
	α	1.545	0.127			
	k	0.557	0.106			
	η	0.728	0.231			

The statistics summary under the GLL–AH model, and other AH models with different baseline distributions are presented in Table 3. Based on the information criteria values, we deduce that the GLL–AH model beats its rival AH models becaue it has the lowest AIC, CAIC, and HQIC values when compared to the other AH models with various baseline distributions.

Table 3. Results from the fitted proposed fully parametric AH regression model with different baseline distributions to gastric cancer dataset.

Models	Parameter(s)	Estimate	SE	AIC	CAIC	HQIC
GLL-AH	β	2.690	0.021	**244.318**	**242.845**	**248.351**
	α	1.505	0.040			
	k	0.542	0.036			
	η	0.133	0.022			
PGW-AH	β	1.930	0.082	251.186	249.878	255.218
	α	1.687	0.142			
	k	0.821	0.066			
	η	2.226	0.102			
GG-AH	β	2.688	0.130	252.645	251.368	256.677
	α	1.821	0.122			
	k	0.482	0.236			
	η	0.737	0.042			
EW-AH	β	2.066	0.110	252.667	251.390	256.699
	α	0.789	0.212			
	k	0.911	0.086			
	η	2.283	0.052			
LL-AH	β	1.097	0.020	247.492	246.686	250.517
	α	1.913	0.052			
	k	1.213	0.019			
LN-AH	β	0.261	0.120	263.830	263.197	266.854
	α	0.065	0.101			
	k	1.260	0.032			
BXII-AH	β	0.923	0.142	249.144	248.359	252.168
	α	0.880	0.119			
	k	1.890	0.120			
W-AH	β	2.581	0.214	256.776	256.078	259.800
	α	1.013	0.049			
	k	1.818	0.112			
G-AH	β	2.367	0.430	255.121	254.406	258.145
	α	1.495	0.039			
	k	1.252	0.123			

7.3. Likelihood Ratio Test

The proposed AH model with the GLL baseline distribution is compared to its submodels, which include the log-logistic AH, Burr-XII AH, and Weibull AH models, by using the likelihood ratio test (LRT). It is required to reduce the number of parameters in a model and evaluate how this affects the model's capacity to match the data in order to draw thorough statistical conclusions about the model. In Table 4, statistics and related P-values demonstrate that the GLL–AH model fits the gastric dataset with crossing survival curves better than its submodels.

Table 4. LRT test for the GH model and its submodels.

Model	Hypothesis	LRT	*p*-Value
GLL-AH vs. BXII-AH	$H_0 : \eta = 1, H_1 : H_0$ is false,	6.999	0.008
GLL-AH vs. LL-AH	$H_0: \eta = k, H_1 : H_0$ is false,	5.347	0.021
GLL-AH vs. W-AH	$H_0: \eta^\alpha \to 0, H_1 : H_0$ is false,	14.533	<0.001

7.4. Bayesian Analysis

We used Bayesian analysis to compare the proposed GLL–AH model with its competing models, such as the GLL–PH, GLL–AH, and GLL–AFT models, and some of its submodels, including the LL–AH, BXII–AH, and W–AH regression models. The baseline distribution parameters $\alpha \sim G(a_1, b_1), \eta \sim G(a_2, b_2)$, and $k \sim G(a_3, b_3)$ with hyperparameter values $(a_1 = b_1 = a_2 = b_2 = a_3 = b_3 = 10)$ are assumed to have separate gamma priors that are independent and noninformative normal prior with a value of $N(0, 100)$ for $\beta's$ (regression coefficients). The Rstan package was utilized for our analysis [45].

7.4.1. Numerical Summary

In this section, we used the McMC sample of posterior properties for the proposed fully parametric AH, PO, AFT, and PH models with the GLL baseline distribution in Table 5 to examine several posterior characteristics of interest and their numerical values. The submodels of the GLL baseline distribution using the AH model are also examined in Table 6 to assess several posterior characteristics of interest and their numerical values.

Table 5. Results for the posterior properties of the GLL–AH, GLL–PO, GLL–PH and GLL–AFT models.

Models	Par (s)	Estimate	SE	SD	2.5%	Medium	97.5%	N_{eff}	$\hat{R}$
GLL–AH	β	1.016	0.009	0.476	0.030	1.027	1.909	2684	1.001
	α	0.836	0.002	0.106	0.648	0.829	1.064	3097	1.002
	k	1.553	0.004	0.196	1.205	1.544	1.969	2714	1.001
	η	0.674	0.003	0.191	0.353	0.653	1.105	3023	1.001
GLL–PO	β	0.565	0.006	0.353	−0.135	0.562	1.268	3617	1.001
	α	1.414	0.003	0.156	1.136	1.405	1.741	3257	1.000
	k	0.804	0.002	0.115	0.600	0.796	1.054	2951	1.001
	η	0.806	0.004	0.214	0.429	0.792	1.262	2918	1.000
GLL–PH	β	0.106	0.004	0.224	−0.330	0.107	0.540	3216	1.000
	α	1.341	0.002	0.146	1.077	1.332	1.646	3588	1.001
	k	0.876	0.002	0.122	0.662	0.869	1.134	3068	1.001
	η	0.837	0.004	0.221	0.452	0.820	1.315	3239	1.001
GLL–AFT	β	0.418	0.005	0.269	−0.116	0.415	0.949	3396	1.000
	α	1.435	0.003	0.177	1.124	1.423	1.804	3373	1.000
	k	0.809	0.002	0.114	0.609	0.801	1.060	2963	1.000
	η	0.850	0.004	0.210	0.479	0.836	1.311	2728	1.000

Table 6. Results for the posterior properties of the submodels of the GLL–AH model including LL–AH, W–AH, and BXII–AH models.

Models	Par (s)	Estimate	SE	SD	2.5%	Medium	97.5%	N_{eff}	$\hat{R}$
LL–AH	β	0.764	0.007	0.385	−0.073	0.800	1.421	3228	1.001
	α	1.636	0.004	0.197	1.261	1.629	2.039	2930	1.000
	k	0.879	0.002	0.107	0.688	0.873	1.109	3681	1.001
W–AH	β	−0.007	0.014	0.949	−1.850	−0.019	1.860	4377	1.000
	α	0.984	0.001	0.085	0.821	0.982	1.152	3521	1.000
	k	0.559	0.001	0.068	0.437	0.554	0.702	3875	1.001
BXII–AH	β	0.678	0.007	0.378	−0.135	0.697	1.345	3291	1.000
	α	1.627	0.004	0.209	1.247	1.620	2.062	3099	1.000
	k	0.949	0.002	0.115	0.740	0.943	1.186	3932	1.000

7.4.2. Visual Summary

Figures 3–9 provide the trace and autocorrelation (AC) plots for the baseline distribution parameters and regression coefficients of the proposed AH model and its submodels, plus other competing survival regression models, including the GLL–PH, GLL–PO, and GLL–AFT models, indicating convergence of the chains.

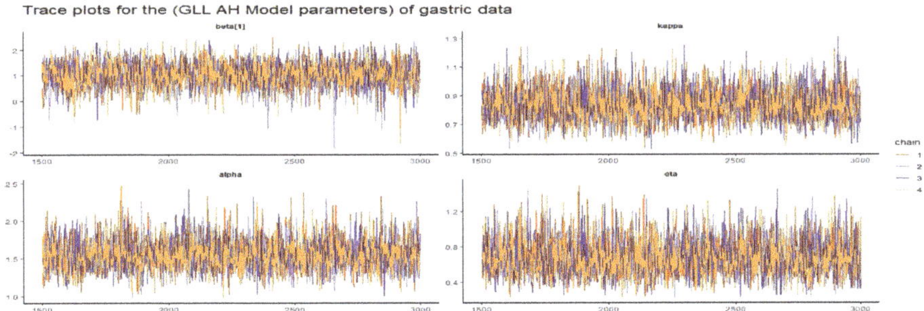

Figure 3. The GLL–AH model posterior parameters trace plots of the gastric cancer data.

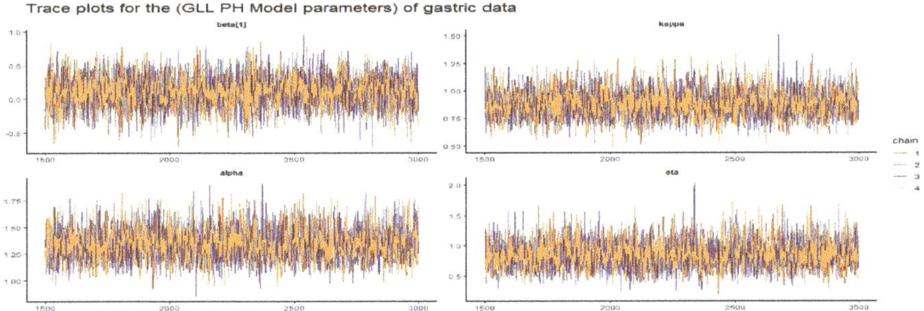

Figure 4. The GLL–PH model posterior parameters trace plots of the gastric cancer data.

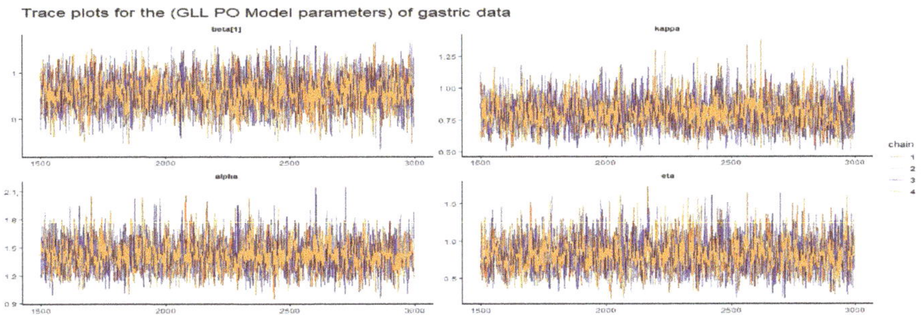

Figure 5. The GLL–PO model posterior parameters trace plots of the gastric cancer data.

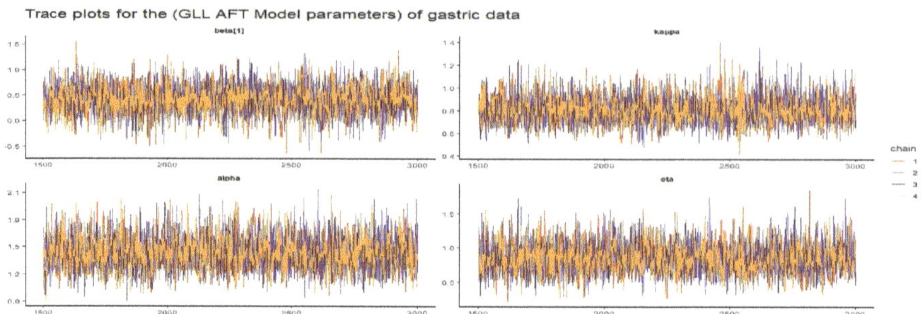

Figure 6. The GLL–AFT model posterior parameters trace plots of the gastric cancer data.

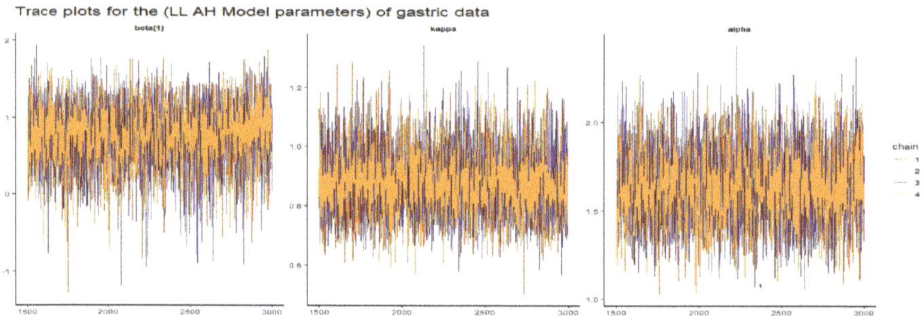

Figure 7. The LL–AH model posterior parameters trace plots of the gastric cancer data.

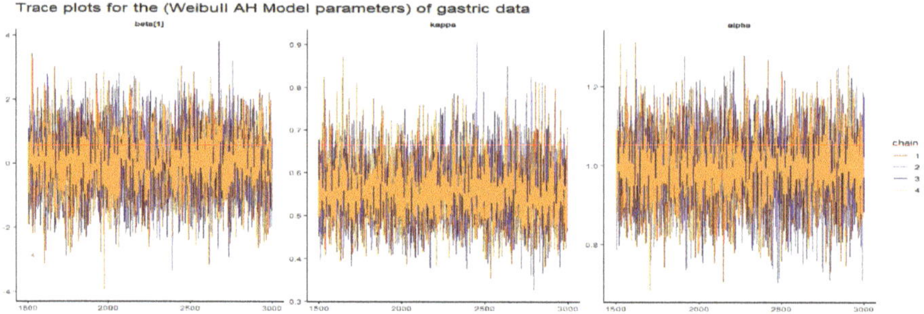

Figure 8. The W–AH model posterior parameters trace plots of the gastric cancer data.

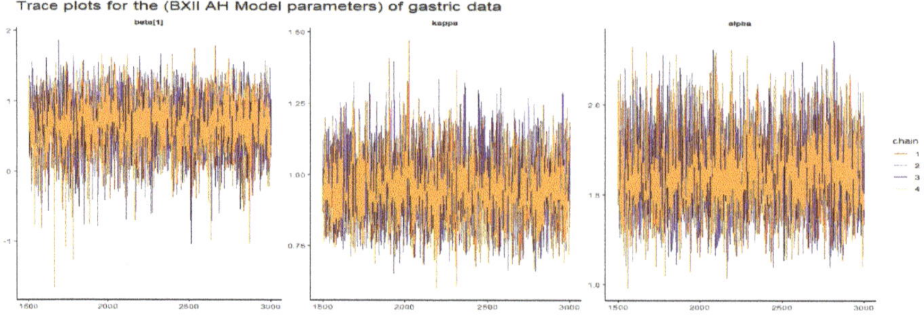

Figure 9. The BXII–AH model posterior parameters trace plots of the gastric cancer data.

7.4.3. Posterior Predictive Checks

If a fitted Bayesian parametric hazard-based regression model predicts future observations that are consistent with the current data, it is considered sufficient or performing well. By using the Bayesplot R package, posterior predictive check (PPC) plots are used to visually evaluate model fit. It can be seen from PPC in Figure 10, that the GLL–AH model fits the data quite well.

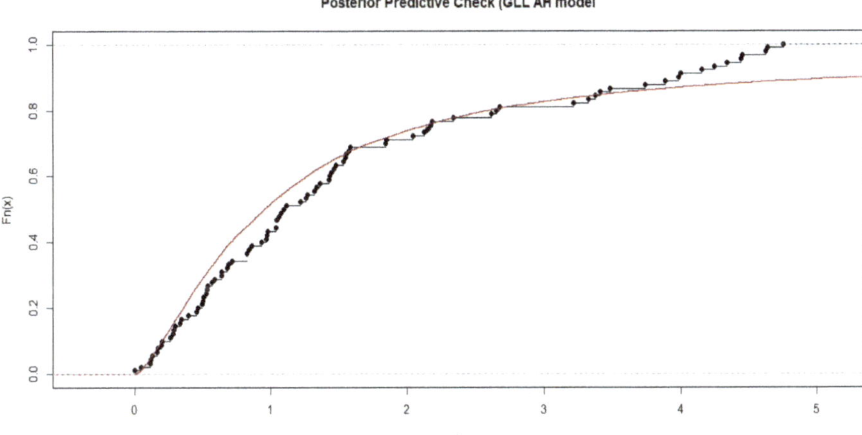

Figure 10. The empirical CDF, the dotted line and the CDF of the fitted model, the smooth curve, show that the fitted GLL–AH model predicts the future observations that are consistent with the current data.

7.4.4. McMC Convergence Diagnostics

We applied both numerical and visual methods to evaluate the convergence of the McMC algorithm for the proposed models and their special cases. The McMC algorithm HMC-NUTS has converged to the joint posterior distribution, as shown by the summary results in the above table, because the potential scale reduction factor $hatR$ is 1, the effective sample size (n_{eff}) is greater than 400, and the MC error (SE) is less than 0.05 of the posterior standard deviations for all parameters.

Visually assessing convergence is often done by using AC and trace graphs [23]. Figures 3–9 show a stationary pattern fluctuating within a band, demonstrating the convergence of the McMC algorithm. Figure 11, showing the AC plot, demonstrates how the AC rapidly decreases to zero as the period of lag increases, indicating good mixing and the convergence of the algorithm to the desired posterior distribution. Finally, Figure 12 indicates the pdf plots for the GLL-AH model posterior parameters.

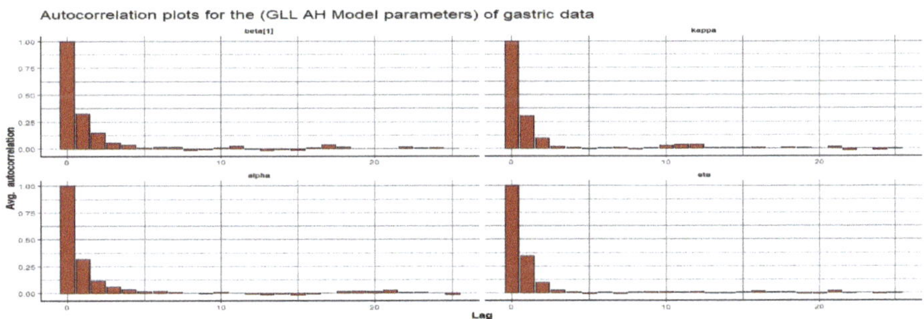

Figure 11. The GLL–AH model posterior parameters AC plots of the gastric cancer data.

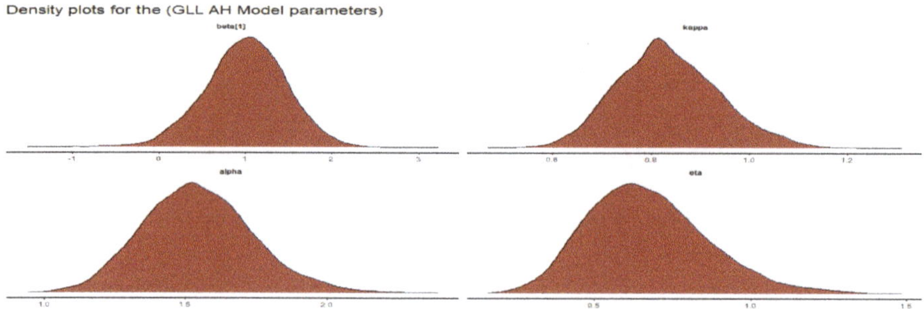

Figure 12. The GLL–AH model posterior parameters PDF plots of the gastric cancer data.

7.4.5. Bayesian Model Selection

We implemented two information criteria, the Watanabe–Akaike information criterion (WAIC), proposed by [46], for the Bayesian model comparison, and the leave-one-out information criterion (LOOIC) proposed by Vehtari et al. [47]. A model may be said to be best suited if it has the lowest WAIC and LOOIC values for both information criteria. In addition to Stan fitting, posterior predictive check (PPC) and determining WAIC and LOOIC are performed by using the R package loo [47]. Table 7 below shows that, when compared to its rival models, the GLL–AH model is the most effective. In addition, Table 8 demonstrates that, when compared to its sub-models, again the GLL–AH model is the superior one.

Table 7. Bayesian model comparison for the GLL–AH, GLL–PO, GLL–AFT, and GLL–PH models.

Model	WAIC	LOOIC
GLL–AH	243.20	243.20
GLL–PO	251.40	251.42
GLL–AFT	251.80	251.90
GLL–PH	254.80	254.82

Table 8. Bayesian model comparison for the GLL–AH and its special cases including LL–AH, W–AH, and BXII–AH models

Model	WAIC	LOOIC
GLL–AH	243.20	243.20
LL–AH	249.30	249.40
W–AH	255.01	255.00
BXII–AH	247.05	247.08

Figure 13 indicates the Kaplan–Meier estimate and the sf estimate for the proposed GLL–AH model parameters.

Figures 14 and 15 demonstrate the Kaplan–Meier estimate and the survival estimate curves for the proposed regression models with GLL baseline distribution and the AH model with various baseline hazards. In Figure 14, the GLL–AH model survival curve is closer to the KM survival curve compared to all other survival regression models. The same thing occurred in Figure 15.

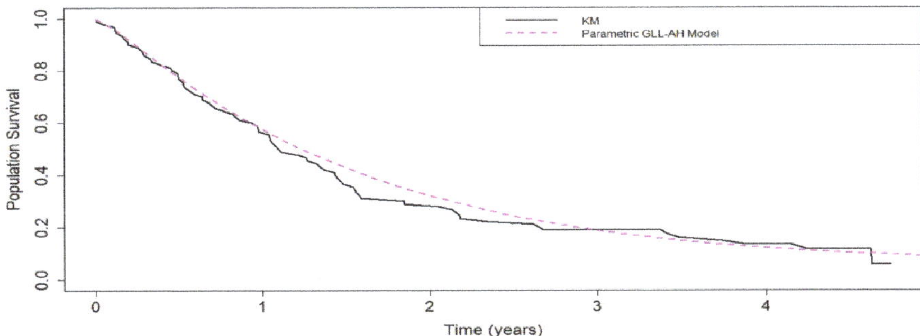

Figure 13. Kaplan–Meier and fitted survival curve for the GLL–AH model of the gastric cancer dataset.

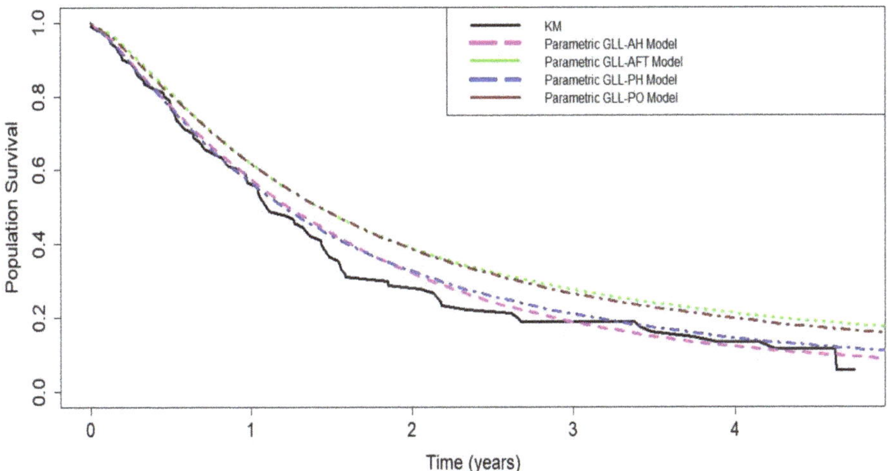

Figure 14. Kaplan–Meier and estimated survival plots for the competitive regression models with the GLL baseline distribution of the gastric cancer dataset.

The main advantage of this study is that, unlike other parametric survival regression models like the PH, PO, and AFT models, the parametric AH model may accommodate survival datasets with crossover survival curves. The proposed parametric model, on the other hand, is inappropriate when the baseline distribution is exponential, which is one of the study's limitations. Another limitation is that when the baseline distribution is the Weibull distribution, the proposed model performs identically to existing parametric hazard-based regression models, such as PH and AFT models.

Extension of the AH model's structure to incorporate survival datasets with or without crossover survival curves is one possible future endeavor. Additionally, this framework may include other parametric survival regression models, such as the additive hazards model.

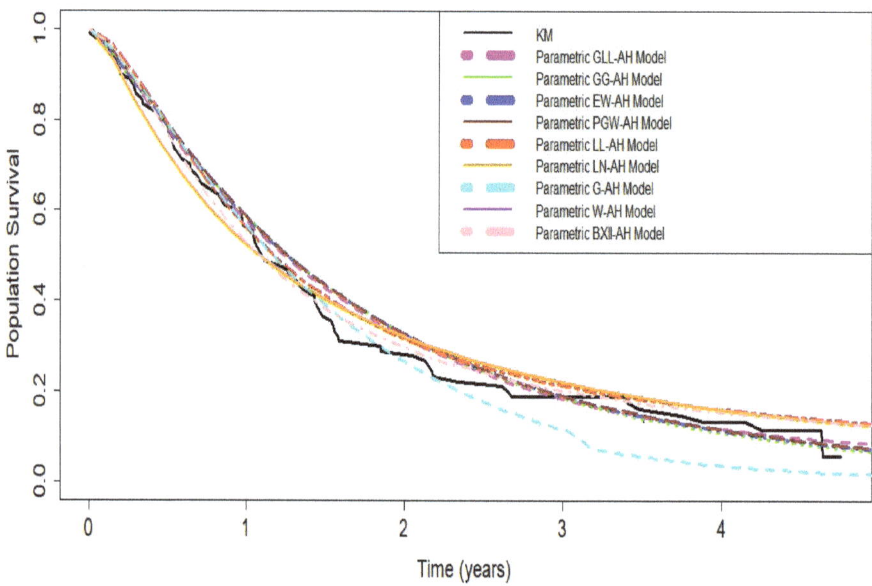

Figure 15. Kaplan–Meier and estimated survival plots for the competitive AH models of the gastric cancer dataset.

8. Conclusions

This article proposes a fully parametric AH model for dealing with censored lifetime data with crossover survival curves as an extension of the semi-parametric AH model [14]. The primary distinction between this modification and others is that we used a modified baseline distribution that can capture different hazard rate shapes to provide a more flexible depiction of the baseline hazard. By adopting a flexible parametric baseline distribution like the GLL distribution, we showed that it is possible to carry out both Bayesian and classical likelihood inference using the rstan package of the R programming language.

This also defines the paper's key contribution, as no other study combining these two characteristics (AH model and a modified baseline distribution) can be found in the time-to-event analysis field. Furthermore, employing both Bayesian and classical inference via MLE will address the semi-parametric AH model's limited use due to a lack of efficient and trustworthy estimation methods. Additionally, using the GLL distribution as a baseline hazard offers several benefits as compared to other parametric baseline distributions that may accept different hazard rate shapes, such as the gamma, GG, Weibull, EW, PGW, LL, Bur-XII, and LN distributions.

Following the simulation study, the paper gave a real-world demonstration involving a well-known dataset with crossover survival curves and was concerned with a clinical study for patients with gastric cancer. In summary, the GLL–AH model outperforms the other competing parametric AH models with various baseline hazards and other survival regression models with the same baseline hazard. Finally, we developed an R package, "AHSurv", to fit the proposed model in this study as an addendum to this paper; the source code is accessible at [48].

Author Contributions: Conceptualization, A.H.M., C.C., O.N. and S.M.; Data curation, A.H.M., C.C., O.N. and S.M.; Formal analysis, A.H.M., C.C., O.N. and S.M.; Investigation, A.H.M., C.C., O.N. and S.M.; Methodology, A.H.M., C.C., O.N. and S.M.; Software, A.H.M., C.C., O.N. and S.M.; Supervision, C.C., O.N. and S.M.; Validation, A.H.M., C.C., O.N. and S.M.; Visualization, A.H.M., C.C., O.N. and S.M.; Writing—original draft, A.H.M. and C.C.; Writing—review & Editing, A.H.M., C.C., O.N. and S.M. All authors have read and agreed to the published version of the manuscript.

Funding: This research received no external funding.

Data Availability Statement: Datasets are mentioned along the paper.

Acknowledgments: The authors would like to thank the academic editors and referees for their valuable suggestions and comments which improved the paper. The first author would like to thank Pan African University for supporting his work.

Conflicts of Interest: The authors declare no conflict of interest.

References

1. Cox, D.R. Regression models and life-tables. *J. R. Stat. Soc. Ser. B (Methodol.)* **1972**, *34*, 187–202. [CrossRef]
2. Kalbfleisch, J.D. Non-parametric Bayesian analysis of survival time data. *J. R. Stat. Soc. Ser. B (Methodol.)* **1978**, *40*, 214–221. [CrossRef]
3. Buckley, J.; James, I. Linear regression with censored data. *Biometrika* **1979**, *66*, 429–436. [CrossRef]
4. Komárek, A.; Lesaffre, E. Bayesian accelerated failure time model with multivariate doubly interval-censored data and flexible distributional assumptions. *J. Am. Stat. Assoc.* **2008**, *103*, 523–533. [CrossRef]
5. Bennett, S. Analysis of survival data by the proportional odds model. *Stat. Med.* **1983**, *2*, 273–277. [CrossRef]
6. Demarqui, F.N.; Mayrink, V.D. Yang and Prentice model with piecewise exponential baseline distribution for modeling lifetime data with crossing survival curves. *Braz. J. Probab. Stat.* **2021**, *35*, 172–186. [CrossRef]
7. Breslow, N.E.; Edler, L.; Berger, J. A two-sample censored-data rank test for acceleration. *Biometrics* **1984**, *40*, 1049–1062. [CrossRef]
8. Diao, G.; Zeng, D.; Yang, S. Efficient semiparametric estimation of short-term and long-term hazard ratios with right-censored data. *Biometrics* **2013**, *69*, 840–849. [CrossRef]
9. Egge, K.; Zahl, P.H. Survival of glaucoma patients. *Acta Ophthalmol. Scand.* **1999**, *77*, 397–401. [CrossRef]
10. Putter, H.; Sasako, M.; Hartgrink, H.; Van de Velde, C.; Van Houwelingen, J. Long-term survival with non-proportional hazards: Results from the Dutch Gastric Cancer Trial. *Stat. Med.* **2005**, *24*, 2807–2821. [CrossRef]
11. Shyur, H.J.; Elsayed, E.; Luxhøj, J.T. A general model for accelerated life testing with time-dependent covariates. *Nav. Res. Logist. (NRL)* **1999**, *46*, 303–321. [CrossRef]
12. Zhang, H.; Wang, P.; Sun, J. Regression analysis of interval-censored failure time data with possibly crossing hazards. *Stat. Med.* **2018**, *37*, 768–775. [CrossRef]
13. Demarqui, F.N.; Mayrink, V.D.; Ghosh, S.K. An Unified Semiparametric Approach to Model Lifetime Data with Crossing Survival Curves. *arXiv* **2019**, arXiv:1910.04475.
14. Chen, Y.Q.; Wang, M.C. Analysis of accelerated hazards models. *J. Am. Stat. Assoc.* **2000**, *95*, 608–618. [CrossRef]
15. Chen, Y.Q.; Jewell, N.P.; Yang, J. Accelerated hazards model: Method, theory and applications. *Handb. Stat.* **2003**, *23*, 431–441.
16. Lee, S.H. Some estimators and tests for accelerated hazards model using weighted cumulative hazard difference. *J. Appl. Stat.* **2009**, *36*, 473–482. [CrossRef]
17. Lee, S.H. On the estimators and tests for the semiparametric hazards regression model. *Lifetime Data Anal.* **2016**, *22*, 531–546. [CrossRef]
18. Rubio, F.J.; Remontet, L.; Jewell, N.P.; Belot, A. On a general structure for hazard-based regression models: An application to population-based cancer research. *Stat. Methods Med. Res.* **2019**, *28*, 2404–2417. [CrossRef]
19. Khan, S.A. Exponentiated Weibull regression for time-to-event data. *Lifetime Data Anal.* **2018**, *24*, 328–354. [CrossRef]
20. Lawless, J.F. *Statistical Models and Methods for Lifetime Data*; John Wiley & Sons: New York, NY, USA, 2011.
21. Muse, A.H.; Mwalili, S.; Ngesa, O.; Chesneau, C.; Alshanbari, H.M.; El-Bagoury, A.A.H. Amoud Class for Hazard-Based and Odds-Based Regression Models: Application to Oncology Studies. *Axioms* **2022**, *11*, 606. [CrossRef]
22. Muse, A.H.; Ngesa, O.; Mwalili, S.; Alshanbari, H.M.; El-Bagoury, A.A.H. A Flexible Bayesian Parametric Proportional Hazard Model: Simulation and Applications to Right-Censored Healthcare Data. *J. Healthc. Eng.* **2022**, *2022*, 2051642. [CrossRef] [PubMed]
23. Ashraf-Ul-Alam, M.; Khan, A.A. Generalized Topp-Leone-Weibull AFT modeling: A Bayesian Analysis with MCMC Tools Using R and Stan. *Austrian J. Stat.* **2021**, *50*, 52–76. [CrossRef]
24. Alvares, D.; Rubio, F.J. A tractable Bayesian joint model for longitudinal and survival data. *Stat. Med.* **2021**, *40*, 4213–4229. [CrossRef] [PubMed]
25. Muse, A.H.; Mwalili, S.; Ngesa, O.; Alshanbari, H.M.; Khosa, S.K.; Hussam, E. Bayesian and frequentist approach for the generalized log-logistic accelerated failure time model with applications to larynx-cancer patients. *Alex. Eng. J.* **2022**, *61*, 7953–7978. [CrossRef]

26. Al-Aziz, S.N.; Muse, A.H.; Jawad, T.M.; Sayed-Ahmed, N.; Aldallal, R.; Yusuf, M. Bayesian inference in a generalized log-logistic proportional hazards model for the analysis of competing risk data: An application to stem-cell transplanted patients data. *Alex. Eng. J.* **2022**, *61*, 13035–13050. [CrossRef]
27. Khan, S.A.; Khosa, S.K. Generalized log-logistic proportional hazard model with applications in survival analysis. *J. Stat. Distrib. Appl.* **2016**, *3*, 16. [CrossRef]
28. Chen, Y.; Hanson, T.; Zhang, J. Accelerated hazards model based on parametric families generalized with Bernstein polynomials. *Biometrics* **2014**, *70*, 192–201. [CrossRef]
29. Zhang, J.; Peng, Y. Crossing hazard functions in common survival models. *Stat. Probab. Lett.* **2009**, *79*, 2124–2130. [CrossRef]
30. Muse, A.H.; Mwalili, S.; Ngesa, O.; Chesneau, C.; Al-Bossly, A.; El-Morshedy, M. Bayesian and Frequentist Approaches for a Tractable Parametric General Class of Hazard-Based Regression Models: An Application to Oncology Data. *Mathematics* **2022**, *10*, 3813. [CrossRef]
31. Chen, Y.Q.; Jewell, N.P. On a general class of semiparametric hazards regression models. *Biometrika* **2001**, *88*, 687–702. [CrossRef]
32. Muse, A.H.; Mwalili, S.; Ngesa, O.; Almalki, S.J.; Abd-Elmougod, G.A. Bayesian and classical inference for the generalized log-logistic distribution with applications to survival data. *Comput. Intell. Neurosci.* **2021**, *2021*, 5820435. [CrossRef]
33. De Santana, T.V.F.; Ortega, E.M.; Cordeiro, G.M.; Silva, G.O. The Kumaraswamy-log-logistic distribution. *J. Stat. Theory Appl.* **2012**, *11*, 265–291.
34. Teamah, A.E.A.; Elbanna, A.A.; Gemeay, A.M. Heavy-tailed log-logistic distribution: Properties, risk measures and applications. *Stat. Optim. Inf. Comput.* **2021**, *9*, 910–941. [CrossRef]
35. Muse, A.H.; Tolba, A.H.; Fayad, E.; Abu Ali, O.A.; Nagy, M.; Yusuf, M. modeling the COVID-19 mortality rate with a new versatile modification of the log-logistic distribution. *Comput. Intell. Neurosci.* **2021**, *2021*, 8640794. [CrossRef]
36. Mansour, M.M.; Ibrahim, M.; Aidi, K.; Shafique Butt, N.; Ali, M.M.; Yousof, H.M.; Hamed, M.S. A new log-logistic lifetime model with mathematical properties, copula, modified goodness-of-fit test for validation and real data modeling. *Mathematics* **2020**, *8*, 1508. [CrossRef]
37. Alkhairy, I.; Nagy, M.; Muse, A.H.; Hussam, E. The Arctan-X family of distributions: Properties, simulation, and applications to actuarial sciences. *Complexity* **2021**, *2021*, 4689010. [CrossRef]
38. Alfaer, N.M.; Gemeay, A.M.; Aljohani, H.M.; Afify, A.Z. The extended log-logistic distribution: Inference and actuarial applications. *Mathematics* **2021**, *9*, 1386. [CrossRef]
39. Muse, A.H.; Mwalili, S.M.; Ngesa, O. On the log-logistic distribution and its generalizations: A survey. *Int. J. Stat. Probab.* **2021**, *10*, 93. [CrossRef]
40. Haghighi, M.N.F. On the power generalized Weibull family: Model for cancer censored data. *Metron* **2009**, *67*, 75–86.
41. Mudholkar, G.S.; Hutson, A.D. The exponentiated Weibull family: Some properties and a flood data application. *Commun. Stat.-Methods* **1996**, *25*, 3059–3083. [CrossRef]
42. Stacy, E.W. A generalization of the gamma distribution. *Ann. Math. Stat.* **1962**, *33*, 1187–1192. [CrossRef]
43. Elshahhat, A.; Muse, A.H.; Egeh, O.M.; Elemary, B.R.; et al. Estimation for Parameters of Life of the Marshall-Olkin Generalized-Exponential Distribution Using Progressive Type-II Censored Data. *Complexity* **2022**, *2022*, 8155929. [CrossRef]
44. Muse, A.H.; Mwalili, S.; Ngesa, O.; Chesneau, C. AmoudSurv: An R Package for Tractable Parametric Odds-Based Regression Models. 2022. Available online: https://cran.r-project.org/web//packages/AmoudSurv/index.html (accessed on 1 November 2022).
45. Carpenter, B.; Gelman, A.; Hoffman, M.D.; Lee, D.; Goodrich, B.; Betancourt, M.; Brubaker, M.; Guo, J.; Li, P.; Riddell, A. Stan: A probabilistic programming language. *J. Stat. Softw.* **2017**, *76*, 1–32. [CrossRef]
46. Watanabe, S. A widely applicable Bayesian information criterion. *J. Mach. Learn. Res.* **2013**, *14*, 867–897.
47. Vehtari, A.; Gelman, A.; Gabry, J. Practical Bayesian model evaluation using leave-one-out cross-validation and WAIC. *Stat. Comput.* **2017**, *27*, 1413–1432. [CrossRef]
48. Muse, A.H.; Mwalili, S.; Ngesa, O.; Kilai, M. AHSurv: An R Package for Flexible Parametric Accelerated Hazards (AH) Regression Models. 2022. Available online: https://cran.r-project.org/web/packages/AHSurv/index.html (accessed on 1 November 2022).

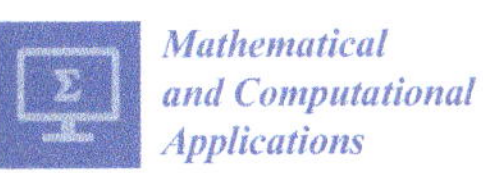

Mathematical and Computational Applications

Article

New Lifetime Distribution for Modeling Data on the Unit Interval: Properties, Applications and Quantile Regression

Suleman Nasiru [1], Abdul Ghaniyyu Abubakari [1] and Christophe Chesneau [2,*]

[1] Department of Statistics, School of Mathematical Sciences, C. K. Tedam University of Technology and Applied Sciences, Kassena-Nankana Navrongo-Kolgo Road, Navrongo, Upper East, Ghana

[2] Department of Mathematics, LMNO, CNRS-Université de Caen, Campus II, Science 3, 14032 Caen, France

* Correspondence: christophe.chesneau@unicaen.fr

Abstract: Probability distributions are very useful in modeling lifetime datasets. However, no specific distribution is suitable for all kinds of datasets. In this study, the bounded truncated Cauchy power exponential distribution is proposed for modeling datasets on the unit interval. The probability density function exhibits desirable shapes, such as left-skewed, right-skewed, reversed J, and bathtub shapes, whereas the hazard rate function displays J and bathtub shapes. For the purpose of modeling dependence between measures in a dataset, a bivariate extension of the proposed distribution is developed. The bivariate probability density function displays monotonic and non-monotonic shapes, making it suitable for modeling complex bivariate relations. Subsequently, the applications of the distribution are illustrated using COVID-19 data. The results revealed that the new distribution provides a better fit to the datasets compared to other existing distributions. Finally, a new quantile regression model is developed and its application demonstrated. The generated quantile regression model offers a decent fit to the data, according to the residual analysis.

Keywords: COVID-19; bounded distribution; estimation methods; Cauchy; regression; bivariate

Citation: Nasiru, S.; Abubakari, A.G.; Chesneau, C. New Lifetime Distribution for Modeling Data on the Unit Interval: Properties, Applications and Quantile Regression. *Math. Comput. Appl.* **2022**, *27*, 105. https://doi.org/10.3390/mca27060105

Received: 31 October 2022
Accepted: 2 December 2022
Published: 3 December 2022

Publisher's Note: MDPI stays neutral with regard to jurisdictional claims in published maps and institutional affiliations.

1. Introduction

Disease modeling and prediction are primary tasks of epidemiologists and researchers interested in the estimation of disease occurrences. To perform these tasks, modeling the variability in disease occurrences using probability distributions is essential. With the emergence of the novel coronavirus disease in late 2019 (COVID-19) and its negative impact on humanity, many researchers have proposed new probability distributions (discrete or continuous) for modeling the number of infections, mortality rate, and recovery rates, among others. Some of the proposed probability distributions or families of distributions include: Marshall–Olkin reduced Kies distribution [1], modified inverse Weibull distribution [2], weighted Weibull distribution [3], type I half logistic Burr X-G family [4], unit power Weibull distribution [5], new extended exponentiated Weibull distribution [6], discrete extended odd Weibull exponential distribution [7], odd Weibull inverse Topp–Leone distribution [8], log-logistic tangent distribution [9], discrete-type half-logistic exponential distribution [10], and unit Johnson S_U distribution [11].

Among these probability distributions used for modeling diseases, those defined on the unit interval play a major role due to their usefulness in areas such as health, psychology, and epidemiology, among others. For instance, researchers may be interested in modeling mortality or recovery rates. Observations measured on these variables are usually proportions, fractions, or rates, which are defined in the unit interval. Although the beta distribution is the oldest for modeling datasets measured on the unit interval, the intractability of its cumulative distribution function (CDF) and quantile function has called for the development of new distributions with tractable CDFs and quantile functions that are also capable of modeling data on the unit interval. Unit distributions proposed recently in literature include: unit Gamma/Gompertz distribution [12], bounded odd inverse Pareto exponential distribution [13], bounded shifted Gompertz distribution [14], unit modified Burr-III distribution [15], unit generalized half normal distribution [16], unit Lindley distribution [17], unit Gompertz distribution [18], logit slash distribution [19], unit Weibull distribution [20] and unit inverse Gaussian distribution [21].

Despite the existence of many unit distributions in the literature, no single distribution is capable of modeling all forms of data since the data generating process produces data with different characteristics such as symmetric, skewed, varied degrees of kurtosis, and monotonic and non-monotonic failure rates. This study thus proposes a new unit distribution called the bounded truncated Cauchy power exponential (BTCPE) distribution. The motivations for developing the new distribution are as follows: to provide a model capable of modeling complex data on unit interval that exhibits platykurtic, leptokurtic, reversed J, left-skewed, right-skewed, bathtub, and J shapes; to develop a bivariate distribution for modeling interdependence between random data on unit interval; and to develop a quantile regression model for understanding the relationship between a response variable and given covariates.

The remainder of the paper is organized in nine sections, described as follows: Section 2 presents the development of the BTCPE distribution, Section 3 describes some of its important properties, Section 4 focuses on a special bivariate extension of the BTCPE distribution, Section 5 is devoted to the parametric estimation methods, Section 6 presents the Monte Carlo simulation of nine frequentist estimation methods, Section 7 contains the univariate applications of the BTCPE distribution, Section 8 is about the quantile regression model and its application, and finally the conclusion of the paper is presented in Section 9.

2. Bounded Truncated Cauchy Power Exponential Distribution

A random variable X follows the truncated Cauchy power exponential (TCPE) distribution if its CDF and probability density function (PDF), respectively, are defined as

$$F_X(x; \alpha, \lambda) = \frac{4}{\pi} \arctan[(1 - e^{-\lambda x})^\alpha], \alpha > 0, \lambda > 0, x > 0, \tag{1}$$

and

$$f_X(x; \alpha, \lambda) = \frac{4\alpha\lambda e^{-\lambda x}(1 - e^{-\lambda x})^{\alpha-1}}{\pi[1 + (1 - e^{-\lambda x})^{2\alpha}]}, x > 0. \tag{2}$$

The TCPE distribution can be presented as a special case of the TCP Weibull distribution proposed by [22]. Now, we define a new unit distribution, called the BTCPE distribution, corresponding to the distribution of $Y = e^{-X}$. The associated CDF is obtained as follows:

$$F_Y(y; \alpha, \lambda) = \mathbb{P}(e^{-X} \leq y) = \mathbb{P}(-X \leq \log(y))$$
$$= 1 - \mathbb{P}(X \leq -\log(y))$$
$$= 1 - F_X(-\log(y); \alpha, \lambda).$$

Hence, the CDF of the BTCPE distribution is expressed as

$$F_Y(y; \alpha, \lambda) = 1 - \frac{4}{\pi} \arctan[(1 - y^\lambda)^\alpha], 0 < y < 1, \tag{3}$$

and $\alpha > 0$ and $\lambda > 0$ are the shape parameters that have to be estimated. The associated PDF of the BTCPE distribution is obtained by differentiating Equation (3), and it is given by

$$f_Y(y; \alpha, \lambda) = \frac{4\alpha\lambda y^{\lambda-1}(1 - y^\lambda)^{\alpha-1}}{\pi[1 + (1 - y^\lambda)^{2\alpha}]}, 0 < y < 1. \tag{4}$$

Often, the PDFs are expressed in expanded form for easy derivation of the statistical properties of the proposed distribution. The expanded form of the PDF of the BTCPE distribution is mainly obtained using the generalized binomial expansion, $(y + a)^{-n} = \sum_{k=0}^{\infty} \binom{-n}{k} y^k a^{-n-k}, |y| < a$, where n is any real number. Thus, it is given by

$$f_Y(y; \alpha, \lambda) = \frac{4\alpha\lambda}{\pi} \sum_{i=0}^{\infty} (-1)^i y^{\lambda-1} (1 - y^\lambda)^{\alpha(2i+1)-1}, 0 < y < 1. \tag{5}$$

The corresponding hazard rate function (HRF) is given by

$$h_Y(y; \alpha, \lambda) = \frac{\alpha\lambda y^{\lambda-1}(1 - y^\lambda)^{\alpha-1}}{[1 + (1 - y^\lambda)^{2\alpha}]\arctan[(1 - y^\lambda)^\alpha]}, 0 < y < 1. \tag{6}$$

The shapes of the PDF and HRF for some given parameter values are shown in Figure 1. The PDF exhibits symmetric, bathtub, left-skewed and right-skewed shapes for the given parameter values. The HRF displays bathtub and increasing failure rates.

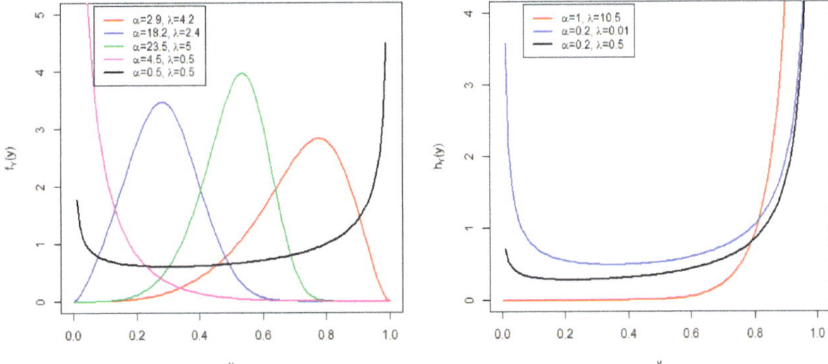

Figure 1. PDF (**left**) and HRF (**right**) of the BTCPE distribution.

3. Some Important Properties

This section presents some relevant properties of the BTCPE distribution.

3.1. Distribution Inequalities

This subsection investigates some desirable inequalities satisfied by the CDF of the BTCPE distribution. These inequalities are very essential in determining the first order stochastic dominance of random variables [23].

Proposition 1. *The CDF of the BTCPE distribution is increasing with respect to the parameter α. The CDF of the BTCPE distribution is decreasing with respect to the parameter λ.*

Proof. For the first point, since $(1 - y^\lambda)^\alpha \log(1 - y^\lambda) < 0$, for $y \in (0, 1)$, we have

$$\frac{\partial F_Y(y; \alpha, \lambda)}{\partial \alpha} = -\frac{4(1 - y^\lambda)^\alpha \log(1 - y^\lambda)}{\pi[1 + (1 - y^\lambda)^{2\alpha}]} \geq 0.$$

This means that $F_Y(y; \alpha, \lambda)$ is increasing with respect to α. For the second point, since $y^\lambda (1 - y^\lambda)^{\alpha-1} \log(y) < 0$, for $y \in (0, 1)$, we have

$$\frac{\partial F_Y(y; \alpha, \lambda)}{\partial \lambda} = \frac{4\alpha y^\lambda (1 - y^\lambda)^{\alpha-1} \log(y)}{\pi[1 + (1 - y^\lambda)^{2\alpha}]} \leq 0.$$

This implies that $F_Y(y; \alpha, \lambda)$ is decreasing with respect to λ. This completes the proof of the proposition. From Proposition 1, the following stochastic ordering property follows immediately: if $\alpha_1 \leq \alpha_2$ then $F_Y(y; \alpha_1, \lambda) \leq F_Y(y; \alpha_2, \lambda)$. Also, if $\lambda_1 \leq \lambda_2$ then $F_Y(y; \alpha, \lambda_2) \leq F_Y(y; \alpha, \lambda_1)$. $\square$

3.2. Quantile Function

The quantile function or the inverse CDF is simply the solution $Q(u; \alpha, \lambda)$ of the following nonlinear equation: $F_Y(Q(u; \alpha, \lambda); \alpha, \lambda) = u$, for all $u \in (0, 1)$. Thus, after some algebraic manipulation, we have

$$Q(u; \alpha, \lambda) = \left\{ 1 - \left(\tan\left[\frac{\pi}{4}(1 - u) \right] \right)^{1/\alpha} \right\}^{1/\lambda}, u \in (0, 1). \tag{7}$$

The median is obtained by substituting $u = 0.5$. The quantile function plays an important role in the generation of random observations from the BTCPE distribution. The quantile function values are also useful in computing measures of skewness and kurtosis. As a classical quantile measure, the MacGillivray measure of skewness [24] is given by

$$\rho(u; \alpha, \lambda) = \frac{Q(1 - u; \alpha, \lambda) + Q(u; \alpha, \lambda) - 2Q(0.5; \alpha, \lambda)}{Q(1 - u; \alpha, \lambda) - Q(u; \alpha, \lambda)}, u \in (0, 1).$$

In particular, the MacGillivray measure of skewness can be used to efficiently describe the effect of the parameters (α, λ) on the skewness. The more the shapes of $\rho(u; \alpha, \lambda)$ vary according to the parameters, the more flexible the skewness is. Figure 2 shows the plot of this skewness measure for a fixed value of λ while α varies and for a fixed value of α while λ varies. From Figure 2, the wider variations seen imply that both parameters have a strong influence on the skewness of the BTCPE distribution. In addition, as the values of α or λ increase, $\rho(u; \alpha, \lambda)$ gets closer to the horizontal line. This shows that utilizing higher values of the parameter can result in a symmetrical distribution.

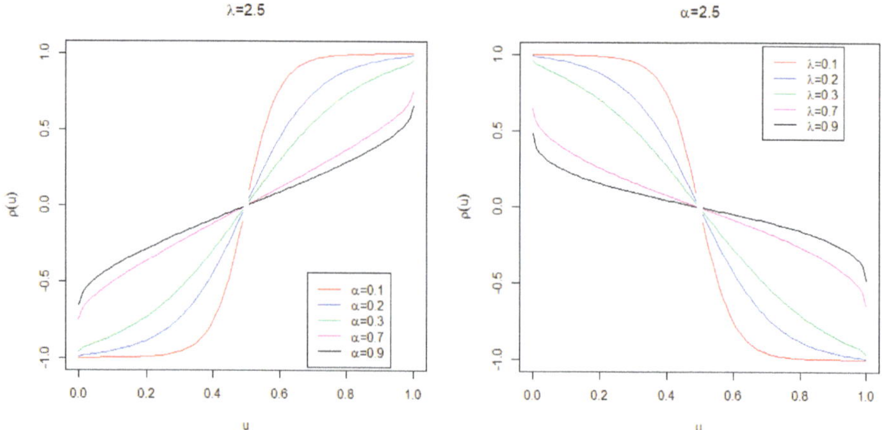

Figure 2. Plots of the MacGillivray skewness.

The kurtosis of the BTCPE distribution can be studied using the Moors kurtosis [25]. The Moors (coefficient of) kurtosis is usually given by

$$K(\alpha, \lambda) = \frac{Q(7/8; \alpha, \lambda) - Q(5/8; \alpha, \lambda) + Q(3/8; \alpha, \lambda) - Q(1/8; \alpha, \lambda)}{Q(3/4; \alpha, \lambda) - Q(1/4; \alpha, \lambda)}.$$

Large values of the Moors kurtosis imply that the distribution has a heavy tail, and small values are indications of a light tail. Figure 3 displays the Moors kurtosis for the BTCPE distribution. It can be observed that the BTCPE distribution exhibits various degrees of kurtosis. When the parameters α and λ are equal, the distribution displays a platykurtic shape. The overall shapes show how flexible the BTCPE distribution is with regards to modeling datasets having different degrees of kurtosis and skewness.

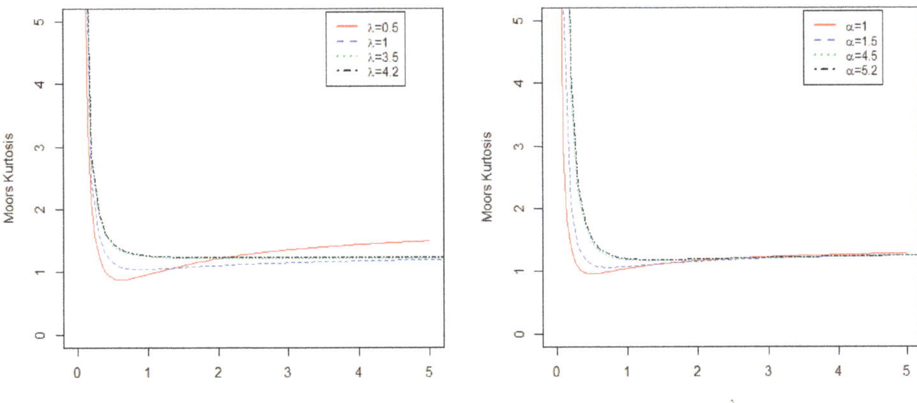

Figure 3. Plots of Moors kurtosis.

3.3. Moments and Moments Generating Function

The rth moments, incomplete moments and moment generating function of the BTCPE distribution are presented in this subsection.

Proposition 2. *If Y is a BTCPE random variable, i.e., a random variable with the BTCPE distribution, then its rth non-central moment is given by*

$$\mu'_r = \frac{4\alpha}{\pi} \sum_{i=0}^{\infty} (-1)^i B\left(\frac{r}{\lambda} + 1, \alpha(1 + 2i)\right), r = 1, 2, \ldots, \tag{8}$$

where $B(a, b) = \int_0^1 z^{a-1}(1 - z)^{b-1} dz$ is the beta integral function.

Proof. The rth non-central moment of the BTCPE random variable is defined as $\mu'_r = E(Y^r) = \int_0^1 y^r f_Y(y; \alpha, \lambda) dy$. Thus, substituting the expanded form of the PDF given in Equation (5) yields

$$\mu'_r = \frac{4\alpha\lambda}{\pi} \sum_{i=0}^{\infty} (-1)^i \int_0^1 y^{r+\lambda-1}(1 - y^\lambda)^{\alpha(1+2i)-1} dy.$$

Letting $z = y^\lambda, y \to 0, z \to 0; y \to 1, z \to 1$ and $dz = \lambda y^{\lambda-1} dy$, we get

$$\mu'_r = \frac{4\alpha}{\pi} \sum_{i=0}^{\infty} (-1)^i \int_0^1 z^{\frac{r}{\lambda}}(1 - z)^{\alpha(1+2i)-1} dz.$$

Hence, several algebraic manipulation yield

$$\mu'_r = \frac{4\alpha}{\pi} \sum_{i=0}^{\infty} (-1)^i B\left(\frac{r}{\lambda} + 1, \alpha(1 + 2i)\right).$$

This completes the proof. $\square$

The non-central moments can be used to derive other important characteristics of the BTCPE distribution such as estimating the variance, coefficient of skewness and kurtosis.

Proposition 3. *The rth incomplete moment of the BTCPE random variable is given by*

$$\varphi_r = \frac{4\alpha}{\pi} \sum_{i=0}^{\infty} (-1)^i B\left(y^\lambda; \frac{r}{\lambda} + 1, \alpha(1 + 2i)\right), r = 1, 2, \ldots, \tag{9}$$

where $B(q; a, b) = \int_0^q z^{a-1}(1 - z)^{b-1} dz$ is the incomplete beta integral function.

Proof. By definition, the rth incomplete moment is given by

$$\varphi_r = E(Y^r 1\{Y < y\}) = \int_0^y x^r f(x; \alpha, \lambda) dx.$$

Hence, substituting the expanded form of the PDF into the definition yields

$$\varphi_r = \frac{4\alpha\lambda}{\pi} \sum_{i=0}^{\infty} (-1)^i \int_0^y x^{r+\lambda-1}(1 - x^\lambda)^{\alpha(1+2i)-1} dx.$$

Letting $z = x^\lambda, x \to 0, z \to 0; x \to y, z \to y^\lambda$ and $dz = \lambda x^{\lambda-1} dx$. Hence, applying similar concepts for proving the incomplete moments yields

$$\varphi_r = \frac{4\alpha}{\pi} \sum_{i=0}^{\infty} (-1)^i B\left(y^\lambda; \frac{r}{\lambda} + 1, \alpha(1 + 2i)\right).$$

This completes the proof. $\square$

The moment generating function is useful for deriving the moments of a random variable if only the moment exists.

Proposition 4. *The moment generating function of the BTCPE random variable is given by*

$$M_Y(t) = \frac{4\alpha}{\pi} \sum_{r=0}^{\infty} \sum_{i=0}^{\infty} \frac{(-1)^i t^r}{r!} B\left(\frac{r}{\lambda} + 1, \alpha(1 + 2i)\right). \tag{10}$$

Proof. By definition and a standard exponential expansion, we have $M_Y(t) = E(e^{tY}) = \sum_{r=0}^{\infty} \frac{t^r}{r!} \mu'_r$. Hence, substituting the rth non-central moment of the BTCPE distribution into the definition completes the proof. $\square$

Table 1 shows the first six moments of the BTCPE distribution and other useful measures, such as the standard deviation (SD), coefficients of variation (CV), skewness (CS) and kurtosis (CK). The SD, CV, CS and CK are, respectively, given by

$$SD = \sqrt{\mu'_2 - \mu^2},$$

$$CV = \frac{\sigma}{\mu} = \sqrt{\frac{\mu'_2}{\mu^2} - 1},$$

$$CS = \frac{\mu'_3 - 3\mu\mu'_2 + 2\mu^3}{(\mu'_2 - \mu^2)^{3/2}}$$

and

$$CK = \frac{\mu_4' - 4\mu\mu_3' + 6\mu^2\mu_2' - 3\mu^4}{(\mu_2' - \mu^2)^2}.$$

Table 1. Values of moment measures, including the SD, CV, CS and CK.

μ_r'	$\alpha = 0.4, \lambda = 2.5$	$\alpha = 4.5, \lambda = 3.1$	$\alpha = 20.0, \lambda = 1.5$
μ_1'	0.8799	0.5602	0.1339
μ_2'	0.8021	0.3401	0.0242
μ_3'	0.7457	0.2185	0.0053
μ_4'	0.7020	0.1465	0.0013
μ_5'	0.6667	0.1017	0.0004
μ_6'	0.6373	0.0726	0.0001
SD	0.1668	0.1619	0.0794
CV	0.1896	0.2890	0.5931
CS	-1.9527	-0.3403	0.7713
CK	6.6850	2.7084	3.5390

From Table 1, the CS is negative for the given parameter values and positive for others. It can be seen that the BTCPE distribution can be leptokurtic or platykurtic depending on the parameter values, since the CK can be lower than 3 or greater than 3, respectively. The coefficient of skewness also reveals that the BTCPE distribution can model both left and right-skewed data.

3.4. Order Statistics

Order statistics play an imperative role in both statistics and industrial reliability analysis. They can be used to estimate the minimum, maximum, and range of observations. They are used in developing control charts that are useful in industrial quality control analyses. Let $Y_{1:n} \leq Y_{2:n} \leq \ldots \leq Y_{n:n}$ be n order statistics from n BTCPE random variables. Then, the PDF of $Y_{k:n}$ is given by

$$f_{k:n}(y;\alpha,\lambda) = \Omega_{k:n}[F_Y(y;\alpha,\lambda)]^{k-1}[1 - F_Y(y;\alpha,\lambda)]^{n-k}f_Y(y;\alpha,\lambda),$$

where

$$\Omega_{k:n} = \frac{n!}{(k-1)!(n-k)!}.$$

Using the binomial expansion $(1-y)^{\lambda-1} = \sum_{i=0}^{\infty}(-1)^i\binom{\lambda-1}{i}y^i, |y| < 1$, we can write

$$f_{k:n}(y;\alpha,\lambda) = \Omega_{k:n}\sum_{i=0}^{k-1}(-1)^i\binom{k-1}{i}[1 - F_Y(y;\alpha,\lambda)]^{n-k+i}f_Y(y;\alpha,\lambda).$$

Thus, we have

$$f_{k:n}(y;\alpha,\lambda) = \frac{\Omega_{k:n}4\alpha\lambda y^{\lambda-1}(1-y^\lambda)^{\alpha-1}}{\pi[1+(1-y^\lambda)^{2\alpha}]}\sum_{i=0}^{k-1}(-1)^i\binom{k-1}{i}\left[\frac{4}{\pi}\arctan[(1-y^\lambda)^\alpha]\right]^{n-k+i}. \tag{11}$$

On the other side, the CDF of $Y_{1:n}$ is simply given by

$$F_{1:n}(y;\alpha,\lambda) = 1 - [1 - F_Y(y;\alpha,\lambda)]^n$$
$$= 1 - \left[\frac{4}{\pi}\arctan[(1-y^\lambda)^\alpha]\right]^n,$$

and the CDF of $Y_{n:n}$ is derived as

$$F_{n:n}(y; \alpha, \lambda) = [F_Y(y; \alpha, \lambda)]^n = \left[1 - \frac{4}{\pi}\arctan[(1 - y^\lambda)^\alpha]\right]^n.$$

The distribution of the smallest order statistic represents the lifetime of a system connected in series, and that of the maximum order statistic denotes the lifetime of a system connected in parallel. Hence, they are vital in studying the minimum and maximum time to failure of components in engineering reliability. The minimum and maximum (min–max) plots of the order statistics can be used to investigate the distributional behavior of observations. The min–max plot captures not only the information in the tails but all the information about the whole distribution. The min–max plots shown in Figure 4 for some parameter values depend on $E(Y_{1:n})$ and $E(Y_{n:n})$. From the min–max plots, the distribution can exhibit symmetrical, left-skewed, and right-skewed shapes.

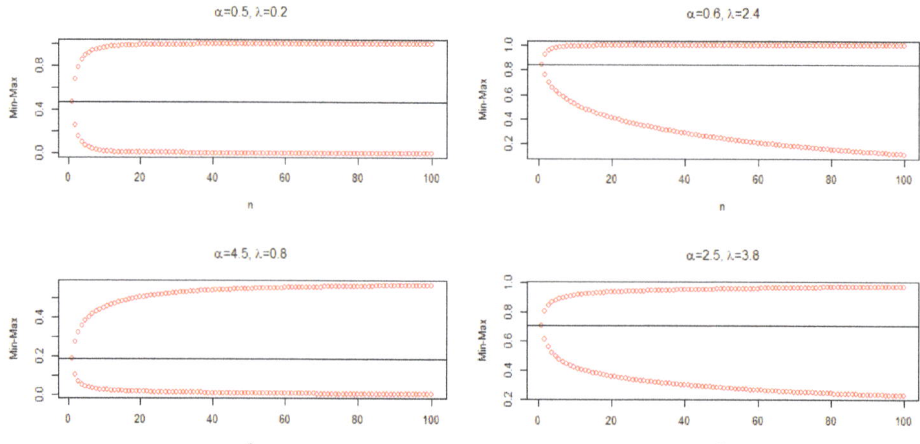

Figure 4. Min–max plots for some parameter values.

4. Bivariate Extension

Researchers may be interested in modeling the dependence between two (quantitative) measures in a dataset. For instance, one may be interested in modeling the relationship between age and the body mass index of individuals. Bivariate distributions have been used in reliability analysis, queuing theory, finance, and insurance risk analysis, among others, to study interdependency (see [26]). In this section, the bivariate extension of the BTCPE (BEBTCPE) distribution is proposed following the strategy developed in [26,27]. Given a bivariate continuous random vector (X, Y), the CDF of the BEBTCPE distribution with parameters $\alpha, \lambda, \delta_1, \delta_2, \delta_3$, where $\alpha > 0, \lambda > 0, -1 < \delta_1 + \delta_3 < 1$, $-1 < \delta_2 + \delta_3 < 1, 0 < x < 1$ and $0 < y < 1$, is given by

$$F_{XY}(x, y; \eta) = \frac{(1 - \frac{4}{\pi}\arctan[(1 - x^\lambda)^\alpha])(1 - \frac{4}{\pi}\arctan[(1 - y^\lambda)^\alpha])}{\left\{1 + (\delta_1 + \delta_3)\frac{4}{\pi}\arctan[(1 - x^\lambda)^\alpha] + (\delta_2 + \delta_3)\frac{4}{\pi}\arctan[(1 - y^\lambda)^\alpha]\right\}^{-1}}, \quad (12)$$

where $\eta = (\alpha, \beta, \delta_1, \delta_2, \delta_3)^{\mathrm{T}}$. The parameters δ_1, δ_2 and δ_3 quantify the dependence between the two variables of a BEBTCPE random vector. The plots of the CDF for the following parameter values are shown in Figure 5:

(a) $\alpha = 3.5, \lambda = 8.2, \delta_1 = 0.3, \delta_2 = 0.1, \delta_3 = 0.3$;
(b) $\alpha = 2.5, \lambda = 0.8, \delta_1 = 0.5, \delta_2 = 0.4, \delta_3 = 0.2$ and
(c) $\alpha = 0.5, \lambda = 4.8, \delta_1 = -0.3, \delta_2 = -0.7, \delta_3 = -0.1$.

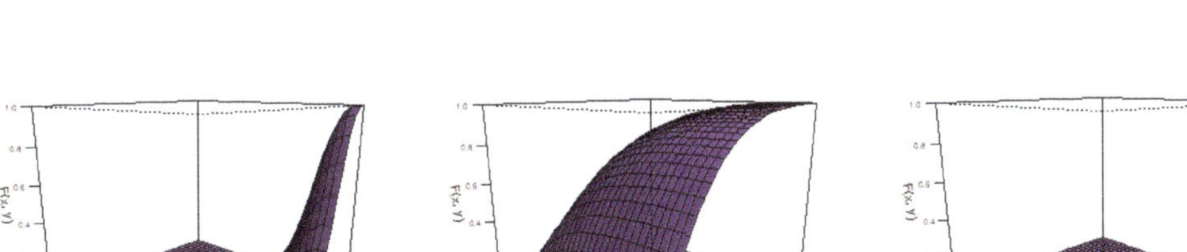

Figure 5. CDF plots of the BEBTCPE distribution.

We notice various concave and convex shapes from these plots.
The corresponding bivariate PDF is given by

$$f_{XY}(x,y;\eta) = \frac{(4\alpha\lambda/\pi)^2(xy)^{\lambda-1}(1-x^\lambda-y^\lambda+(xy)^\lambda)^{\alpha-1}\left[1+(1-x^\lambda)^{2\alpha}\right]^{-1}\left[1+(1-y^\lambda)^{2\alpha}\right]^{-1}}{\left\{1+(\delta_1+\delta_3)\frac{8}{\pi}\arctan[(1-x^\lambda)^\alpha]+(\delta_2+\delta_3)\frac{8}{\pi}\arctan[(1-y^\lambda)^\alpha]\right\}^{-1}}. \tag{13}$$

Figure 6 shows the bivariate PDF plots of the BEBTCPE distribution for the following parameter values:

(a) $\alpha = 3.5, \lambda = 8.2, \delta_1 = 0.3, \delta_2 = 0.1, \delta_3 = 0.3$;
(b) $\alpha = 2.5, \lambda = 0.8, \delta_1 = 0.5, \delta_2 = 0.4, \delta_3 = 0.2$ and
(c) $\alpha = 0.5, \lambda = 4.8, \delta_1 = -0.3, \delta_2 = -0.7, \delta_3 = -0.1$.

The first graph displays a non-monotonic shape whereas the other two exhibit monotonic shapes, illustrating the versatility in the bivariate modeling sense.

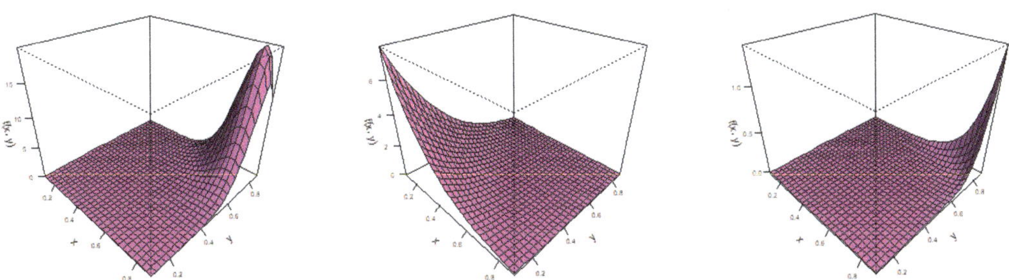

Figure 6. PDF plots of the BEBTCPE distribution.

5. Parameter Estimation Methods

This section presents nine estimation methods for estimating the parameters of the BTCPE distribution. These include the maximum likelihood (ML) estimation (MLE), ordinary least squares (OLS), weighted least squares (WLS), Cramér–von Mises (CVM), percentile (PC) estimation, Anderson–Darling (AD) methods, and maximum and minimum product spacing methods.

5.1. Maximum Likelihood Estimation

One of the most common methods used for estimating the parameters of a developed model is the MLE method. Suppose that Y follows the BTCPE distribution, with $\vartheta = (\alpha, \lambda)^{\mathrm{T}}$ as the parameter vector. For a single observation y of Y, the log-likelihood function $\ell = \ell(\vartheta)$ is

$$\ell = \log\left(\frac{4\alpha\lambda}{\pi}\right) + (\lambda - 1)\log(y) + (\alpha - 1)\log(1 - y^\lambda) - \log(1 + (1 - y^\lambda)^{2\alpha}). \tag{14}$$

To obtain the estimates of the parameters for the single observation, the first partial derivative of Equation (14) with respect to the parameters needs to be derive. Here, we obtain

$$\frac{\partial \ell}{\partial \alpha} = \frac{1}{\alpha} + \log(1 - y^\lambda) - \frac{2(1 - y^\lambda)^{2\alpha}\log(1 - y^\lambda)}{1 + (1 - y^\lambda)^{2\alpha}}, \tag{15}$$

and

$$\frac{\partial \ell}{\partial \lambda} = \frac{1}{\lambda} + \log(y) - \frac{y^\lambda(\alpha - 1)\log(y)}{1 - y^\lambda} + \frac{2\alpha y^\lambda(1 - y^\lambda)^{2\alpha - 1}\log(y)}{1 + (1 - y^\lambda)^{2\alpha}}. \tag{16}$$

Given that $y_1, y_2, \ldots, y_n$ are (independent and identically) observations from n BTCPE random variables, then the total log-likelihood function is given by $\ell_n^* = \sum_{i=1}^{n} \ell_i(\vartheta)$, where $\ell_i(\vartheta), i = 1, 2, \ldots, n$ is defined in Equation (14) with $y = y_i$. The estimates of the parameters can be obtained by maximizing the total log-likelihood function directly using MATLAB, MATHEMATICA and R software. In this study, the R software is used [28]. Alternatively, the estimates of the parameters can be obtained by equating the first partial derivatives with respect to the parameters to zero and solving the resulting system of equations simultaneously. However, since the resulting system of equations does not have a closed form, the nonlinear system of equations $\left(\frac{\partial \ell_n^*}{\partial \alpha}, \frac{\partial \ell_n^*}{\partial \lambda}\right)^{\mathrm{T}} = (0, 0)^{\mathrm{T}}$ is solved numerically to obtain the estimates of the parameters.

5.2. Ordinary and Weighted Least Squares Estimation

Suppose that $y_{(1)}, y_{(2)}, \ldots, y_{(n)}$ are ordered observations from n BTCPE random variables. The OLS estimates of the parameters $\hat{\alpha}_{LSE}$ and $\hat{\lambda}_{LSE}$ are obtained by minimizing the following function:

$$LSE(\alpha, \lambda) = \sum_{i=1}^{n}\left[\left(1 - \frac{4}{\pi}\arctan[(1 - y_{(i)}^\lambda)^\alpha]\right) - \frac{i}{n+1}\right]^2, \tag{17}$$

with respect to the parameters α and λ. On the other hand, the OLS estimates can be obtained by numerically solving the following nonlinear equations:

$$\sum_{i=1}^{n}\left[\left(1 - \frac{4}{\pi}\arctan[(1 - y_{(i)}^\lambda)^\alpha]\right) - \frac{i}{n+1}\right]\Delta_s(y_{(i)}|\alpha, \lambda) = 0, \, s = 1, 2, \tag{18}$$

where

$$\Delta_1(y_{(i)}|\alpha, \lambda) = -\frac{8(1 - y_{(i)}^\lambda)^\alpha \log(1 - y_{(i)}^\lambda)}{\pi[1 + (1 - y_{(i)}^\lambda)^{2\alpha}]}$$

and

$$\Delta_2(y_{(i)}|\alpha, \lambda) = \frac{8\alpha y_{(i)}^\lambda(1 - y_{(i)}^\lambda)^{\alpha - 1}\log(y_{(i)})}{\pi[1 + (1 - y_{(i)}^\lambda)^{2\alpha}]}.$$

The WLS estimates $\hat{\alpha}_{WLS}$ and $\hat{\lambda}_{WLS}$ are obtained by minimizing the following function:

$$WLS(\alpha, \lambda) = \sum_{i=1}^{n} \frac{(n+1)^2(n+2)}{i(n-i+1)} \left[\left(1 - \frac{4}{\pi}\arctan[(1-y_{(i)}^{\lambda})^{\alpha}] \right) - \frac{i}{n+1} \right]^2, \quad (19)$$

with respect to the parameters α and λ. Alternatively, the WLS estimates can be obtained by numerically solving the following nonlinear equations:

$$\sum_{i=1}^{n} \frac{(n+1)^2(n+2)}{i(n-i+1)} \left[\left(1 - \frac{4}{\pi}\arctan[(1-y_{(i)}^{\lambda})^{\alpha}] \right) - \frac{i}{n+1} \right] \Delta_s(y_{(i)}|\alpha, \lambda) = 0, \, s = 1, 2, \quad (20)$$

where $\Delta_s(x_{(i)}|\alpha, \lambda), s = 1, 2$ are defined above.

5.3. Cramér–Von Mises Estimation

Let $y_{(1)}, y_{(2)}, \ldots, y_{(n)}$ be ordered observations from n BTCPE random variables. The CVM estimates of the parameters $\hat{\alpha}_{CVM}$ and $\hat{\lambda}_{CVM}$ are obtained by minimizing the following function:

$$CVM(\alpha, \lambda) = \frac{1}{12n} + \sum_{i=1}^{n} \left[\left(1 - \frac{4}{\pi}\arctan[(1-y_{(i)}^{\lambda})^{\alpha}] \right) - \frac{2i-1}{2n} \right]^2, \quad (21)$$

with respect to the parameters α and λ. The estimates of the parameters can also be obtained by numerically solving the following equations:

$$\sum_{i=1}^{n} \left[\left(1 - \frac{4}{\pi}\arctan[(1-y_{(i)}^{\lambda})^{\alpha}] \right) - \frac{2i-1}{2n} \right] \Delta_s(y_{(i)}|\alpha, \lambda) = 0, \, s = 1, 2, \quad (22)$$

where $\Delta_s(y_{(i)}|\alpha, \lambda), s = 1, 2$ are given above.

5.4. Anderson–Darling Estimation

Another minimum distance estimation method is the AD estimation technique. Let $y_{(1)}, y_{(2)}, \ldots, y_{(n)}$ be ordered observations from n BTCPE random variables. The AD estimates for the parameters of the BTCPE distribution are obtained by minimizing the following function:

$$AD(\alpha, \lambda) = -n - \frac{1}{n}\sum_{i=1}^{n} (2i-1) \left[\log\left(1 - \frac{4}{\pi}\arctan[(1-y_{(i)}^{\lambda})] \right) + \log\left(\frac{4}{\pi}\arctan[(1-y_{(i)}^{\lambda})] \right) \right], \quad (23)$$

with respect to the parameters α and λ.

5.5. Percentile Estimation

The PC estimation approach is another method of estimating the parameters of a given model. Let $y_{(1)}, y_{(2)}, \ldots, y_{(n)}$ be ordered observations from n BTCPE random variables and $u_i = i/(n+1)$ be an unbiased estimate of $F_Y(y_{(i)}; \alpha, \lambda)$. The PC estimates of the parameters of the BTCPE distribution are obtained by minimizing the following function:

$$PC(\alpha, \lambda) = \sum_{i=1}^{n} \left[y_{(i)} - \left\{ 1 - \left(\tan\left[\frac{\pi}{4}(1-u_i) \right] \right)^{1/\alpha} \right\}^{1/\lambda} \right]^2, \quad (24)$$

with respect to the parameters α and λ.

5.6. Maximum and Minimum Product Spacing Estimation

An alternative parameter estimation technique which is based on the Kullback–Leibler information measure is the maximum product spacing (MPS). Let $y_{(1)}, y_{(2)}, \ldots, y_{(n)}$ be ordered observations from n BTCPE random variables. Consider the uniform spacing

$$D_i = F_Y(y_{(i)}; \alpha, \lambda) - F_Y(y_{(i-1)}; \alpha, \lambda)$$
$$= \tfrac{4}{\pi} \arctan[(1 - y_{(i-1)}^\lambda)] - \tfrac{4}{\pi} \arctan[(1 - y_{(i)}^\lambda)]$$

where $F_Y(y_{(0)}; \alpha, \lambda) = 0$, $F_Y(y_{(n+1)}; \alpha, \lambda) = 1$ and $D_0(\alpha, \lambda) + D_1(\alpha, \lambda) + \ldots + D_{n+1}(\alpha, \lambda) = 1$. The estimates of the parameters are obtained via the MPS approach by maximizing the logarithm of the geometric mean of the spacing defined by

$$MPS(\alpha, \lambda) = \frac{1}{n+1} \sum_{i=1}^{n+1} \log D_i(\alpha, \lambda), \tag{25}$$

with respect to the parameters α and λ.

Additionally, the minimum spacing distance (MSD) estimates for the parameters α and λ are obtained by minimizing the following function:

$$MSD(\alpha, \lambda) = \sum_{i=1}^{n+1} \vartheta\left(D_i(\alpha, \lambda), \frac{1}{n+1}\right), \tag{26}$$

where $\vartheta(x, y)$ is an appropriate distance, with respect to the parameters α and λ. Although different choices of $\vartheta(x, y)$ exist, in this study the absolute distance $|x - y|$ and the absolute-log distance $|\log x - \log y|$ are utilized. Thus, the minimum spacing absolute distance (MSAD) and minimum spacing absolute-log distance (MSALD) estimates are, respectively, obtained by minimizing the following functions:

$$MSAD(\alpha, \lambda) = \sum_{i=1}^{n+1} \left| D_i(\alpha, \lambda) - \frac{1}{n+1} \right| \tag{27}$$

and

$$MSALD(\alpha, \lambda) = \sum_{i=1}^{n+1} \left| \log D_i(\alpha, \lambda) - \log \frac{1}{n+1} \right|, \tag{28}$$

where $D_i(\alpha, \lambda) \neq \frac{1}{n+1}$ and $\log D_i(\alpha, \lambda) \neq \log \frac{1}{n+1}$.

6. Simulation

In this section, simulation experiments are carried out to assess how well the proposed parameters of the BTCPE distribution have been estimated. The experiments are carried out with the following two different parameter combinations: $\alpha = 4.1, \lambda = 2.5$ and $\alpha = 3.1, \lambda = 8.5$. The experiments are replicated 5000 times with the following different sample sizes: $n = 25, 75, 125, 175$ and 225. The bias (AB) and root mean square error (RMSE) of the estimates are then computed and compared.

The AB and RMSE are, respectively, computed using

$$AB = \frac{1}{R} \sum_{i=1}^{R} (\hat{\vartheta}_i - \vartheta)$$

and

$$RMSE = \sqrt{\frac{1}{R} \sum_{i=1}^{R} (\hat{\vartheta}_i - \vartheta)^2},$$

where $\hat{\vartheta}$ is either $\hat{\alpha}$ or $\hat{\lambda}$ and $R = 5000$ is used in this study.

From Tables 2 and 3, most of the estimates have their ABs and RMSEs decreasing as the sample size increases. This is an indication that most of the estimates exhibit the consistency property. From Table 2, it can be observed that for sample sizes 25, 75 and 125 the PC estimate is the best for α and, for the sample sizes 175 and 225, the MLE is the best for α. For the parameter λ, the PC estimate is the best for the sample size 25 and the MLE is the best for 75, 125, 175 and 225. In Table 3, for sample sizes 25 and 75 the AD estimate is the best for the parameter α and the MLE is the best for 125, 175 and 225. For the parameter λ, the MLE is the best for sample sizes 25, 125, 175 and 225. The AD estimate is best for λ when the sample size is 75.

Table 2. AB and RMSE for $\alpha = 4.1$ and $\lambda = 2.5$.

Parameter	n	AB									RMSE								
		MLE	MPS	MADS	MALDS	OLS	WLS	CVM	AD	PC	MLE	MPS	MADS	MALDS	OLS	WLS	CVM	AD	PC
α	25	0.7980	2.2327	−2.3189	0.531	0.3530	−2.6423	1.1477	0.4713	−0.3155	2.2233	3.5728	2.9322	3.3391	2.4457	2.6729	3.5377	1.9964	1.4972
	75	0.2140	0.7443	−1.8442	0.0634	0.1365	−3.1632	0.3415	0.1180	−0.1694	0.9157	1.3139	2.4241	1.1330	1.1489	3.1657	1.2820	0.9535	0.8506
	125	0.1342	0.4372	−1.3088	−0.0031	0.0472	−3.3268	0.1337	0.0713	−0.0783	0.6843	0.8313	1.9860	0.7795	0.8149	3.3279	0.8159	0.7025	0.6641
	175	0.0914	0.2987	−0.8738	−0.0272	0.0460	−2.1721	0.1164	0.0657	−0.0484	0.5365	0.6791	1.5544	0.6323	0.6845	2.1990	0.6955	0.5941	0.5393
	225	0.0677	0.2509	−0.6976	0.0062	0.0301	3.0791	0.1096	0.0365	−0.0623	0.4841	0.5505	1.3266	0.5926	0.5906	3.3377	0.6147	0.5240	0.4860
λ	25	0.1871	0.5436	−1.1017	0.0300	−0.0075	−1.3344	0.2060	0.0670	−0.1737	0.6038	0.8201	1.3862	0.7382	0.6538	1.3687	0.7340	0.5749	0.5401
	75	0.0478	0.2197	−0.8939	−0.0079	0.0078	−2.0003	0.0802	0.1185	−0.0672	0.3089	0.4026	1.2090	0.3996	0.3692	2.0029	0.3886	0.3379	0.3175
	125	0.0378	0.1293	−0.6271	−0.0160	−0.0055	−2.1461	0.0305	0.0146	−0.0448	0.2407	0.2740	0.9681	0.2987	0.2752	2.1472	0.2798	0.2560	0.2466
	175	0.0233	0.0866	−0.3986	−0.0139	0.0034	−1.6394	0.0280	0.0114	−0.0267	0.1959	0.2314	0.7264	0.2315	0.2372	1.6455	0.2383	0.2134	0.2008
	225	0.0208	0.0820	−0.3101	0.0021	0.0003	0.1937	0.0230	0.0007	−0.0225	0.1810	0.1939	0.6057	0.2196	0.2079	0.3066	0.2124	0.1874	0.1835

Table 3. AB and RMSE for $\alpha = 3.1$ and $\lambda = 8.5$.

Parameter	n	AB									RMSE								
		MLE	MPS	MADS	MALDS	OLS	WLS	CVM	AD	PC	MLE	MPS	MADS	MALDS	OLS	WLS	CVM	AD	PC
α	25	0.5120	1.5281	−1.3158	0.3712	0.2359	−1.9121	0.7701	0.2725	−0.6748	1.4793	2.5597	2.0817	1.9689	1.6086	1.9366	2.4325	1.3231	1.4204
	75	0.2081	0.5190	−0.9456	0.0477	0.0264	−2.2924	0.1973	0.0989	−0.3302	0.6848	0.8671	1.5913	0.7865	0.7088	2.2949	0.8122	0.6294	0.8145
	125	0.0994	0.3218	−0.6895	0.0570	0.0153	−2.4255	0.1020	0.0757	−0.2704	0.4778	0.6228	1.2854	0.5657	0.5532	2.4266	0.5644	0.5109	0.6262
	175	0.0867	0.2259	−0.5478	0.0107	0.0240	−1.3857	0.0882	0.0554	−0.2187	0.4077	0.4554	1.0538	0.4821	0.4813	1.4810	0.4940	0.4163	0.5242
	225	0.0461	0.1719	−0.4192	0.0007	0.0195	2.2166	0.0555	0.0200	−0.1735	0.3331	0.3880	0.8460	0.4021	0.4133	2.3785	0.4184	0.3477	0.4768
λ	25	0.5725	1.9602	−3.5282	0.1675	0.1107	−4.6443	0.7293	0.2368	−1.6432	2.0951	3.1555	4.8358	2.6753	2.3883	4.7703	2.6517	2.1203	2.6346
	75	0.2957	0.8211	−2.4563	−0.0292	−0.0416	−6.9243	0.2449	0.0825	−0.7220	1.1627	1.4130	3.9050	1.3676	1.2655	6.9330	1.3246	1.1047	1.4877
	125	0.1098	0.4964	−1.6975	0.0128	−0.0435	−7.3960	0.1015	0.0784	−0.5259	0.8837	1.0321	3.0631	1.0694	0.9899	7.3994	0.9670	0.9327	1.1482
	175	0.1182	0.3504	−1.2348	−0.0232	0.0022	−5.5678	0.1061	0.0622	−0.4011	0.7361	0.7786	2.4368	0.8864	0.8608	5.5937	0.8687	0.7734	0.9365
	225	0.0631	0.2843	−0.9196	−0.0034	0.0015	0.6715	0.043	0.0249	−0.3171	0.6223	0.7091	1.9241	0.7604	0.7452	1.0898	0.7515	0.6590	0.8305

7. Applications

Three applications of the BTCPE distribution are illustrated in this section, and its performance is compared to other competitive distributions defined in the unit interval. The performance of the BTCPE distribution was compared with that of the beta, unit Burr-III (UBIII) [29], bounded M-O extended exponential (BMOEE) [30], unit Gompertz (UG) [18], unit Lindley (UL) [17], unit Weibull (UW) [20] and unit-improved second-degree Lindley (UISDL) [31] distributions. The Akaike information criterion (AIC), Bayesian information criterion (BIC), Anderson–Darling (AD) test, and Cramér–von Mises (CVM) are the model selection techniques employed in arriving at the best model. For these selection techniques, the best model is the one with the smallest test statistic. The datasets represent the mortality rate of COVID-19 patients in Canada and the United Kingdom (UK), and the recovery rate of COVID-19 patients in Spain. The first two datasets were recently reported by [8].

The first dataset is the mortality rate for UK from 1 December 2020 to 29 January 2021. The data are: 0.1292, 0.3805, 0.4049, 0.2564, 0.3091, 0.2413, 0.1390, 0.1127, 0.3547, 0.3126, 0.2991, 0.2428, 0.2942, 0.0807, 0.1285, 0.2775, 0.3311, 0.2825, 0.2559, 0.2756, 0.1652, 0.1072, 0.3383, 0.3575, 0.2708, 0.2649, 0.0961, 0.1565, 0.1580, 0.1981, 0.4154, 0.3990, 0.2483, 0.1762, 0.1760, 0.1543, 0.3238, 0.3771, 0.4132, 0.4602, 0.352, 0.1882, 0.1742, 0.4033, 0.4999, 0.3930, 0.3963, 0.3960, 0.2029, 0.1791, 0.4768, 0.5331, 0.3739, 0.4015, 0.3828, 0.1718, 0.1657, 0.4542, 0.4772, 0.3402.

The second dataset denotes the mortality rate for Canada from 1 November to 26 December 2020. The data are: 0.1622, 0.1159, 0.1897, 0.1260, 0.3025, 0.2190, 0.2075, 0.2241, 0.2163, 0.1262, 0.1627, 0.2591, 0.1989, 0.3053, 0.2170, 0.2241, 0.2174, 0.2541, 0.1997, 0.3333, 0.2594, 0.2230, 0.2290, 0.1536, 0.2024, 0.2931, 0.2739, 0.2607, 0.2736, 0.2323, 0.1563, 0.2677, 0.2181, 0.3019, 0.2136, 0.2281, 0.2346, 0.1888, 0.2729, 0.2162, 0.2746, 0.2936, 0.3259, 0.2242, 0.1810, 0.2679, 0.2296, 0.2992, 0.2464, 0.2576, 0.2338, 0.1499, 0.2075, 0.1834, 0.3347, 0.2362.

The third dataset constitutes the recovery rates of COVID-19 patients in Spain from 3 March to 7 May 2020. The dataset can be found in [1] and are: 0.6670, 0.5000, 0.5000, 0.4286, 0.7500, 0.6531, 0.5161, 0.7895, 0.7689, 0.6873, 0.5200, 0.7251, 0.6375, 0.6078, 0.6289, 0.5712, 0.5923, 0.6061, 0.5924, 0.5921, 0.5592, 0.5954, 0.6164, 0.6455, 0.6725, 0.6838, 0.6850, 0.6947, 0.7210, 0.7315, 0.7412, 0.7508, 0.7519, 0.7547, 0.7645, 0.7715, 0.7759, 0.7807, 0.7838, 0.7847, 0.7871, 0.7902, 0.7934, 0.7913, 0.7962, 0.7971, 0.7977, 0.8007, 0.8038, 0.8289, 0.8322, 0.8354, 0.8371, 0.8387, 0.8456, 0.8490,0.8535, 0.8547, 0.8564, 0.8580, 0.8604, 0.8628, 0.6586, 0.7070, 0.7963, 0.8516.

The ML estimates of the parameters are estimated using the *bbmle* package in R [32]. The initial values of the parameters of the fitted distributions used for the optimization are obtained using the *GenSA* package in R [33]. Table 4 displays the descriptive statistics for COVID-19 mortality for the UK and Canada, as well as the recovery rate for Spain. The datasets are platykurtic due to the negative excess kurtosis. The UK mortality is right-skewed and that of Canada is left-skewed. The recovery rate for Spain is also left-skewed. This is affirmed by the boxplot of the datasets shown in Figure 7.

Table 4. Descriptive statistics for datasets.

Country	Minimum	Maximum	Mean	Skewness	Kurtosis
UK	0.0807	0.5331	0.2888	0.0476	−1.1034
Canada	0.1159	0.3347	0.2305	−0.0850	−0.4402
Spain	0.4286	0.8628	0.7240	−0.6890	−0.4761

7.1. UK COVID-19 Mortality

Table 5 presents ML estimates of the parameters and their corresponding standard errors in brackets, the log-likelihood (ℓ), AIC, BIC, AD, and CVM for the fitted distributions. Given that it has the lowest values for the AIC, BIC, AD, and CVM and the maximum log-likelihood, the BTCPE distribution offers the best fit to the UK mortality dataset.

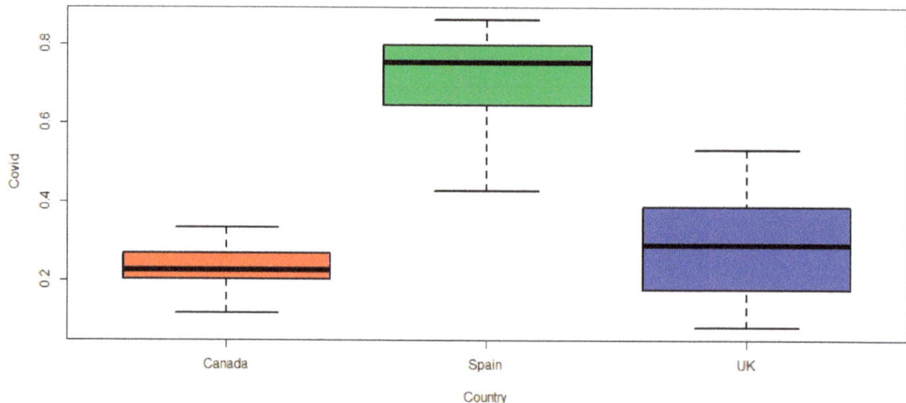

Figure 7. Boxplots of COVID-19 datasets.

Table 5. Parameter estimates and model selection criteria for UK.

Model	Parameter	ℓ	AIC	BIC	AD	CVM
BTCPE	$\alpha = 16.6904(5.2798)$ $\lambda = 2.3884(0.2865)$	45.4400	-86.8726	-82.6840	0.6494	0.1049
Beta	$\alpha = 4.0502(0.7128)$ $\beta = 10.0132(1.8287)$	45.4000	-86.7958	-82.6071	0.7356	0.1280
UBIII	$\alpha = 0.0757(0.0383)$ $\beta = 13.3804(6.5631)$	38.9000	-73.8075	-69.6188	2.8948	0.5248
BMOEE	$\alpha = 105.2655(59.9004)$ $\beta = 3.5949(0.4092)$	40.7200	-77.4396	-73.2509	1.1465	0.1698
UW	$\alpha = 0.2834(0.0602)$ $\beta = 3.1228(0.3047)$	42.5600	-81.1208	-76.9322	1.0656	0.1820
UG	$\alpha = 686.3600(2.2295 \times 10^{-10})$ $\beta = 0.0011(1.4051 \times 10^{-4})$	2.8400	-1.6760	2.5127	12.2290	2.4707
UL	$\alpha = 2.8293(0.3029)$	32.3800	-62.7533	-60.6590	4.4878	0.7574
UISDL	$\alpha = 3.4259(0.3151)$	33.6100	-65.2142	-63.1198	3.9972	0.6545

Figure 8 displays the empirical and fitted PDFs and CDFs of the various distributions used to model the UK mortality dataset. The figure gives an indication that the BTCPE distribution provides a good fit to the dataset compared to the other models.

Figure 9 is the probability–probability (P-P) plots of the fitted distributions. Figure 9 once more shows that the BTCPE distribution fits the UK drought mortality well because the points cluster along the diagonal.

The profile log-likelihood plots for the estimated parameter values of the BTCPE distribution for the UK mortality data are shown in Figure 10. From the plots, it can be observed that the estimated values are the true maxima.

7.2. Canada COVID-19 Mortality

Table 6 presents ML estimates of the parameters and their corresponding standard errors in brackets and model selection criteria for the fitted distributions. The BTCPE distribution again provides the best fit to the Canada mortality dataset since it has the highest log-likelihood and the lowest values of the AIC, BIC, AD, and CVM.

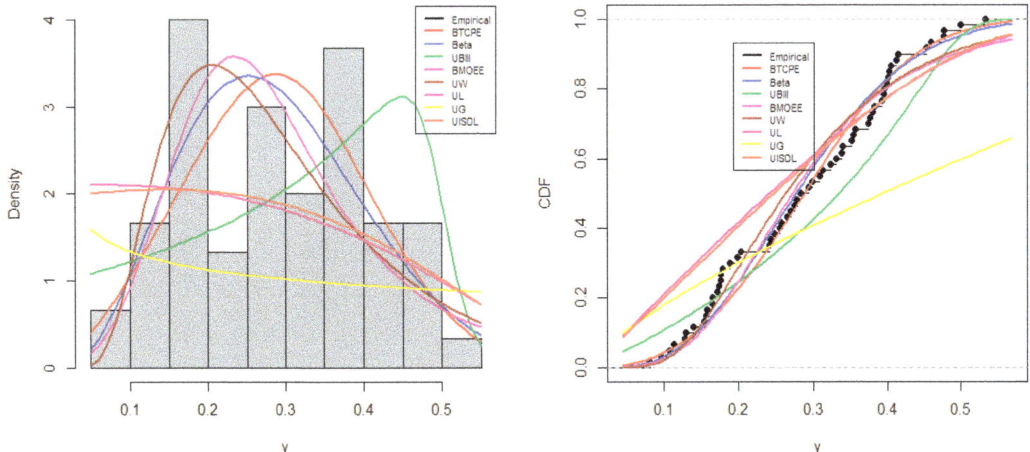

Figure 8. Empirical and fitted PDFs (**left**) and CDFs (**right**) of UK dataset.

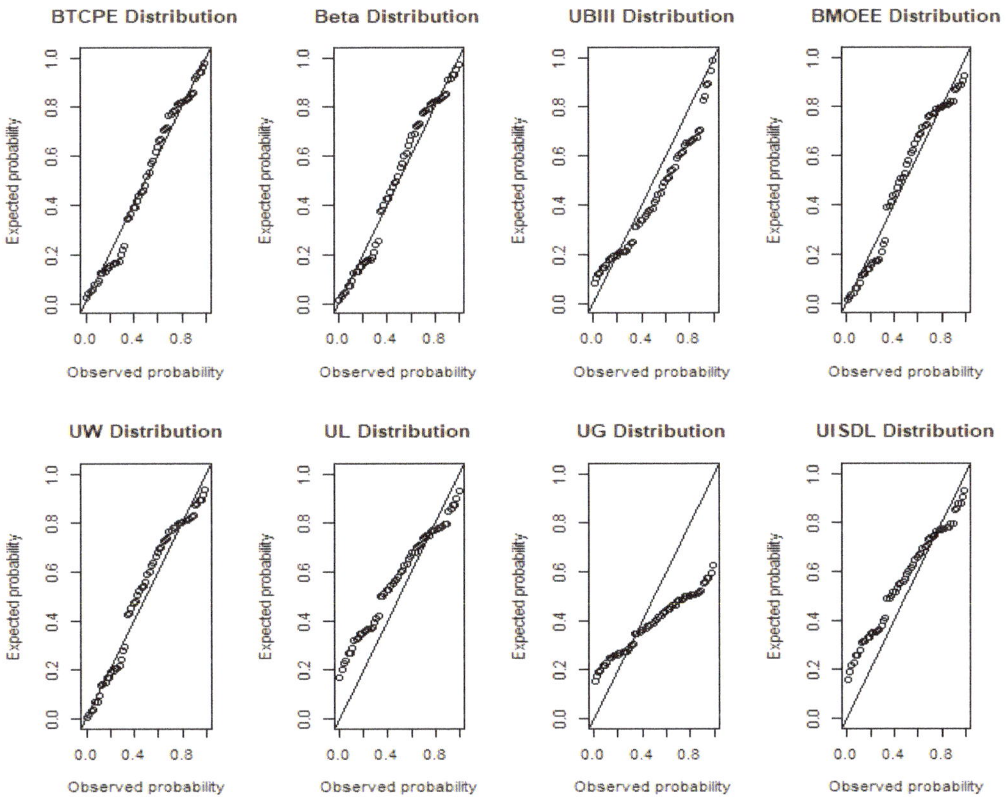

Figure 9. P-P plots for UK drought mortality.

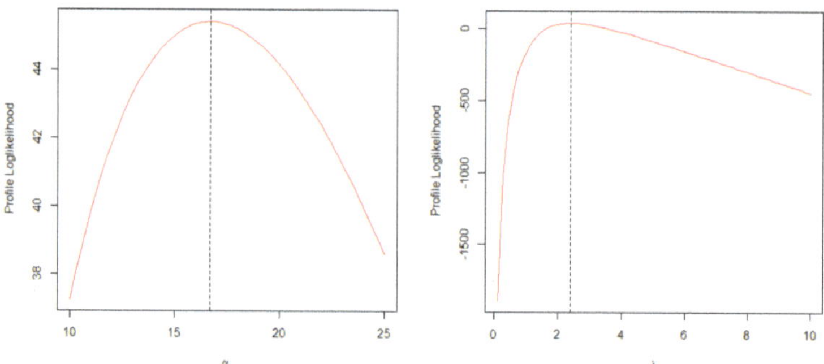

Figure 10. Profile log-likelihood plots for estimated parameters of BTCPE for UK.

Table 6. Parameter estimates and model selection criteria for Canada.

Model	Parameter	ℓ	AIC	BIC	AD	CVM
BTCPE	$\alpha = 622.2064(399.8188)$ $\lambda = 4.5085(0.4837)$	86.4400	-168.8806	-164.8299	0.3767	0.0689
Beta	$\alpha = 14.5128(2.7128)$ $\beta = 48.4900(9.1745)$	85.9400	-167.8800	-163.8293	0.4398	0.0692
UBIII	$\alpha = 0.0080(0.0011)$ $\beta = 101.7700(8.4127 \times 10^{-8})$	30.8900	-57.7749	-53.7242	14.8770	3.1113
BMOEE	$\alpha = 2822.9776(3.3087 \times 10^{-5})$ $\beta = 5.4444(0.1439)$	80.6700	-157.3394	-153.2887	1.5514	0.2327
UW	$\alpha = 0.0552(0.0193)$ $\beta = 6.1602(0.5868)$	79.9500	-155.9080	-151.8573	1.4890	0.2389
UG	$\alpha = 628.3885(2.4072 \times 10^{-10})$ $\beta = 0.0011(1.4212 \times 10^{-4})$	5.2500	-6.4901	-2.4393	18.5180	3.9712
UL	$\alpha = 3.9381(0.4506)$	41.1400	-80.2707	-78.2453	12.7090	2.5936
UISDL	$\alpha = 3.4259(0.3151)$	42.2000	-82.3913	-80.3660	12.3010	2.4925

Figure 11 shows the empirical and fitted PDFs and CDFs of the various distributions used to model the Canada drought mortality dataset. The figure gives an indication that the BTCPE distribution provides a better fit to the drought mortality for Canada than the other models, as it mimics the empirical PDF and CDF of the dataset better than the other models.

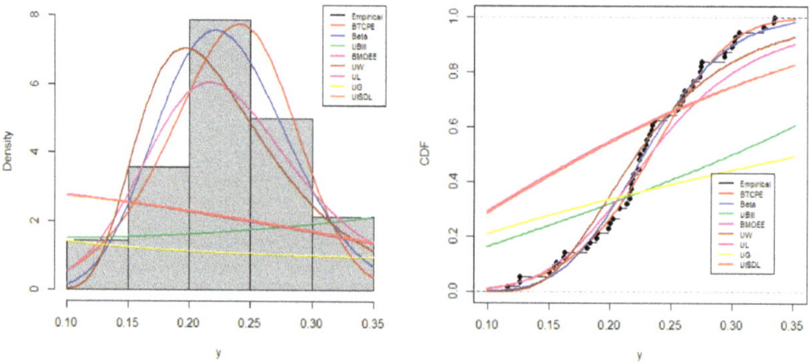

Figure 11. Empirical and fitted PDFs (**left**) and CDFs (**right**) of Canada dataset.

Figure 12 shows the P-P plots of the fitted models. Figure 12 gives an indication that the BTCPE distribution provides a good fit to the Canada mortality as the points cluster along the diagonal.

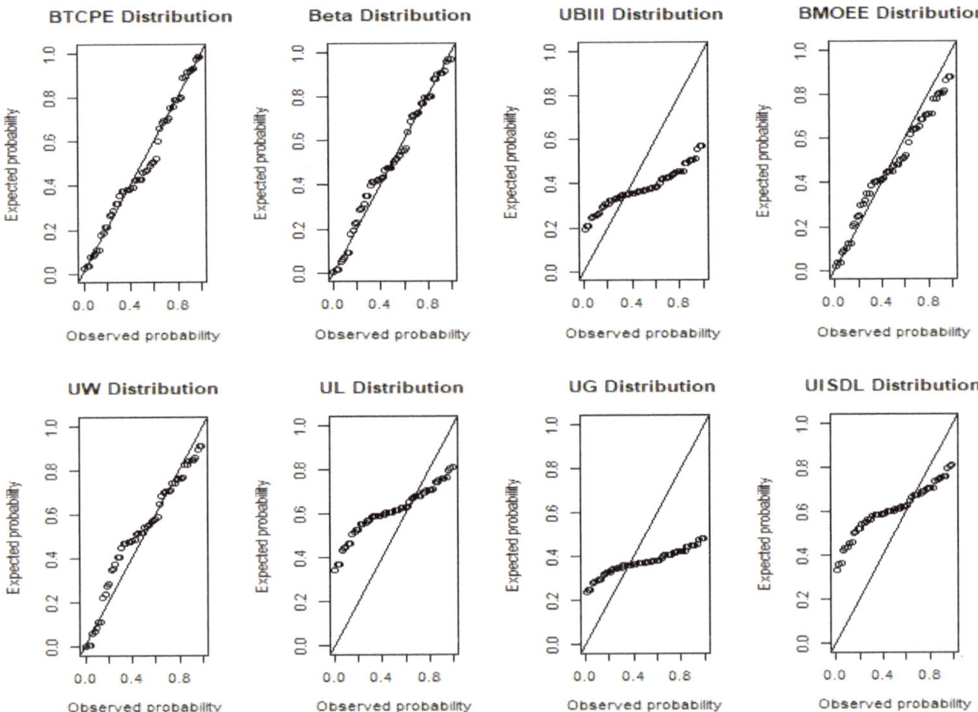

Figure 12. P-P plots for Canada mortality.

Figure 13 displays the profile log-likelihood plots for the estimated parameter values of the BTCPE distribution for the Canada mortality data. It can be observed from the plots that the estimates are unique and represent the true maxima.

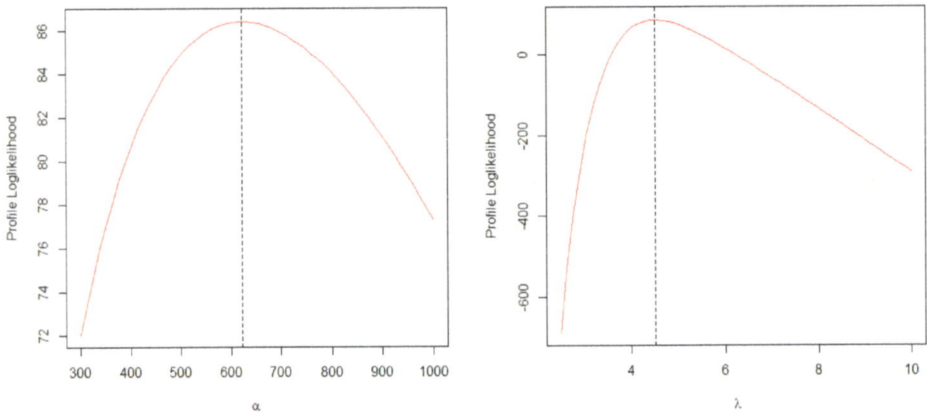

Figure 13. Profile log-likelihood plots for estimated parameters of BTCPE for Canada.

7.3. Spain COVID-19 Recovery Rate

The ML estimates of the parameters and their corresponding standard errors in brackets and model selection criteria for the fitted distributions are shown in Table 7. Because it has the lowest values for the AIC, BIC, AD, and CVM and the maximum log-likelihood, the BTCPE distribution again offers the best fit to the Spain recovery rate dataset.

Table 7. Parameter estimates and model selection criteria for Canada.

Model	Parameter	ℓ	AIC	BIC	AD	CVM
BTCPE	$\alpha = 7.1385(1.7764)$ $\lambda = 7.1961(0.9033)$	58.7500	-113.4953	-109.1160	0.8770	0.1363
Beta	$\alpha = 12.7943(2.2291)$ $\beta = 4.8994(0.8270)$	57.5700	-111.1489	-106.7692	1.0520	0.1783
UBIII	$\alpha = 5.4398(0.7948)$ $\beta = 2.0613(0.1723)$	53.8000	-103.5927	-99.2134	1.3725	0.2209
BMOEE	$\alpha = 22.1286(9.9041)$ $\beta = 10.0043(1.2381)$	51.4600	-98.9276	-94.5483	1.4958	0.2100
UW	$\alpha = 8.6445(1.6973)$ $\beta = 2.2320(0.2036)$	53.9700	-103.9316	-99.5523	1.3830	0.2238
UG	$\alpha = 0.2792(0.1059)$ $\beta = 3.8482(0.6025)$	46.0300	-88.0569	-83.6776	2.4709	0.3691
UL	$\alpha = 0.5200(0.0466)$	46.1100	-90.2298	-88.0402	4.2480	0.6736
UISDL	$\alpha = 0.7403(0.0539)$	52.0400	-102.0717	-99.8820	2.3450	0.3194

The empirical and fitted PDFs and CDFs of the various distributions used to model the Spain recovery rate dataset are shown in Figure 14. It can be seen that the BTCPE distribution provides a better fit to the recovery rate data than the other models.

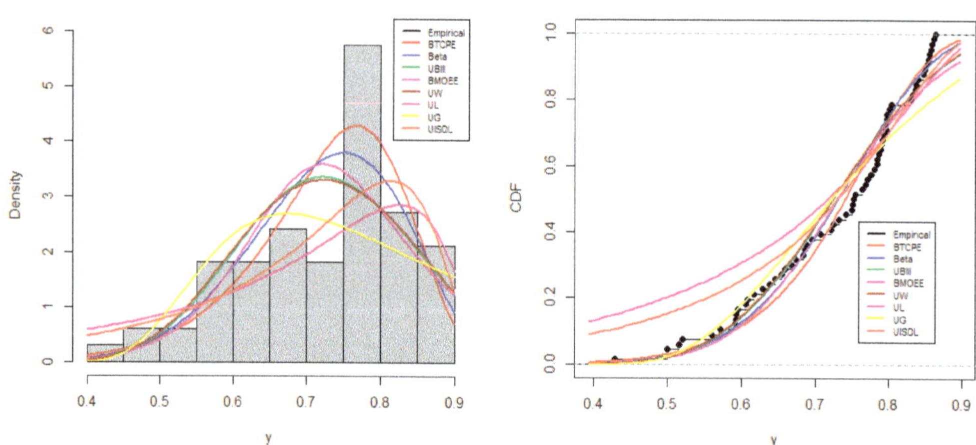

Figure 14. Empirical and fitted PDFs (**left**) and CDFs (**right**) of Spain dataset.

The P-P plots of the fitted models for the recovery rate data are displayed in Figure 15. The plots indicate that the BTCPE distribution provides a good fit to the recovery rate data as the points cluster along the diagonal.

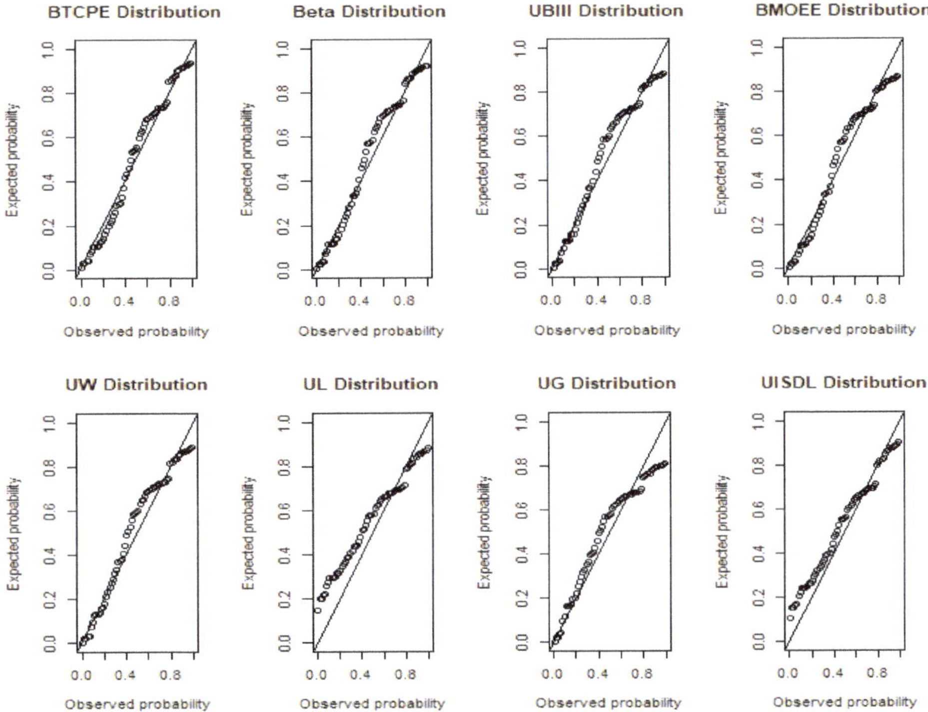

Figure 15. P-P plots for Spain recovery data.

The profile log-likelihood plots for the estimated parameter values of the BTCPE distribution for the recovery rate data are shown in Figure 16. The plots suggest that the estimates are unique and represent the true maxima.

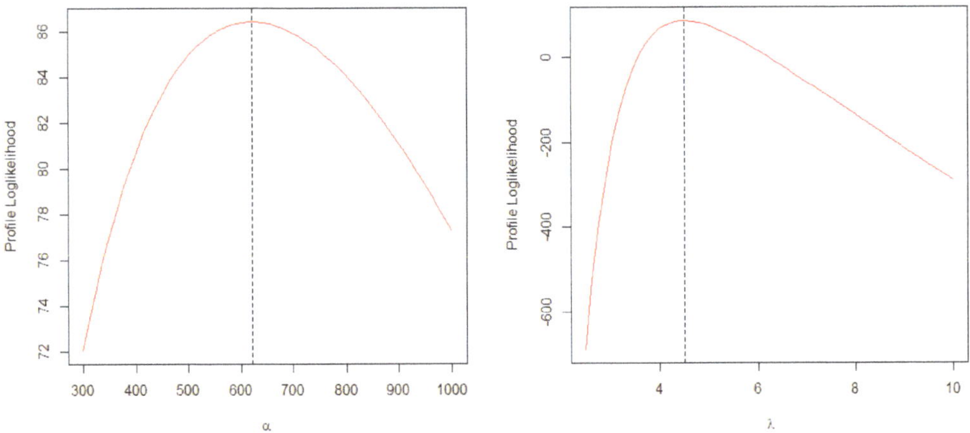

Figure 16. Profile log-likelihood plots for estimated parameters of BTCPE for Spain.

8. Quantile Regression

When the response variable defined in the unit interval is skewed or contaminated with outliers, the beta regression model, which models the conditional mean of the response variable, is no longer reliable. A robust regression model is needed to model the effects of the covariates on the response variable. In this study, a quantile regression model is proposed for modeling the conditional quantile of the response variable. Given the quantile function of the BTCPE distribution, the PDF of the BTCPE distribution can be re-parameterized in terms of its uth quantile as $\rho = Q(u; \alpha, \lambda), \rho \in [0,1]$. If $\lambda = \log(1 - (\tan[\pi(1-u)/4])^{1/\alpha})/\log(\rho)$, then the re-parameterized PDF is

$$f_Y(y; \alpha, \lambda) =$$
$$\frac{4\alpha(\log(1-(\tan[\pi(1-u)/4])^{1/\alpha})/\log(\rho))y^{(\log(1-(\tan[\pi(1-u)/4])^{1/\alpha})/\log(\rho))-1}(1-y^{(\log(1-(\tan[\pi(1-u)/4])^{1/\alpha})/\log(\rho))})^{\alpha-1}}{\pi[1+(1-y^{(\log(1-(\tan[\pi(1-u)/4])^{1/\alpha})/\log(\rho))})^{2\alpha}]}. \tag{29}$$

The parameter ρ is the quantile parameter. The BTCPE quantile regression is defined as

$$g(\rho_i) = z'_i\boldsymbol{\theta},$$

where $\boldsymbol{\theta} = (\theta_0, \theta_1, \ldots, \theta_p)'$ is the vector of unknown parameters, ρ_i is the ith quantile parameter and $z'_i = (1, z_{i1}, z_{i2}, \ldots, z_{ip})$ are the known i^{th} vector of covariates. The link function $g(\cdot)$ is used to link the covariates to the conditional median of the dependent variable Y. The logit link function is used to link the covariates to the conditional quantile since $y \in (0, 1)$. Hence, we have

$$g(\rho_i) = \text{logit}(\rho_i) = \log(\frac{\rho_i}{1 - \rho_i}).$$

Further, we can write

$$\rho_i = \frac{\exp(z'_i\boldsymbol{\theta})}{1 + \exp(z'_i\boldsymbol{\theta})}.$$

Substituting ρ_i into the re-parameterized PDF, the log-likelihood for estimating the parameters of the BTCPE quantile regression is given by

$$\begin{aligned}
\ell = &\sum_{i=1}^n \log\Big((4\alpha/\pi)(\log(1 - (\tan[\pi(1-u)/4])^{1/\alpha})/\log(\rho_i))\Big) - \sum_{i=1}^n \log(1 + (1 - z_i)^{2\alpha}) + \\
&\sum_{i=1}^n [(\log(1 - (\tan[\pi(1-u)/4])^{1/\alpha})/\log(\rho_i)) - 1]\log(y_i) + (\alpha - 1)\sum_{i=1}^n \log(1 - z_i),
\end{aligned} \tag{30}$$

where $z_i = y_i^{(\log(1-(\tan[\pi(1-u)/4])^{1/\alpha})/\log(\rho_i))}$. The estimates of the parameters of the regression equation are obtained by directly maximizing the log-likelihood function. They will be denoted as $\hat{\alpha}$ and $\hat{\boldsymbol{\theta}} = (\hat{\theta}_0, \ldots, \hat{\theta}_p)'$ of α and $\boldsymbol{\theta}$, respectively.

8.1. Residual Analysis

Model diagnostics are very essential when fitting a model to a dataset. Often, the behavior of the model residuals is examined to see if the model really provides a good fit to the data. In this study, the randomized quantile residuals are used to assess the adequacy of the regression model. The randomized quantile residuals are defined as

$$r_i = \Phi^{-1}(F_Y(y_i; \hat{\alpha}, \hat{\boldsymbol{\theta}})), i = 1, 2, \ldots, n,$$

where $\Phi^{-1}(\cdot)$ is the quantile of the standard normal distribution. The randomized quantile residuals are expected to be distributed as the standard normal distribution if the models provide a good fit to the data.

8.2. Monte Carlo Simulation for Quantile Regression

Monte Carlo simulations are carried out in this section to examine the performance of the ML estimates of the parameters of the BTCPE regression model. The exercise is performed with two covariates. The following regression structure is adopted for the simulation:

$$\rho_i = \frac{\exp(\theta_0 + \theta_1 z_{i1} + \theta_2 z_{i2})}{1 + \exp(\theta_0 + \theta_1 z_{i1} + \theta_2 z_{i2})}.$$

The observations for the response variable are generated from the BTCPE distribution using sample sizes $n = 50, 100, 250, 350, 500, 600$ and 700. The experiments were repeated 5000 times for each sample size. The performance of the ML estimates is examined using AB and RMSE. The simulations were carried out using the median, $u = 0.5$. The following parameter combinations were used in the simulation: $I : (\alpha, \theta_0, \theta_1, \theta_2) = (0.7, 0.2, 0.8, 0.3)$, $II : (\alpha, \theta_0, \theta_1, \theta_2) = (0.6, 0.5, 0.4, 1.8)$ and $III : (\alpha, \theta_0, \theta_1, \theta_2) = (0.8, 0.4, 0.9, 0.6)$. From the simulation results shown in Table 8, the ABs and RMSEs of the estimates' decrease as the sample size increases. Hence, the ML estimates for the BTCPE regression parameters are consistent.

Table 8. Simulation results for the quantile regression.

		I		II		III	
Parameter	n	AB	RMSE	AB	RMSE	AB	RMSE
θ_0	50	0.1949	0.2235	0.3599	0.3753	0.2609	0.2969
	100	0.1946	0.1961	0.3551	0.3726	0.2178	0.2579
	250	0.1919	0.1941	0.3465	0.3673	0.1525	0.1926
	350	0.1898	0.1928	0.3271	0.3544	0.1320	0.1700
	500	0.1838	0.1927	0.3109	0.3482	0.1101	0.1431
	600	0.1779	0.1886	0.3051	0.3434	0.0998	0.1318
	700	0.1761	0.1850	0.2908	0.3333	0.0908	0.1196
θ_1	50	0.2826	0.3067	0.3485	0.3807	0.8194	0.8276
	100	0.2605	0.2904	0.3181	0.3486	0.8142	0.8238
	250	0.2290	0.2651	0.3171	0.3363	0.8013	0.8134
	350	0.2176	0.2539	0.3138	0.3342	0.7872	0.8041
	500	0.2097	0.2454	0.3083	0.3305	0.7727	0.7945
	600	0.2079	0.2433	0.3020	0.3272	0.7188	0.7610
	700	0.2053	0.2389	0.2978	0.3253	0.6862	0.7447
θ_2	50	1.5889	1.5959	1.7104	1.7153	0.5212	0.5338
	100	1.5835	1.5913	1.7046	1.7102	0.5140	0.5291
	250	1.5818	1.5910	1.6938	1.7006	0.5073	0.5250
	350	1.5698	1.5815	1.6751	1.6845	0.4893	0.5130
	500	1.5566	1.5723	1.6432	1.6578	0.4753	0.5046
	600	1.4749	1.5132	1.5559	1.5917	0.4601	0.4999
	700	1.3803	1.4520	1.4593	1.5264	0.4535	0.4921
α	50	0.0792	0.0998	0.0842	0.1110	0.1091	0.1520
	100	0.0577	0.0745	0.0570	0.0747	0.0872	0.1382
	250	0.0352	0.0463	0.0339	0.0437	0.0523	0.0859
	350	0.0295	0.0378	0.0287	0.0366	0.0427	0.0650
	500	0.0246	0.0316	0.0239	0.0317	0.0340	0.0467
	600	0.0227	0.0287	0.0217	0.0290	0.0317	0.0449
	700	0.0210	0.0267	0.0201	0.0259	0.0287	0.0375

8.3. Application

The application of the quantile regression model is demonstrated in this section using a real dataset. The data are taken from [34] and are also available at http://www.leg.ufpr.br/doku.php/publications:papercompanions:multquasibeta (accessed on 30 August 2022). The data consist of body fat percentage (response variable) measured in five regions: android, arms, gynoids, legs and trunk. The data are comprised of 298 observations and the independent variables are: age (in years), body mass index (in kg/m^2), sex (female or male) and IPAQ (sedentary (S), insufficiently active (I), or active (A)). In this study, the response variable body fat percentage at arms is regressed on age (z_{i1}), body mass index (z_{i2}) and sex

(z_{i3}, 0 for female and 1 for male). The response variable is regressed on the covariates using the relationship $\mathrm{logit}(\rho_i) = \theta_0 + \theta_1 z_{i1} + \theta_2 z_{i2} + \theta_3 z_{i3}, i = 1, 2, \ldots, 298$. Table 9 presents ML estimates, standard errors, and p-values for the parameters of the fitted models for the different quantiles. The estimates are all significant at the 5% level of significance.

Table 9. ML estimates for quantile regression.

u		$\hat{\theta}_0$	$\hat{\theta}_1$	$\hat{\theta}_2$	$\hat{\theta}_3$	$\hat{\alpha}$
	Estimates	−3.6699	0.0076	0.0905	−1.004	308.7724
0.10	Standard error	0.1681	1.1670×10^{-3}	7.5355×10^{-3}	4.3797×10^{-2}	9.3305×10^{-5}
	p-value	< 0.0001	< 0.0001	< 0.0001	< 0.0001	< 0.0001
	Estimates	−3.2544	0.0071	0.0845	−0.9326	325.4705
0.25	Standard error	0.1545	1.0687×10^{-3}	6.9379×10^{-3}	4.0103×10^{-2}	4.6137×10^{-5}
	p-value	< 0.0001	< 0.0001	< 0.0001	< 0.0001	< 0.0001
	Estimates	−2.8977	0.0067	0.0792	−0.8732	340.4285
0.50	Standard error	0.1436	9.9065×10^{4}	6.4570×10^{-3}	3.7166×10^{-2}	1.3990×10^{-5}
	p-value	< 0.0001	< 0.0001	< 0.0001	< 0.0001	< 0.0001
	Estimates	−2.6424	0.0064	0.0766	−0.8384	281.1611
0.75	Standard error	0.1405	9.7128×10^{-4}	6.3363×10^{-3}	3.6303×10^{-2}	6.4012×10^{-6}
	p-value	< 0.0001	< 0.0001	< 0.0001	< 0.0001	< 0.0001
	Estimates	−2.4030	0.0061	0.0731	−0.7987	273.9968
0.90	Standard error	0.1353	9.3470×10^{-4}	6.1047×10^{-3}	3.4900×10^{-2}	2.9792×10^{-5}
	p-value	< 0.0001	< 0.0001	< 0.0001	< 0.0001	< 0.0001

Table 10 presents the model selection criteria for the different quantiles. It is observed that the 0.90th quantile provides the best fit for the data as it has the least values of the model selection criteria.

Table 10. Model selection criteria for quantile regression.

u	-2ℓ	AIC	BIC
0.10	−885.3517	−875.3517	−856.8663
0.25	−887.4067	−877.4067	−858.9212
0.50	−889.1990	−879.1990	−860.7136
0.75	−889.8634	−879.8634	−861.3779
0.90	−890.8307	−880.8307	−862.3453

Figure 17 shows the rate of change of the regression coefficients for the different quantile levels and the corresponding 95% confidence interval (CI). It can be observed that all the coefficients approach zero as the quantile level increases, suggesting that they are more important in explaining smaller quantiles.

Figures 18 and 19 show the P-P plots and half-normal plots with simulated envelopes, respectively, for the randomized quantile residuals. These figures display good fits of the BTCPE quantile regression model to the uth percentage of body fat in arms for $u \in (0.10, 0.25, 0.50, 0.75, 0.90)$.

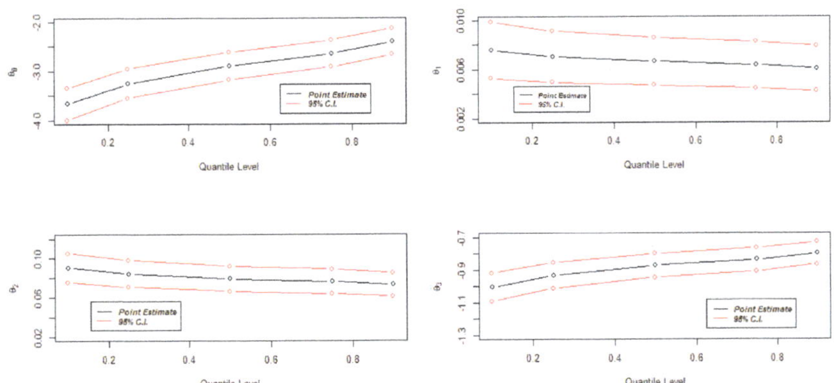

Figure 17. Rate of change of regression coefficients for different quantiles.

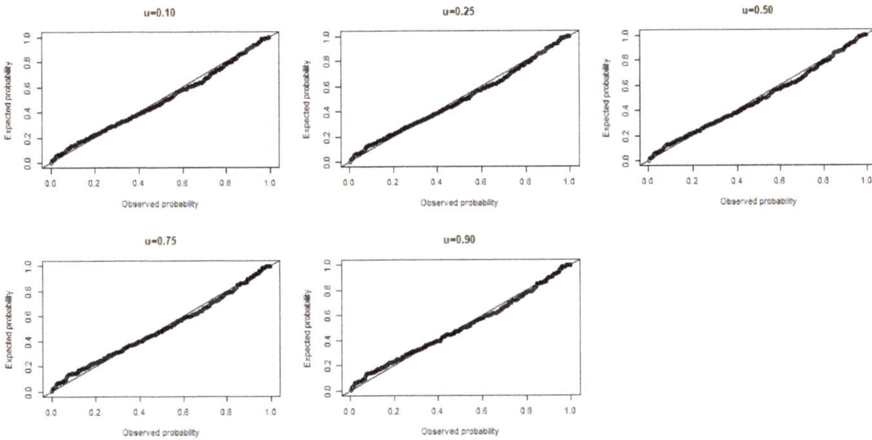

Figure 18. P-P plots for randomized quantile residuals.

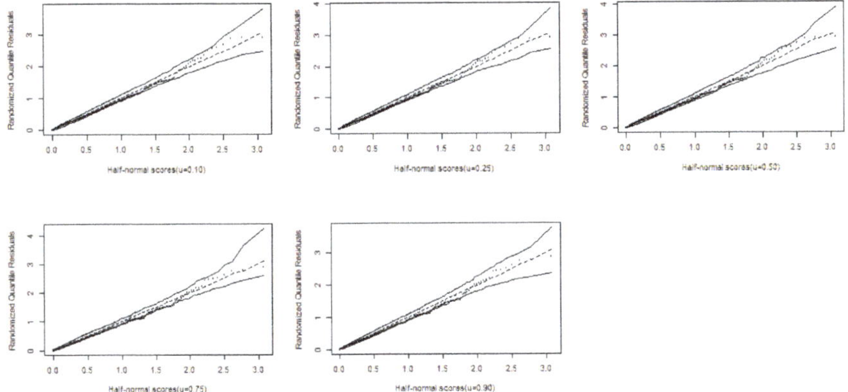

Figure 19. Half-normal plots with simulated envelopes for randomized quantile residuals.

9. Conclusions

In this study, the BTCPE distribution is proposed for modeling datasets that are defined on the unit interval. The PDF of this distribution exhibits left-skewed, right-skewed, reversed J, and approximately symmetric shapes. The HRF displays increasing and bathtub shapes. This makes the distribution a suitable candidate for modeling datasets that exhibit these traits. Nine estimation methods were proposed for estimating the parameters of the distribution, and simulation results revealed that most of these estimates were consistent when it came to the estimation of the parameters of the distribution. The applications of the BTCPE distribution were illustrated using datasets on the mortality rate and recovery rates of COVID-19. The results revealed that for the three datasets, the BTCPE model provided a better fit than the other competing models. A quantile regression model for studying the relationship between the conditional quantiles of a bounded response variable and a set of covariates was proposed. The application of the regression model was illustrated using real data. The study only defined the cumulative distribution and probability density functions of the bivariate distribution. Our future research will study the detailed properties of the bivariate distribution, estimate its parameters, and illustrate its applications.

Author Contributions: Conceptualization, S.N., A.G.A., and C.C.; Data curation, S.N., A.G.A., and C.C.; Methodology, S.N., A.G.A., and C.C.; Supervision, S.N., and C.C.; Validation, S.N., and C.C.; Visualization, S.N., and A.G.A.; Writing, S.N., and A.G.A.; Review & editing, S.N., and C.C. All authors read and agreed to the published version of the manuscript.

Funding: This research received no external funding.

Conflicts of Interest: The authors declare no conflict of interest.

References

1. Afify, A.Z.; Nassar, M.; Kumar, D.; Cordeiro, G.M. A new unit distribution: Properties and applications. *Electron. J. Appl. Stat. Anal.* **2022**, *15*, 460–484.
2. Almazah, M.M.A.; Ullah, K.; Hussam, E.; Hossain, M.; Aldallal, R.; Riad, F.H. New Statistical Approaches for Modeling the COVID-19 Data Set: A Case Study in the Medical Sector. *Complexity* **2022**, *2022*, 1325825. [CrossRef]
3. Alahmadi, A.A.; Alqawba, M.; Almutiry, W.; Shawki, A.W.; Alrajhi, S.; Al-Marzouki, S.; Elgarhy, M. A New Version of Weighted Weibull Distribution: Modelling to COVID-19 Data. *Discret. Dyn. Nat. Soc.* **2022**, *2022*, 3994361. [CrossRef]
4. Algarni, A.; Almarashi, A.M.; Elbatal, I.; Hassan, A.S.; Almetwally, E.M.; Daghistani, A.M.; Elgarhy, M. Type I Half Logistic Burr X-G Family: Properties, Bayesian, and Non-Bayesian Estimation under Censored Samples and Applications to COVID-19 Data. *Math. Probl. Eng.* **2021**, *2021*, 5461130. [CrossRef]
5. Bantan, R.A.R.; Shafiq, S.; Tahir, M.H.; Elhassanein, A.; Jamal, F.; Almutiry, W.; Elgarhy, M. Statistical Analysis of COVID-19 Data: Using a New Univariate and Bivariate Statistical Model. *J. Funct. Spaces* **2022**, *2022*, 2851352. [CrossRef]
6. Arif, M.; Khan, D.M.; Aamir, M.; Khalil, U.; Bantan, R.A.R.; Elgarhy, M. Modeling COVID-19 Data with a Novel Extended Exponentiated Class of Distributions. *J. Math.* **2022**, *2022*, 1908161. [CrossRef]
7. Nagy, M.; Almetwally, E.M.; Gemeay, A.M.; Mohammed, H.S.; Jawa, T.M.; Sayed-Ahmed, N.; Muse, A.H. The New Novel Discrete Distribution with Application on COVID-19 Mortality Numbers in Kingdom of Saudi Arabia and Latvia. *Complexity* **2021**, *2021*, 7192833. [CrossRef]
8. Almetwally, E.M. The Odd Weibull Inverse Topp–Leone Distribution with Applications to COVID-19 Data. *Ann. Data Sci.* **2021**, *9*, 121–140. [CrossRef]
9. Muse, A.H.; Tolba, A.H.; Fayad, E.; Abu Ali, O.A.; Nagy, M.; Yusuf, M. Modelling the COVID-19 Mortality Rate with a New Versatile Modification of the Log-Logistic Distribution. *Comput. Intell. Neurosci.* **2021**, *2021*, 8640794. [CrossRef]
10. Haq, M.A.U.; Babar, A.; Hashmi, S.; Alghamdi, A.S.; Afify, A.Z. The Discrete Type-II Half-Logistic Exponential Distribution with Applications to COVID-19 Data. *Pak. J. Stat. Oper. Res.* **2021**, *17*, 921–932. [CrossRef]
11. Gündüz, S.; Korkmaz, M. A New Unit Distribution Based on the Unbounded Johnson Distribution Rule: The Unit Johnson SU Distribution. *Pak. J. Stat. Oper. Res.* **2020**, *16*, 471–490. [CrossRef]
12. Bantan, R.; Jamal, F.; Chesneau, C.; Elgarhy, M. Theory and Applications of the Unit Gamma/Gompertz Distribution. *Mathematics* **2021**, *9*, 1850. [CrossRef]
13. Nasiru, S.; Abubakari, A.G.; Angbing, I.D. Bounded Odd Inverse Pareto Exponential Distribution: Properties, Estimation, and Regression. *Int. J. Math. Math. Sci.* **2021**, *2021*, 9955657. [CrossRef]
14. Jodrá, P. A bounded distribution derived from the shifted Gompertz law. *J. King Saud Univ.-Sci.* **2020**, *32*, 523–536. [CrossRef]

15. Haq, M.A.U.; Hashmi, S.; Aidi, K.; Ramos, P.L.; Louzada, F. Unit Modified Burr-III Distribution: Estimation, Characterizations and Validation Test. *Ann. Data Sci.* **2020**, *99*, 1–26. [CrossRef]
16. Korkmaz, M.Ç. The unit generalized half normal distribution: A new bounded distribution with inference and application. *U. P. B. Sci. Bull. Ser. A* **2020**, *82*, 133–140.
17. Mazucheli, J.; Menezes, A.F.B.; Chakraborty, S. On the one parameter unit-Lindley distribution and its associated regression model for proportion data. *J. Appl. Stat.* **2019**, *46*, 700–714. [CrossRef]
18. Mazucheli, J.; Menezes, A.F.; Dey, S. Unit-Gompertz distribution with applications. *Statistica* **2019**, *79*, 25–43.
19. Korkmaz, M. A new heavy-tailed distribution defined on the bounded interval. *J. Appl. Stat.* **2019**, *47*, 2097–2119. [CrossRef]
20. Mazucheli, J.; Menezes, A.F.; Ghitany, M.E. The unit Weibull distribution and associated inference. *J. Appl. Probab. Stat.* **2018**, *13*, 1–22.
21. Ghitany, M.E.; Mazucheli, J.; Menezes, A.F.B.; Alqallaf, F. The unit-inverse Gaussian distribution: A new alternative to two-parameter distributions on the unit interval. *Commun. Stat.-Theory Methods* **2018**, *48*, 3423–3438. [CrossRef]
22. Aldahlan, M.A.; Jamal, F.; Chesneau, C.; Elgarhy, M.; Elbatal, I. The Truncated Cauchy Power Family of Distributions with Inference and Applications. *Entropy* **2020**, *22*, 346. [CrossRef] [PubMed]
23. Shaked, M.; Shanthikumar, J.G. *Stochastic Orders*; Wiley: New York, NY, USA, 2007.
24. MacGillivray, H.L. Skewness and Asymmetry: Measures and Orderings. *Ann. Stat.* **1986**, *14*, 994–1011. [CrossRef]
25. Moors, J.J. A quantile alternative for kurtosis. *J. R. Stat. Soc. Ser. D* **1988**, *37*, 25–32. [CrossRef]
26. Elhassanein, A. On Statistical Properties of a New Bivariate Modified Lindley Distribution with an Application to Financial Data. *Complexity* **2022**, *2022*, 2328831. [CrossRef]
27. Ganji, M.; Bevrani, H.; Hami, N. A New Method For Generating Continuous Bivariate Families. *J. Iran. Stat. Soc.* **2018**, *17*, 109–129. [CrossRef]
28. R Core Team. *R: A Language and Environment for Statistical Computing*; R Foundation for Statistical Computing: Vienna, Austria, 2019; Available online: https://www.R-project.org/ (accessed on 30 October 2022).
29. Modi, K.; Gill, V. Unit Burr-III distribution with application. *J. Stat. Manag. Syst.* **2019**, *23*, 579–592. [CrossRef]
30. Ghosh, I.; Dey, S.; Kumar, D. Bounded M-O Extended Exponential Distribution with Applications. *Stoch. Qual. Control* **2019**, *34*, 35–51. [CrossRef]
31. Altun, E.; Cordeiro, G.M. The unit-improved second-degree Lindley distribution: Inference and regression modeling. *Comput. Stat.* **2019**, *35*, 259–279. [CrossRef]
32. Bolker, B. *Tools for General Maximum Likelihood Estimation*; R Development Core Team: Vienna, Austria, 2014. Available online: https://github.com/bbolker/bbmle (accessed on 30 October 2022).
33. Xiang, Y.; Gubian, S.; Suomela, B.; Hoeng, J. Generalized simulated annealing: GenSA package. *R J.* **2013**, *5*, 13–29. [CrossRef]
34. Petterle, R.R.; Bonat, W.H.; Scarpin, C.T.; Jonasson, T.; Borba, V.Z.C. Multivariate quasi-beta regression models for continuous bounded data. *Int. J. Biostat.* **2020**, *17*, 39–53. [CrossRef] [PubMed]

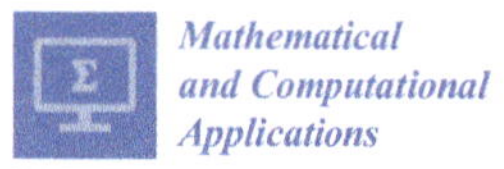
Mathematical and Computational Applications

Article

Pseudo-Poisson Distributions with Concomitant Variables

Barry C. Arnold [1,*] **and Bangalore G. Manjunath** [2,*]

1   Department of Statistics, University of California, Riverside, CA 92521, USA
2   School of Mathematics and Statistics, University of Hyderabad, Hyderabad 500046, India
*   Correspondence: barnold@ucr.edu (B.C.A.); bgmanjunath@gmail.com (B.G.M.)

Abstract: It has been argued in Arnold and Manjunath (2021) that the bivariate pseudo-Poisson distribution will be the model of choice for bivariate data with one equidispersed marginal and the other marginal over-dispersed. This is due to its simple structure, straightforward parameter estimation and fast computation. In the current note, we introduce the effects of concomitant variables on the bivariate pseudo-Poisson parameters and explore the distributional and inferential aspects of the augmented models. We also include a small simulation study and an example of application to real-life data.

Keywords: correlation; likelihood ratio test; maximum likelihood estimators; pseudo-Poisson; regression

Citation: Arnold, B.C.; Manjunath, B.G. Pseudo-Poisson Distributions with Concomitant Variables. *Math. Comput. Appl.* **2023**, *28*, 11. https://doi.org/10.3390/mca28010011

Academic Editor: Sandra Ferreira

Received: 15 December 2022
Revised: 8 January 2023
Accepted: 10 January 2023
Published: 12 January 2023

1. Introduction

The classical "one-dimensional" Poisson distribution has historically been found to be useful in modeling a wide variety of "integer-valued" phenomena, such as the number of accidents and associated fatalities, disease advances, rate of rare event occurrences and so on. With regard to the Poisson model with concomitants, i.e., Poisson regression or count regression, its best known applications are in (i) modeling counts of bacteria exposed to various environmental conditions and dilutions; (ii) modeling counts of infant mortality among groups with different demographics. All these examples are typically modeled under the assumption of equi-dispersion. However, count data also exhibits over and under dispersion. In this context, the one-dimensional Conway–Maxwell–Poisson model or its regression version fills the bill precisely by allowing us to model over, equi- and under-dispersion data.

In general, bivariate count data, along with having marginal over-, under- or equi-dispersion, will also exhibit a variety of dependence structures. In particular, for linear dependence, the possible relations are positive or negative correlation. The classical bivariate Poisson model is appropriate for data having equi-dispersed marginals with positive correlation. Here again, the bivariate Conway–Maxwell–Poisson is more flexible in that it can adapt to both under and over dispersed data, see Sellers et al. [1]. Concerning bivariate Poisson regression models, the first version involving explanatory variables acting on the marginal means was introduced in Kocherlakota and Kocherlakota [2] based on the classical bivariate Poisson model. In addition, the derivation of Wald, score and likelihood ratio test statistics for testing a single coefficient parameter vector are discussed in Riggs et al. [3]. Zamani et al. [4] proposed a bivariate Poisson model which can be fitted to both positive and negatively correlated data. Recently, Chowdhury et al. [5] considered the Poisson–Poisson regression model (which is the particular case of the bivariate pseudo-Poisson model) to analyze the impact of covariates on the daily new cases and fatalities associated with the COVID-19 pandemic. Finally, we refer to Karlis and Ntzoufras [6] and the R package *bivpois* for maximum likelihood estimation, using an Expectation-Maximization (EM) algorithm, for diagonally inflated bivariate Poisson regression models.

In recent work, Arnold and Manjunath [7] recommended the bivariate pseudo-Poisson model to fit data which have one marginal and other conditional of the Poisson form. Due to its straightforward structure with no restrictions on the conditional mean function, it allows us to include a variety of dependence structures, including positive and negative correlation. In the following, we introduce explanatory variables acting on the pseudo-Poisson parameters. Thanks to the simple structure, the concomitant effects can be introduced into each of the parameters to generate a family of models with a variety of dependence structures. We refer to Arnold et al. [8] and Veeranna et al. [9] on Bayesian and goodness-of-fit tests for the bivariate pseudo-Poisson model, respectively, which can also be adapted to accommodate the presence of concomitant variables. We refer to Arnold et al. [10] and Filus et al. [11] for further reading on conditional specified models and the triangular transformations, respectively. Finally, we refer Ghosh et al. [12] on the recent results on bivariate count model which has both conditionals with a Poisson structure.

We next review the concept of multivariate pseudo-Poisson distributions, as discussed in Arnold and Manjunath [7].

Definition 1. *A k-dimensional random variable $\underline{X} = (X_1, X_2, \ldots, X_k)$ is said to have a k-dimensional pseudo-Poisson distribution if there exists a positive constant λ_1 such that*

$$X_1 \sim \mathscr{P}(\lambda_1)$$

and $k - 1$ functions $\{\lambda_\ell : \ell = 2, 3, \ldots, k\}$ where, for each ℓ, $\lambda_\ell : \{0, 1, 2, \ldots\}^{(\ell-1)} \to (0, \infty)$ such that

$$X_\ell | \underline{X}_{(\ell-1)} = \underline{x}_{(\ell-1)} \sim \mathscr{P}(\lambda_\ell(\underline{x}_{(\ell-1)})),$$

where $\underline{X}_{(\ell-1)} = (X_1, \ldots, X_{l-1})^\top$. Note that there are no constraints on the forms of the functions λ_ℓ, $\ell = 2, 3, \ldots, k$ that appear in the definition, save for measurability. In applications, it would typically be the case that the λ_ℓ's would be chosen to be relatively simple functions depending on a limited number of parameters.

Definition 2. *A random pair of variables (X_1, X_2) is said to have a bivariate pseudo-Poisson distribution if there exists a positive constant λ_1 such that*

$$X_1 \sim \mathscr{P}(\lambda_1)$$

and a function $\lambda_2 : \{0, 1, 2, \ldots\} \to (0, \infty)$ such that, for every non-negative integer x_1,

$$X_2 | X_1 = x_1 \sim \mathscr{P}(\lambda_2(x_1)).$$

The fact that there are no constraints on the $\lambda_2(x_1)$ allows us to adapt to a variety of dependence structures including positive or negative correlation.

Example 1. *A judicious choice of a parametric family for $\lambda_2(x_1)$ will admit positive and negative correlation between X_1 and X_2. For example, if we consider*

$$\lambda_2(x_1; \gamma, \delta) = 1 + (2\gamma - 1)(1 - e^{-\delta x_1}). \tag{1}$$

For $\delta > 0$, the above function will be increasing if $\gamma > 1/2$, decreasing if $\gamma < 1/2$ and constant if $\gamma = 1/2$. Consequently, X_1 and X_2 will have a positive correlation if $\gamma > 1/2$, negative correlation if $\gamma < 1/2$ and will be uncorrelated if $\gamma = 1/2$. A more general model with the same properties can be obtained by replacing $1 - e^{-\delta x_1}$ by $F(x_1; \underline{\theta})$, a parameterized family of distribution functions with support $(0, \infty)$.

2. Incorporating Concomitant Variables

In many (perhaps, most) applications, in addition to the observed values of the $(X_{1,i}, X_{2,i})$'s pairs, there will be available values of arrays of concomitant variables which are expected to influence the stochastic behavior of the observed data points. A straightforward manner in which to incorporate vectors of concomitant variables $\underline{u}_i = (u_{1i}, \ldots, u_{di})^\top$ into the model is as follows:

$$X_1 \sim \mathscr{P}\left(\lambda_1 \exp\left(\underline{\alpha}^\top \underline{u}\right)\right) \tag{2}$$

and

$$X_2 | X_1 = x_1 \sim \mathscr{P}\left(\lambda_2 \exp\left(\underline{\beta}^\top \underline{u}\right) + \lambda_3 \exp\left(\underline{\gamma}^\top \underline{u}\right) x_1\right) \tag{3}$$

where $\lambda_1 > 0$, $\lambda_2 \geq 0$, $\lambda_3 > 0$, $\underline{\alpha} = (\alpha_1, \ldots, \alpha_d)^\top$, $\underline{\beta} = (\beta_1, \ldots, \beta_d)^\top$ and $\underline{\gamma} = (\gamma_1, \ldots, \gamma_d)^\top$ are d-dimensional unknown parameters.

There are certainly many other manners in which one can model the influence of concomitant variables. If there is scientific justification for alternative models that do not introduce the concomitants via log-linear adjustments of the form specified in (2) and (3), then one should certainly utilize the scientifically appropriate link functions.

Just as in classical multiple regression, it is worthwhile to determine whether a simple linear dependence assumption for the effect of concomitants will be adequate to fit the data. In the remainder of this paper, we will focus on the simple model (2) and (3).

3. Moments

In the following, we derive some population moments for the model specified in (2) and (3).

$$
\begin{aligned}
E(X_1) = Var(X_1) &= \lambda_1 \exp\left(\underline{\alpha}^\top \underline{u}\right) \\
E(X_2) &= \lambda_2 \exp\left(\underline{\beta}^\top \underline{u}\right) + \lambda_1 \lambda_3 \exp\left((\underline{\gamma} + \underline{\alpha})^\top \underline{u}\right) \\
V(X_2) &= \lambda_2 \exp\left(\underline{\beta}^\top \underline{u}\right) + \lambda_1 \lambda_3 \exp\left((\underline{\gamma} + \underline{\alpha})^\top \underline{u}\right) \\
&\quad + \lambda_1 \lambda_3^2 \exp\left((2\underline{\gamma} + \underline{\alpha})^\top \underline{u}\right) \\
Cov(X_1, X_2) &= \lambda_1 \lambda_3 \exp\left((\underline{\gamma} + \underline{\alpha})^\top \underline{u}\right).
\end{aligned}
$$

The marginal dispersion indices are

$$
\begin{aligned}
DI(X_1) &= \frac{Var(X_1)}{E(X_1)} = 1. \\
DI(X_2) &= \frac{Var(X_2)}{E(X_2)} = 1 + \frac{\lambda_1 \lambda_3^2 \exp\left((2\underline{\gamma} + \underline{\alpha})^\top \underline{u}\right)}{\lambda_2 \exp\left(\underline{\beta}^\top \underline{u}\right) + \lambda_1 \lambda_3 \exp\left((\underline{\gamma} + \underline{\alpha})^\top \underline{u}\right)}.
\end{aligned}
$$

For $\lambda_2 = 0$

$$DI(X_2) = 1 + \lambda_3 \exp\left(\underline{\gamma}^\top \underline{u}\right).$$

Define

$$
\begin{aligned}
E(\underline{X}) &= (E(X_1), E(X_2))^\top \\
&= \left(\lambda_1 \exp\left(\underline{\alpha}^\top \underline{u}\right), \lambda_2 \exp\left(\underline{\beta}^\top \underline{u}\right) + \lambda_1 \lambda_3 \exp\left((\underline{\gamma} + \underline{\alpha})^\top \underline{u}\right)\right)^\top
\end{aligned}
$$

and

$$Cov(\underline{X}) = \begin{bmatrix} \lambda_1 \exp\left(\underline{\alpha}^\top \underline{u}\right) & \lambda_1\lambda_3 \exp\left((\underline{\gamma}+\underline{\alpha})^\top \underline{u}\right) \\ \lambda_1\lambda_3 \exp\left((\underline{\gamma}+\underline{\alpha})^\top \underline{u}\right) & \lambda_2 \exp\left(\underline{\beta}^\top \underline{u}\right) + \lambda_1\lambda_3 \exp\left((\underline{\gamma}+\underline{\alpha})^\top \underline{u}\right) + \\ & +\lambda_1\lambda_3^2 \exp\left((2\underline{\gamma}+\underline{\alpha})^\top \underline{u}\right) \end{bmatrix}$$

By using the definition given in the paper by Kokonendji and Puig [13] page 183, the bivariate Fisher index of dispersion is given by

$$GDI(\underline{X}) = \frac{\sqrt{E(\underline{X})}^\top Cov(\underline{X}) \sqrt{E(\underline{X})}}{E(\underline{X})^\top E(\underline{X})}$$

which is a case of over-disperson, cf. Arnold and Manjunath [7] page 2311 for the dispersion index proof for the bivariate pseudo-Poisson distribution.

4. Statistical Inference

In this section, we obtain maximum likelihood estimators (m.l.e.) of parameters λ_1, λ_2, λ_3, $\underline{\alpha}$, $\underline{\beta}$ and $\underline{\gamma}$. In addition, we construct the likelihood ratio test for the possible parallelism, coincidence and significance of each of the regression coefficients.

4.1. Estimation

Let $(X_{1i}, X_{2i})^\top$, $i = 1, 2, \ldots, n$ be a bivariate count sample from the pseudo-Poisson distribution (in Section 2) and let $\underline{u}_1, \ldots, \underline{u}_n$ be d-dimensional known covariates. Then, the log-likelihood function is

$$\begin{aligned} \log L \;=\; & -\lambda_1 \sum_{i=1}^{n} \exp\left(\underline{\alpha}^\top \underline{u}_i\right) + \sum_{i=1}^{n} x_{1i} \log\left(\lambda_1 \exp\left(\underline{\alpha}^\top \underline{u}_i\right)\right) \\ & - \sum_{i=1}^{n}\left(\lambda_2 \exp\left(\underline{\beta}^\top \underline{u}_i\right) + \lambda_3 \exp\left(\underline{\gamma}^\top \underline{u}_i\right)x_{1i}\right) \\ & + \sum_{i=1}^{n} x_{2i} \log\left(\lambda_2 \exp\left(\underline{\beta}^\top \underline{u}_i\right) + \lambda_3 \exp\left(\underline{\gamma}^\top \underline{u}_i\right)x_{1i}\right) \\ & - \sum_{i=1}^{n} \log(x_{1i}!x_{2i}!). \end{aligned} \tag{4}$$

Partial differentiation with respect to each parameters λ_1, λ_2 and λ_3 and equating to zero gives

$$-\sum_{i=1}^{n} \exp\left(\underline{\alpha}^\top \underline{u}_i\right) + \sum_{i=1}^{n} x_{1i} \frac{\exp\left(\underline{\alpha}^\top \underline{u}_i\right)}{\lambda_1 \exp\left(\underline{\alpha}^\top \underline{u}_i\right)} = 0 \tag{5}$$

$$-\sum_{i=1}^{n} \exp\left(\underline{\beta}^\top \underline{u}_i\right) + \sum_{i=1}^{n} x_{2i} \frac{\exp\left(\underline{\beta}^\top \underline{u}_i\right)}{\lambda_2 \exp\left(\underline{\beta}^\top \underline{u}_i\right) + \lambda_3 \exp\left(\underline{\gamma}^\top \underline{u}_i\right)x_{1i}} = 0 \tag{6}$$

$$-\sum_{i=1}^{n} x_{1i}\exp\left(\underline{\gamma}^\top \underline{u}_i\right) + \sum_{i=1}^{n} x_{1i}x_{2i} \frac{\exp\left(\underline{\gamma}^\top \underline{u}_i\right)}{\lambda_2 \exp\left(\underline{\beta}^\top \underline{u}_i\right) + \lambda_3 \exp\left(\underline{\gamma}^\top \underline{u}_i\right)x_{1i}} = 0. \tag{7}$$

Now, taking partial derivatives of $\log L$ with respect to α_j, β_j and γ_j for $j \in \{1, \ldots, d\}$ and equating to zero yields

$$-\lambda_1 \sum_{i=1}^{n} u_{ji} \exp\left(\underline{\alpha}^{\top}\underline{u}_i\right) + \sum_{i=1}^{n} x_{1i} u_{ji} = 0 \tag{8}$$

$$-\sum_{i=1}^{n} u_{ji} \exp\left(\underline{\beta}^{\top}\underline{u}_i\right) + \sum_{i=1}^{n} x_{2i} u_{ji} \frac{\exp\left(\underline{\beta}^{\top}\underline{u}_i\right)}{\lambda_2 \exp\left(\underline{\beta}^{\top}\underline{u}_i\right) + \lambda_3 \exp\left(\underline{\gamma}^{\top}\underline{u}_i\right) x_{1i}} = 0 \tag{9}$$

$$-\sum_{i=1}^{n} x_{1i} u_{ji} \exp\left(\underline{\gamma}^{\top}\underline{u}_i\right) + \sum_{i=1}^{n} x_{1i} x_{2i} u_{ji} \frac{\exp\left(\underline{\gamma}^{\top}\underline{u}_i\right)}{\lambda_2 \exp\left(\underline{\beta}^{\top}\underline{u}_i\right) + \lambda_3 \exp\left(\underline{\gamma}^{\top}\underline{u}_i\right) x_{1i}} = 0. \tag{10}$$

In particular, consider $d = 1$ and let $u_1, \ldots, u_n$ be the observed covariates. The likelihood equations from (5) to (10) simplify to become (with notation $\alpha_1 = \alpha$, $\beta_1 = \beta$ and $\gamma_1 = \gamma$)

$$\lambda_1 \sum_{i=1}^{n} \exp(\alpha u_i) = \sum_{i=1}^{n} x_{1i} \tag{11}$$

$$\sum_{i=1}^{n} \exp(u_i \beta) = \sum_{i=1}^{n} x_{2i} \frac{1}{\lambda_2 + \lambda_3 \exp(u_i(\gamma - \beta)) x_{1i}} \tag{12}$$

$$\sum_{i=1}^{n} x_{1i} \exp(u_i \gamma) = \sum_{i=1}^{n} x_{1i} x_{2i} \frac{1}{\lambda_2 \exp(u_i(\beta - \gamma)) + \lambda_3 x_{1i}}. \tag{13}$$

In the same way,

$$\lambda_1 \sum_{i=1}^{n} u_i \exp\left(\alpha u_i\right) = \sum_{i=1}^{n} x_{1i} u_i \tag{14}$$

$$\sum_{i=1}^{n} u_i \exp\left(\beta u_i\right) = \sum_{i=1}^{n} x_{2i} u_i \frac{1}{\lambda_2 + \lambda_3 x_{1i} \exp(u_i(\gamma - \beta))} \tag{15}$$

$$\sum_{i=1}^{n} x_{1i} u_i \exp\left(\gamma u_i\right) = \sum_{i=1}^{n} x_{1i} x_{2i} u_i \frac{1}{\lambda_2 \exp(u_i(\beta - \gamma)) + \lambda_3 x_{1i}}. \tag{16}$$

Note that the equations from (11) to (16) do not yield explicit expressions for the maximum likelihood estimates. However, one can use numerical methods to solve the system of six equations with six unknown parameters.

4.2. Likelihood Ratio Test

The general form of a generalized likelihood ratio test statistic is of the form

$$\Lambda = \frac{\sup_{\theta \in \Theta_0} L(\theta)}{\sup_{\theta \in \Theta} L(\theta)} \tag{17}$$

Here, Θ_0 is a subset of Θ and we envision testing $H_0 : \theta \in \Theta_0$. We reject the null hypothesis for small values of Λ.

Now, for the bivariate pseudo-Poisson model, the natural parameter space under the full model is $\Theta = \{(\lambda_1, \lambda_2, \lambda_3, \underline{\alpha}, \underline{\beta}, \underline{\gamma})^{\top} : \lambda_1 \geq 0, \lambda_2 \geq 0, \lambda_3 \geq 0, \underline{\alpha} \in \mathbb{R}^d, \underline{\beta} \in \mathbb{R}^d, \underline{\gamma} \in \mathbb{R}^d\}$. The m.l.e.'s under the complete parameter space are obtained by taking partial differentiation of Equation (4) with respect to $\lambda_1, \lambda_2, \lambda_3, \underline{\alpha}, \underline{\beta}, \underline{\gamma}$ and equating to zero. We denote the obtained numerical solution m.l.e.'s by $\hat{\lambda}_1, \hat{\lambda}_2, \hat{\lambda}_3$ and $\underline{\hat{\alpha}}, \underline{\hat{\beta}}, \underline{\hat{\gamma}}$.

Remark 1. *We used the "maxLik" optimization function in R (in the package "maxLik") to obtain the m.l.e.'s by numerical solution. This function also allows us to use a different methods of optimization using algorithms such as Newton–Raphson, Broyden–Fletcher–Goldfarb–Shanno, Berndt–Hall–Hall–Hausman, Berndt–Hall–Hall–Hausman, Simulated Annealing, Conjugate Gradients and Nelder–Mead methods. In the current paper, we use the Newton–Raphson method to estimate parameters and to compute their standard errors.*

4.2.1. Testing $H_0 : \underline{\alpha} = \underline{\beta} = \underline{\gamma} = \underline{0}$

In the following, we will construct a likelihood ratio test for testing whether the observed concomitant does not affect the distribution of (X_1, X_2). Under the null hypothesis, the natural parameter space is $\Theta_0 = \{(\lambda_1, \lambda_2, \lambda_3, \underline{\alpha}, \underline{\beta}, \underline{\gamma})^\top : \lambda_1 > 0, \lambda_2 \geq 0, \lambda_3 \geq 0, \underline{\alpha} = \underline{0}, \underline{\beta} = \underline{0}, \underline{\gamma} = \underline{0}\}$. Now, taking partial derivatives of Equation (4) with respect to each parameters $\lambda_1, \lambda_2, \lambda_3$ and equating to zero yields

$$-n + \frac{1}{\lambda_1} \sum_{i=1}^{n} X_{1i} = 0 \tag{18}$$

$$-n + \sum_{i=1}^{n} \frac{X_{2i}}{\lambda_2 + \lambda_3 X_{1i}} = 0 \tag{19}$$

$$-\sum_{i=1}^{n} X_{1i} + \sum_{i=1}^{n} \frac{X_{1i} X_{2i}}{\lambda_2 + \lambda_3 X_{1i}} = 0. \tag{20}$$

Equation (18) is readily solved, to obtain the m.l.e. for λ_1, namely, $\hat{\lambda}_1^* = \overline{X}_1$. The remaining two Equations (19) and (20) must be solved numerically to obtain $\hat{\lambda}_2^*, \hat{\lambda}_3^*$.

Now let $\hat{\lambda}_1, \hat{\lambda}_2, \hat{\lambda}_3$ and $\underline{\hat{\alpha}}, \underline{\hat{\beta}}, \underline{\hat{\gamma}}$ be the m.l.e. estimates on unrestricted space. Then, the likelihood (as defined in Equation (4)) ratio test statistic is

$$\Lambda_1 = \frac{L(\hat{\lambda}_1^*, \hat{\lambda}_2^*, \hat{\lambda}_3^*, \underline{0}, \underline{0}, \underline{0})}{L(\hat{\lambda}_1, \hat{\lambda}_2, \hat{\lambda}_3, \underline{\hat{\alpha}}, \underline{\hat{\beta}}, \underline{\hat{\gamma}})}. \tag{21}$$

If n is large, then $-2\log(\Lambda_1)$ may be compared with a suitable χ^2_{3d} percentile in order to decide whether H_0 should be rejected or not.

4.2.2. Testing $H_0 : \underline{\alpha} = \underline{0}$

Here, we are testing that the observed concomitant does not affect the marginal distribution of X_1. Note that under the null hypothesis, the natural parameter space is $\Theta_0 = \{(\lambda_1, \lambda_2, \lambda_3, \underline{\alpha}, \underline{\beta}, \underline{\gamma})^\top : \lambda_1 > 0, \lambda_2 \geq 0, \lambda_3 \geq 0, \underline{\alpha} = \underline{0}, \underline{\beta} \in \mathbb{R}^d, \underline{\gamma} \in \mathbb{R}^d\}$. Now, again taking partial derivatives of Equation (4) with respect to parameters $\lambda_1, \lambda_2, \lambda_3$ & $\underline{\beta}, \underline{\gamma}$ and equating to zero gives m.l.e.'s, denoted by $\hat{\lambda}_1^*, \hat{\lambda}_2^*, \hat{\lambda}_3^*, \underline{\hat{\beta}}^*$ and $\underline{\hat{\gamma}}^*$, respectively. The likelihood ratio test statistic is

$$\Lambda_2 = \frac{L(\hat{\lambda}_1^*, \hat{\lambda}_2^*, \hat{\lambda}_3^*, \underline{0}, \underline{\hat{\beta}}^*, \underline{\hat{\gamma}}^*)}{L(\hat{\lambda}_1, \hat{\lambda}_2, \hat{\lambda}_3, \underline{\hat{\alpha}}, \underline{\hat{\beta}}, \underline{\hat{\gamma}})} \tag{22}$$

If n is large, then $-2\log(\Lambda_2)$ may be compared with a suitable χ^2_d percentile in order to decide whether H_0 should be rejected or not.

4.2.3. Testing $H_0 : \underline{\beta} = \underline{\gamma} = \underline{0}$

In this case, we are testing whether the observed concomitant does not affect the conditional distribution of X_2 given X_1. Under the null hypothesis, the natural parameter space is $\Theta_0 = \{(\lambda_1, \lambda_2, \lambda_3, \underline{\alpha}, \underline{\beta}, \underline{\gamma})^\top : \lambda_1 > 0, \lambda_2 \geq 0, \lambda_3 \geq 0, \underline{\alpha} \in \mathbb{R}^d, \underline{\beta} = \underline{0}, \underline{\gamma} = \underline{0}\}$. Again, taking partial derivatives of Equation (4) with respect to each parameters $\lambda_1, \lambda_2, \lambda_3$ & $\underline{\alpha}$

and equating to zero gives to m.l.e.'s denoted by $\hat{\lambda}_1^*, \hat{\lambda}_2^*, \hat{\lambda}_3^*, \hat{\underline{\alpha}}^*$. The likelihood ratio test statistic is

$$\Lambda_3 = \frac{L(\hat{\lambda}_1^*, \hat{\lambda}_2^*, \hat{\lambda}_3^*, \hat{\underline{\alpha}}^*, \underline{0}, \underline{0})}{L(\hat{\lambda}_1, \hat{\lambda}_2, \hat{\lambda}_3, \hat{\underline{\alpha}}, \hat{\underline{\beta}}, \hat{\underline{\gamma}})} \tag{23}$$

If n is large, then $-2\log(\Lambda_3)$ may be compared with a suitable χ^2_{2d} percentile in order to decide whether H_0 should be rejected or not.

4.2.4. Testing $H_0 : \underline{\beta} = \underline{0}$

Here, we are interested in testing whether the observed concomitant does not affect the intercept term of the pseudo-Poisson model. Now, under the null hypothesis, the natural parameter space is $\Theta_0 = \{(\lambda_1, \lambda_2, \lambda_3, \underline{\alpha}, \underline{\beta}, \underline{\gamma})^\top : \lambda_1 > 0, \lambda_2 \geq 0, \lambda_3 \geq 0, \underline{\alpha} \in \mathbb{R}^d, \underline{\beta} = \underline{0}, \underline{\gamma} \in \mathbb{R}^d\}$. Again, taking partial derivatives of Equation (4) with respect to each parameters $\lambda_1, \lambda_2, \lambda_3$ & $\underline{\alpha}, \underline{\gamma}$ and equating to zero gives to m.l.e.'s denoted by $\hat{\lambda}_1^*, \hat{\lambda}_2^*, \hat{\lambda}_3^*, \hat{\underline{\alpha}}^*, \hat{\underline{\gamma}}^*$. The likelihood ratio test statistic is

$$\Lambda_4 = \frac{L(\hat{\lambda}_1^*, \hat{\lambda}_2^*, \hat{\lambda}_3^*, \hat{\underline{\alpha}}^*, \underline{0}, \hat{\underline{\gamma}}^*)}{L(\hat{\lambda}_1, \hat{\lambda}_2, \hat{\lambda}_3, \hat{\underline{\alpha}}, \hat{\underline{\beta}}, \hat{\underline{\gamma}})} \tag{24}$$

If n is large, then $-2\log(\Lambda_4)$ may be compared with a suitable χ^2_d percentile in order to decide whether H_0 should be rejected or not.

4.2.5. Testing $H_0 : \underline{\gamma} = \underline{0}$

In this case, we wish to determine whether the concomitant does not affect the dependence structure of the pseudo-Poisson model. Thus, under the null hypothesis, parameter space is $\Theta_0 = \{(\lambda_1, \lambda_2, \lambda_3, \underline{\alpha}, \underline{\beta}, \underline{\gamma})^\top : \lambda_1 > 0, \lambda_2 \geq 0, \lambda_3 \geq 0, \underline{\alpha} \in \mathbb{R}^d, \underline{\beta} \in \mathbb{R}^d, \underline{\gamma} = \underline{0}\}$. Now, taking partial derivatives of Equation (4) with respect to the parameters $\lambda_1, \lambda_2, \lambda_3$ & $\underline{\alpha}, \underline{\beta}$ and equating to zero gives, m.l.e.'s denoted by $\hat{\lambda}_1^*, \hat{\lambda}_2^*, \hat{\lambda}_3^*, \hat{\underline{\alpha}}^*, \hat{\underline{\beta}}^*$. The likelihood ratio test statistic is

$$\Lambda_5 = \frac{L(\hat{\lambda}_1^*, \hat{\lambda}_2^*, \hat{\lambda}_3^*, \hat{\underline{\alpha}}^*, \hat{\underline{\beta}}^*, \underline{0})}{L(\hat{\lambda}_1, \hat{\lambda}_2, \hat{\lambda}_3, \hat{\underline{\alpha}}, \hat{\underline{\beta}}, \hat{\underline{\gamma}})} \tag{25}$$

If n is large, then $-2\log(\Lambda_5)$ may be compared with a suitable χ^2_d percentile in order to decide whether H_0 should be rejected or not.

In the next examples, we are interested in testing some hypotheses concerning the relationship between the explanatory and response variables. In particular, we are interested in testing whether the regression planes are parallel or if they are coincident. We illustrate the testing procedure using the simple sub-model given by

$$X_1 \sim \mathscr{P}\left(\exp\left(\sum_{j=1}^{d} u_{ij}\alpha_j\right)\right) \tag{26}$$

and

$$X_2 | X_1 = x_1 \sim \mathscr{P}\left(\exp\left(\sum_{j=1}^{d} u_{ij}\gamma_j\right)x_1\right). \tag{27}$$

4.2.6. Testing for Parallelism

In the following, we are interested in testing whether the planes on which the means lie are parallel. If we set $u_{1i} = 1$ for $i \in \{1, \ldots, n\}$ then the two marginal means are

$$log(E(X_1)) \;=\; \alpha_1 + \sum_{j=2}^{d} u_{ij}\alpha_j \tag{28}$$

$$log(E(X_2)) \;=\; \alpha_1 + \gamma_1 + \sum_{j=2}^{d} u_{ij}(\alpha_j + \gamma_j). \tag{29}$$

For the bivariate pseudo-Poisson regression model specified in (26) and (27), now it is interesting to examine the hypothesis that the planes on which the mean lies are parallel. This is equivalent to testing for the hypothesis $H_0 : \gamma_j = 0$, for $j \in \{1, \ldots, d\}$. Under the null hypothesis, the pseudo-Poisson regression model will be

$$X_1 \sim \mathscr{P}\left(\exp\left(\alpha_1 + \sum_{j=2}^{d} u_{ij}\alpha_j \right) \right) \tag{30}$$

and

$$X_2 | X_1 = x_1 \sim \mathscr{P}\left(\exp(\gamma_1)x_1 \right). \tag{31}$$

The log-likelihood is

$$
\begin{aligned}
log(L_{PH}) \;=\; & -\sum_{i=1}^{n} \exp\left(\alpha_1 + \sum_{j=2}^{d} u_{ij}\alpha_j \right) + \sum_{i=1}^{n} x_{1i} \log\left(\exp\left(\alpha_1 + \sum_{j=2}^{d} u_{ij}\alpha_j \right) \right) \\
& -\sum_{i=1}^{n} \exp(\gamma_1)x_{1i} + \sum_{i=1}^{n} x_{2i} \log\left(\exp(\gamma_1)x_{1i} \right) - \sum_{i=1}^{n} \log(x_{1i}!x_{2i}!).
\end{aligned}
\tag{32}
$$

Note that testing for parallelism for the model specified in (30) and (31) is equivalent to testing for the observed concomitant and has no effect on the conditional distribution of X_2 given X_1. Now, partial differentiation with respect to γ_1 and α_j, $j \in \{1, \ldots, d\}$ and equating to zero gives us

$$log(\bar{X}_1) \;=\; \alpha_1 + \sum_{j=2}^{d} u_{ij}\alpha_j$$

$$\sum_{i=1}^{n} x_{1i}u_{ij} \;=\; \sum_{i=1}^{n} \exp\left(\alpha_1 + \sum_{j=2}^{d} \alpha_j u_{ij} \right), \; j \in \{2, \ldots, d\} \tag{33}$$

Solving the above d equations leads us to the m.l.e. of α_i denoted by α_{Pj}^{*}, $j \in \{1, \ldots, d\}$ and the m.l.e. of γ_1 is

$$\gamma_{P1}^{*} = log\left(\sum_{i=1}^{n} x_{2i} \log(x_{1i}) - \sum_{i=1}^{n} x_{1i} \right). \tag{34}$$

Now, we denote the obtained m.l.e.'s under the complete parameter space by $\hat{\alpha}_{Pj}$ and $\hat{\gamma}_{Pj}$, $j \in \{1, \ldots, d\}$. The likelihood ratio test statistic is

$$\Lambda_P = \frac{L_{PH}(\hat{\alpha}_{P}^{*}, \underline{0})}{L_P(\hat{\underline{\alpha}}_P, \hat{\underline{\gamma}}_P)},$$

where $L_P(.,.)$ is the likelihood of the model in (26) & (27) and (30) & (31). If n is large, then $-2log(\Lambda_P)$ may be compared with a suitable χ^2_{d-1} percentile in order to decide whether H_0 should be rejected or not.

4.2.7. Testing for Coincidence

Here, we assume that the regression relationship does not change from time 1 to time 2 which will occur if the planes on which means lies are coincident. Now, for the given model in (26) and (27), the two marginal means are

$$log(E(X_1)) \;=\; \sum_{j=1}^{d} u_{ij}\alpha_j \tag{35}$$

$$log(E(X_2)) \;=\; \sum_{j=1}^{d} u_{ij}(\alpha_j + \gamma_j). \tag{36}$$

The assumption of coincidence leads us to test $H_0 : \gamma_j = 0$, for $j \in \{1,\ldots,d\}$. Denote by $\hat{\alpha}^*_{Cj}$ for $j \in \{1,\ldots,d\}$ are m.l.e.'s under the null hypothesis and by $\hat{\alpha}_{Cj}$ and $\hat{\gamma}_{Cj}$ for $j \in \{1,\ldots,d\}$ are m.l.e.'s under complete parameter space, for . Now, the likelihood ratio test statistic is

$$\Lambda_C = \frac{L_{CH}(\hat{\underline{\alpha}}^*_P, \underline{0})}{L_C(\hat{\underline{\alpha}}_P, \hat{\underline{\gamma}}_P)},$$

where $L_{CH}(.,)$ and $L_C(.,.)$ are likelihood under null and complete parameter space, respectively. If n is large, then $-2log(\Lambda_C)$ may be compared with a suitable χ^2_d percentile to decide whether H_0 should be rejected or not.

5. Applications

In the following two subsections, we illustrate a simulation study and give examples of real-life applications of the bivariate pseudo-Poisson regression model.

5.1. Simulation

We have simulated 2000 data sets of sample size $n = 20, 30, 50, 100, 200, 500, 1000$ for the parameter values $\lambda_1 = 1, \lambda_2 = 1, \lambda_3 = 4, \alpha_1 = 1, \alpha_2 = 0, \alpha_3 = -1, \beta_1 = 0, \beta_2 = 1, \beta_3 = 1, \gamma_1 = 0, \gamma_2 = 0$ and $\gamma_3 = 1$ from the pseudo-Poisson regression model. We refer to Figures 1–4 for the bootstrapped distribution of each of the parameters. The numerical evidence suggests that as sample size increases, m.l.e.'s approach the true parameter values with standard errors that are decreasing as the sample size increases.

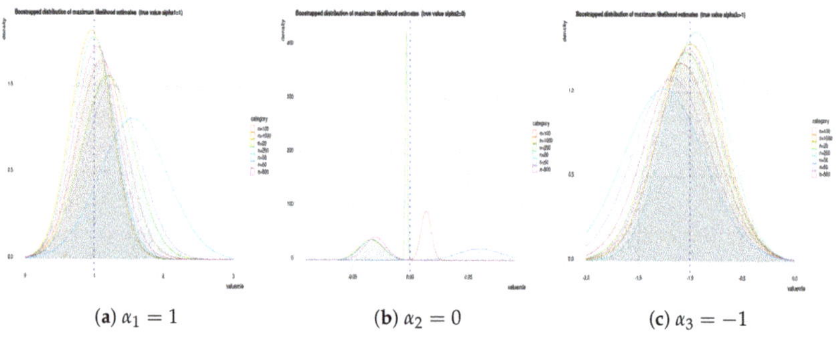

      (a) $\alpha_1 = 1$ (b) $\alpha_2 = 0$ (c) $\alpha_3 = -1$

Figure 1. Boostrapped distribution of $\underline{\alpha} = (\alpha_1, \alpha_2, \alpha_3)^\top$.

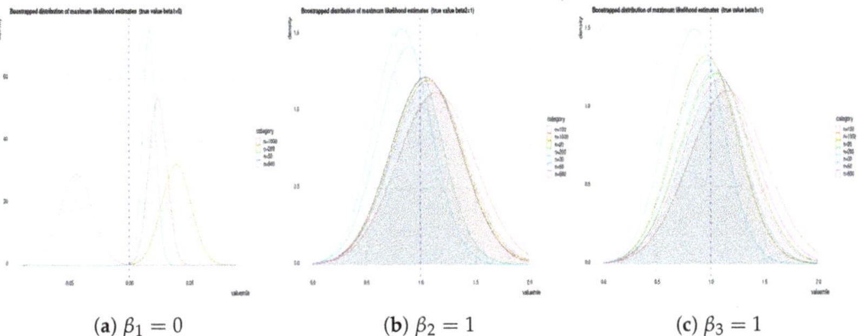

(a) $\beta_1 = 0$　　　　(b) $\beta_2 = 1$　　　　(c) $\beta_3 = 1$

Figure 2. Boostrapped distribution of $\underline{\beta} = (\beta_1, \beta_2, \beta_3)^\top$.

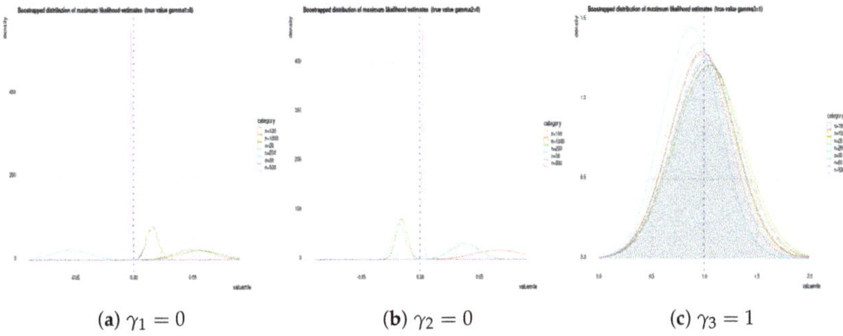

(a) $\gamma_1 = 0$　　　　(b) $\gamma_2 = 0$　　　　(c) $\gamma_3 = 1$

Figure 3. Boostrapped distribution of $\underline{\gamma} = (\gamma_1, \gamma_2, \gamma_3)^\top$.

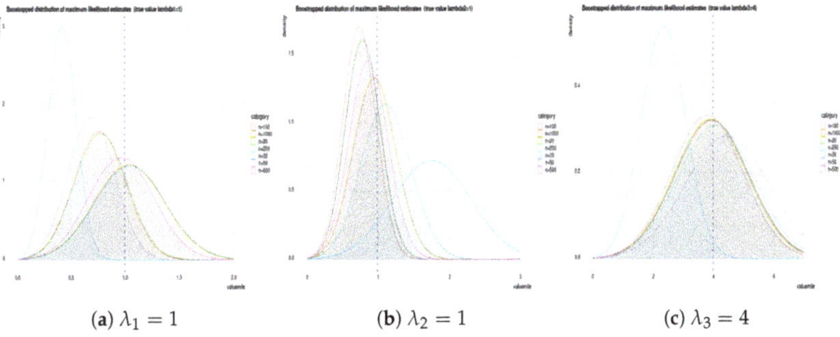

(a) $\lambda_1 = 1$　　　　(b) $\lambda_2 = 1$　　　　(c) $\lambda_3 = 4$

Figure 4. Boostrapped distribution of $(\lambda_1, \lambda_2, \lambda_3)$.

5.2. Real-Life Data

5.2.1. Australian Health Service Utilization Data: 1977–1978

We consider a data set which is mentioned in Islam and Chowdhury [14] that is part of the Health and Retirement Study (HRS). The data represent the number of conditions ever had (X_1) as mentioned by the doctors and utilization of healthcare services (say, hospital, nursing home, doctor and home care) (X_2). The concomitant variables are Gender, Age, Hispanic, and Veteran.

The marginal estimated dispersion indices are 0.779 and 1.029. The sample Pearson correlation coefficient between X_1 and X_2 is 0.063. We can conclude that marginal X_1 is approximately equi-dispersed and marginal X_2 is slightly over-dispersed. Further, the data were also tested for independence and it was concluded that the assumption was rejected, cf. Arnold and Manjunath [7] pages 2321–2322.

We refer to the Table 1 for the log-likelihood values for the following models:

Full Model: The parameters are $\lambda_1, \lambda_2, \lambda_3, \underline{\alpha} = (\alpha_1, \alpha_2, \alpha_3, \alpha_4)^\top$, $\underline{\beta} = (\beta_1, \beta_2, \beta_3, \beta_4)^\top, \underline{\gamma} = (\gamma_1, \gamma_2, \gamma_3, \gamma_4)^\top$

Mirrored, Model   (in which X_1 and X_2 are interchanged): The parameters are $\lambda_1, \lambda_2, \lambda_3,$ $\underline{\alpha} = (\alpha_1, \alpha_2, \alpha_3, \alpha_4)^\top, \underline{\beta} = (\beta_1, \beta_2, \beta_3, \beta_4)^\top, \underline{\gamma} = (\gamma_1, \gamma_2, \gamma_3, \gamma_4)^\top$

Sub-Model I   ($\lambda_2 = 0$): The parameters are $\lambda_1, \lambda_3, \underline{\alpha} = (\alpha_1, \alpha_2, \alpha_3, \alpha_4)^\top$, $\underline{\gamma} = (\gamma_1, \gamma_2, \gamma_3, \gamma_4)^\top$

Sub-Model II   ($\lambda_2 = \lambda_3$): The parameters are $\lambda_1, \lambda_3, \underline{\alpha} = (\alpha_1, \alpha_2, \alpha_3, \alpha_4)^\top$, $\underline{\gamma} = (\gamma_1, \gamma_2, \gamma_3, \gamma_4)^\top$

Sub-Model II   (Mirrored): The parameters are $\lambda_1, \lambda_3, \underline{\alpha} = (\alpha_1, \alpha_2, \alpha_3, \alpha_4)^\top$, $\underline{\gamma} = (\gamma_1, \gamma_2, \gamma_3, \gamma_4)^\top$.

Table 1. Models for the Australian Health Service Utilization Data.

Models	No. Parameters	Log-Likelihood
Full Model	15	−52,654.13
Mirrored, Full Model	15	−16,229.24
Sub-Model I ($\lambda_2 = 0$)	11	−16,586.07
Sub-Model II ($\lambda_2 = \lambda_3$)	11	−16,371.95
Mirrored Sub-Model II ($\lambda_2 = \lambda_3$)	11	−17,585.37

The mirrored Full Model fits the data best. For the detailed discussion on the mirrored model, see Arnold and Manjunath [7] page 2323. In Islam and Chowdhury [14], page 122, the authors fitted the Poisson–Poisson regression model for the same data set. Note that the Poisson–Poisson regression model is a sub-model of the pseudo-Poisson regression model when $\lambda_2 = 0$. Hence, we conclude that our generalized pseudo-Poisson mirrored model fits the data better than the Poisson–Poisson regression model. The parameter estimates for the pseudo-Poisson mirrored model and their standard errors are displayed in Table 2.

Further, we tested for the significance of the regression coefficients. With reference to Table 3, the computed $-2 \log \lambda$ and compared with χ^2 table values with respective degrees of freedom and the size of 0.05 or 0.10 and concluded that there is not enough evidence to accept the null hypotheses.

Table 2. Final model estimates and its standard error (s.e.) for the Australian Health Service Utilization Data.

Parameter	m.l.e.	s.e.
α_1	0.292	0.039
α_2	−0.008	0.004
α_3	−0.186	0.058
α_4	0.140	0.042
β_1	−0.132	0.0273
β_2	0.016	0.0036
β_3	0.038	0.0277
β_4	0.053	0.035
γ_1	1.636	0.656
γ_2	−0.025	0.039
γ_3	−0.996	−
γ_4	−0.148	0.273
λ_1	1.172	0.385
λ_2	0.824	0.224
λ_3	0.313	1.037

Table 3. Hypothesis testing for the Australian Health Service Utilization Data.

Hypothesis	$\log \Lambda^* - \log \Lambda$	d.f.
$\underline{\alpha} = \underline{\beta} = \underline{\gamma} = \underline{0}$	-113.7227	12
$\underline{\alpha} = \underline{0}$	-103.483	4
$\underline{\beta} = \underline{\gamma} = \underline{0}$	-28.26604	8
$\underline{\beta} = \underline{0}$	-24.26104	4
$\underline{\gamma} = \underline{0}$	-6.857175	4

5.2.2. Road Safety Data

The second data set is on road safety, published by the Department for Transport, United Kingdom. The data comprise information about personal injury road accidents in Great Britain and the consequent casualties on public roads. The concomitant variables are Gender of the driver (Male = 1, Female = 0), Area (Urban = 0, Rural = 1), Accident Severity (Fatal Severity = 1 else = 0), Accident Severity (Serious Severity = 1, else = 0), and Light condition (Daylight = 1, Others = 0).

We refer to Table 4 for the log-likelihood values for the following:

Full Model: parameters are $\lambda_1, \lambda_2, \lambda_3, \underline{\alpha} = (\alpha_1, \alpha_2, \alpha_3, \alpha_4, \alpha_5)^\top,$
$\underline{\beta} = (\beta_1, \beta_2, \beta_3, \beta_4, \beta_5)^\top, \underline{\gamma} = (\gamma_1, \gamma_2, \gamma_3, \gamma_4, \gamma_5)^\top$

Mirrored, Model   (X_1 and X_2 are interchanged): parameters are $\lambda_1, \lambda_2, \lambda_3,$
$\underline{\alpha} = (\alpha_1, \alpha_2, \alpha_3, \alpha_4, \alpha_5)^\top, \underline{\beta} = (\beta_1, \beta_2, \beta_3, \beta_4, \beta_5)^\top, \underline{\gamma} = (\gamma_1, \gamma_2, \gamma_3, \gamma_4, \gamma_5)^\top$

Sub-Model I   ($\lambda_2 = 0$): parameters are $\lambda_1, \lambda_3, \underline{\alpha} = (\alpha_1, \alpha_2, \alpha_3, \alpha_4, \alpha_5)^\top,$
$\underline{\gamma} = (\gamma_1, \gamma_2, \gamma_3, \gamma_4, \gamma_5)^\top$

Sub-Model I   (Mirrored): parameters are $\lambda_1, \lambda_3, \underline{\alpha} = (\alpha_1, \alpha_2, \alpha_3, \alpha_4, \alpha_5)^\top,$
$\underline{\gamma} = (\gamma_1, \gamma_2, \gamma_3, \gamma_4, \gamma_5)^\top.$

Sub-Model II   ($\lambda_2 = \lambda_3$): parameters are $\lambda_1, \lambda_3, \underline{\alpha} = (\alpha_1, \alpha_2, \alpha_3, \alpha_4, \alpha_5)^\top,$
$\underline{\gamma} = (\gamma_1, \gamma_2, \gamma_3, \gamma_4, \gamma_5)^\top$

Sub-Model II   (Mirrored): parameters are $\lambda_1, \lambda_3, \underline{\alpha} = (\alpha_1, \alpha_2, \alpha_3, \alpha_4, \alpha_5)^\top,$
$\underline{\gamma} = (\gamma_1, \gamma_2, \gamma_3, \gamma_4, \gamma_5)^\top.$

Table 4. Models for the Road safety data.

Models	No. Parameters	Log-Likelihood
Full Model	18	$-223{,}743.3$
Mirrored Full Model	18	$-243{,}538.7$
Sub-Model I($\lambda_2 = 0$)	11	$-251{,}937.1$
Mirrored Sub-Model I($\lambda_2 = 0$)	11	$-37{,}599.63$
Sub-Model II($\lambda_2 = \lambda_3$)	11	$-36{,}201.52$
Mirrored Sub-Model II($\lambda_2 = \lambda_3$)	11	$-36{,}516.22$

We refer to Table 4 and conclude that the Full Model fits the road safety data and refer to Table 5 for the estimates and their standard errors.

Table 5. Final model estimates and its standard error (s.e.) Road safety data.

Parameter	m.l.e.	s.e.
α_1	1.002	0.006
α_2	0.999	0.005
α_3	0.999	0.017
α_4	0.999	0.005
β_1	1.000	0.005
β_2	1.004	0.005
β_3	1.003	0.005
β_4	1.000	0.015
γ_1	0.999	0.0036
γ_2	1.005	0.004
γ_3	1.355	–
γ_4	1.319	–
λ_1	1.010	–
λ_2	1.105	–
λ_3	-0.078	0.007

6. Concluding Remarks

The bivariate pseudo-Poisson model with its straightforward structure with no restrictions on the conditional mean function allows us to model a variety of dependence structures, including positive and negative correlation. Introducing explanatory variables in such models will be a useful additional to the toolkit for modelers dealing with bivariate count data which have positive or negative correlation. In the current note, we explored distributional and inferential aspects of such models and also included a simulation and real-life data applications. We emphasize the advantage of considering the current model over other available count regression models in Section 5.2. The bivariate pseudo-Poisson regression model has a simple structure, straightforward parameter estimation and fast computation, and will deserve a place in the analysis of count data sets with concomitants.

Author Contributions: Both authors equally contributed. All authors have read and agreed to the published version of the manuscript.

Funding: The second author's research was sponsored by the Institution of Eminence (IoE), University of Hyderabad (UoH-IoE-RC2-21-013).

Institutional Review Board Statement: Not applicable.

Informed Consent Statement: Not applicable.

Data Availability Statement: The data sets used in the current article are available at bpglm: R package for Bivariate Poisson GLM with Covariates.

Conflicts of Interest: The authors declare no conflict of interest.

References

1. Sellers, K.F.; Morris, D.M.; Balakrishnan, N. Bivariate Conway-Maxwell-Poisson distribution: Formulation, properties, and inference. *J. Multi. Anal.* **2016**, *150*, 152–168. [CrossRef]
2. Kocherlakota, S.; Kocherlakota, K. Regression in the bivariate Poisson distribution. *Comm.-Stat.-Theory Methods* **2001**, *30*, 815–825. [CrossRef]
3. Riggs, K.; Young, D.M.; Stamey, J.D. Statistical Inference for a bivariate Poisson regression model. *Adv. Appl. Stat.* **2008**, *10*, 55–73.
4. Zamani, H.; Faroughi, P.; Ismail, N. Bivariate generalized Poisson regression model: Applications on health care data. *Empir. Econ.* **2016**, *15*, 1607–1621. [CrossRef]
5. Chowdhury, R.; Sneddon, G.; Hasan, T.M. Analyzing the effect of duration on the daily new cases of COVID-19 infections and deaths using bivariate Poisson regression: A marginal conditional approach. *Math. Biosci. Eng.* **2020**, *17*, 6085–6097. [CrossRef]
6. Karlis, D.; Ntzoufras, I. Bivariate Poisson and diagonal inflated bivariate Poisson regression models in R. *J. Statist. Softw.* **2005**, *14*, 1–36. [CrossRef]
7. Arnold, B.C.; Manjunath, B.G. Statistical inference for distributions with one Poisson conditional. *J. Appl. Stat.* **2021**, *48*, 2306–2325. [CrossRef] [PubMed]

8. Arnold, B.C.; Veeranna, B.; Manjunath, B.G.; Shobha, B. Bayesian inference for pseudo-Poisson data. *J. Comput. Stat. Simul.* **2022**, 1–28. [CrossRef]
9. Veeranna, B.; Manjunath, B.G.; Shobha, B. Goodness-of-fit test for pseudo-Poisson distribution. 2022, *under review*.
10. Arnold, B.C.; Castillo, E.; Sarabia, J.M. *Conditional Specification of Statistical Models*; Springer Series in Statistics; Springer: New York, NY, USA, 1999.
11. Filus, J.K.; Filus, L.Z.; Arnold, B.C. Families of multivariate distributions involving "Triangular" transformations. *Comm.-Stat.-Theory Methods* **2009**, *39*, 107–116. [CrossRef]
12. Ghosh, I.; Marques, F.; Chakraborty, S. A new bivariate Poisson distribution via conditional specification: Properties and applications. *J. Appl. Stat.* **2021**, *48*, 3025–3047. [CrossRef] [PubMed]
13. Kokonendji, C.C.; Puig, P. Fisher dispersion index for multivariate count distributions: A review and a new proposal. *J. Multivar. Anal.* **2018**, *165*, 180–193. [CrossRef]
14. Islam, M.A.; Chowdhury, R.I. *Analysis of Repeated Measures Data*; Springer Nature: Singapore, 2017.

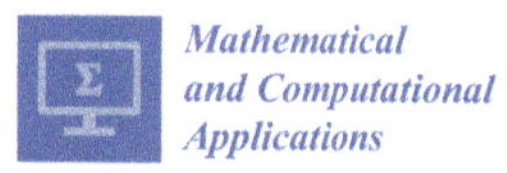 *Mathematical and Computational Applications*

Article

Forecasting Financial and Macroeconomic Variables Using an Adaptive Parameter VAR-KF Model

Nat Promma [1] and Nawinda Chutsagulprom [1,2]*

[1] Advanced Research Center for Computational Simulation (ARCCoS), Department of Mathematics, Faculty of Science, Chiang Mai University, Chiang Mai 50200, Thailand
[2] Centre of Excellence in Mathematics, CHE, Si Ayutthaya Road, Bangkok 10400, Thailand
* Correspondence: nawin.cp@gmail.com

Abstract: The primary objective of this article is to present an adaptive parameter VAR-KF technique (APVAR-KF) to forecast stock market performance and macroeconomic factors. The method exploits a vector autoregressive model as a system identification technique, and the Kalman filter is served as a recursive state parameter estimation tool. A further development was designed by incorporating the GARCH model to quantify an automatic observation covariance matrix in the Kalman filter step. To verify the efficiency of our proposed method, we conducted an experimental simulation applied to the main stock exchange index, real effective exchange rate and consumer price index of Thailand and Indonesia from January 1997 to May 2021. The APVAR-KF method is generally shown to have a superior performance relative to the conventional VAR(1) model and the VAR-KF model with constant parameters.

Keywords: Kalman filter; VAR; GARCH

Citation: Promma, N.; Chutsagulprom, N. Forecasting Financial and Macroeconomic Variables Using an Adaptive Parameter VAR-KF Model. *Math. Comput. Appl.* **2023**, *28*, 19. https://doi.org/10.3390/mca28010019

Academic Editor: Sandra Ferreira

Received: 29 September 2022
Revised: 29 January 2023
Accepted: 1 February 2023
Published: 2 February 2023

1. Introduction

1.1. Motivation and Related Work

Due to an unprecedented increase in the uncertainty of economic and financial market activities, independent investors and policy makers require effective forecasting tools in order to facilitate more accurate decision plans. Numerous forecasting methods, ranging from univariate to multivariate time series models, have been developed to forecast stock market pricing and macroeconomic variables. Some of the most notable univariate techniques include autoregressive integrated moving average (ARIMA) models [1,2], artificial neural networks (ANNs) [3,4] and support vector machines (SVMs) [5,6]. In practice, economics and finance are correlated disciplines in which a change in one activity can cause uncertainty in the other. Macroeconomic fundamentals reflect the general economic environment and can influence the degree of variation in future cash flow in a stock market. Conversely, stock prices are often used as leading indicators that aggregate information about the economy's direction. The existence of an association between macroeconomic indicators and stock prices has been extensively verified by several research studies [7–9]. Therefore, instead of using univariate time series forecasting techniques, multivariate time-series models are more suitable approaches for the predictability of macroeconomic variables, as well as stock indices.

Vector autoregressive (VAR) [10] models are multivariate time series techniques in which the dynamics of state variables can be expressed as a linear combination of past realizations. They are predominantly utilized for structural analysis and macroeconomic forecasting purposes because of their implementation's simplicity and flexibility. Some studies that used VAR models for time series prediction include Suhasono et al. [11], who compared the forecasting performance between vector error correction modeling (VECM) and VAR models for ASEAN stock price indices. Öğünç [12] forecasted the inflation, nominal exchange rate and interest rate in Turkey through VAR variants. However, despite all of their advantages, the linearity assumption underlying VAR models can potentially lead to biased estimates, especially for a highly volatile time series. Several improvements for VAR models have been put forward to handle the inherent nonlinearity structure in data. One extensively used technique is to introduce drifting autoregressive coefficients to capture the presence of nonlinear effects in lagged dependent models. A time-varying parameter vector autoregression (TVP-VAR) model with stochastic volatility is a VAR-based approach in which the parameter estimation is calculated via the Markov chain Monte Carlo (MCMC) sampling algorithm [13]. D'Agostino et al. [14] compared the forecasting accuracy of nine time series methods in US inflation, unemployment and interest rates over the period of 1970–2007. They concluded that the TVP-VAR is the only method that can forecast all three variables accurately. Bekiros et al. [15] reported that using the TVP-VAR technique leads to a better forecasting ability than benchmark autoregression and random walk models when predicting the oil price movement with economic policy uncertainty being included. Kumar [16] examined the forecast ability of the ARIMA, VAR and TVP-VAR methods to predict the daily exchange rates of the Indian rupee against the U.S. dollar. The empirical results show that the TVP-VAR model outperforms other competing approaches. However, in the process of computing the posterior distribution of parameters of the traditional TVP-VAR model, the MCMC sampling algorithm requires a heavy computational burden in high-dimensional cases. To attenuate the curse of dimensionality, Korobilis [17] adopted a stochastic search algorithm using the Gibbs sampler to select potential variables that have a larger contribution to the forecasting accuracy. Many researchers make use of Bayesian data assimilation techniques, preferably the Kalman filtering (KF) [18], for the parameter estimation problem. Bekiros [19] exploited the KF algorithm and an extension of the univariate methodology framework for the parameter estimation in the TVP-VAR model to predict the monthly macroeconomic factors of the EU economy. Koop and Korobilis [20] introduced forgetting factors in the TVP-VAR model with parameters being recursively updated through the KF approach. Their purposed method leads to the scalability of the state-space estimator, and ultimately aids in a dimensionality reduction.

As far as the relationship between stock prices and macroeconomic fundamentals is concerned, it is accordingly plausible to include financial factors in macroeconomic forecasting and vice versa. Nevertheless, researchers tend to not forecast these variables simultaneously via multivariate time series models due to their different observed frequencies. In this work, we will present the hybrid VAR and KF method for the economic and financial trend prediction based on the monthly data. Motivated by Bekiros [19] and Koop and Korobilis [20], the model coefficients were sequentially updated through the joint state-parameter KF procedure rather than employing the filtering technique, particularly for the parameter estimation. The use of the KF model also involves the predetermination of noise covariances, where they are mostly constructed in an ad hoc manner that cannot accurately quantify model uncertainties under complex circumstances. Meanwhile, economic and financial time series are typically characterized by volatility clustering properties, or heteroscedasticity. We therefore enhanced our model with heteroscedastic noise by using a statistical technique to model an observation error covariance matrix and an average of sample covariances for a process error covariance matrix.

1.2. Contribution

The objectives of this paper are:

1. We present a forecasting technique, the adaptive parameter VAR-KF (APVAR-KF) method, in which the state-space equations are constructed through the VAR model and the optimal state and parameter estimates are achieved using the KF approach.
2. A generalized autoregressive conditional heteroskedasticity (GARCH) model was used to generate a measurement noise covariance matrix in the KF step in case of the presence of heteroscedasticity.

3. The estimation and prediction performance of the APVAR-KF method was conducted and compared with VAR-based models with time-invariant parameters for the main stock exchange index and macroeconomic indicators in two selected emerging market economies: Thailand and Indonesia.

1.3. Article Structure

The remainder of this paper is organized as follows. Section 2 presents a detailed description of the proposed model, the APVAR-KF method, where a measurement noise covariance matrix was constructed through the multivariate GARCH with BEKK specification. Section 3 provides a comparative investigation of the estimation and prediction performances of the APVAR-KF and benchmark models for stock exchange index, real effective exchange rate and consumer price index of Thailand and Indonesia. Conclusions and discussion are drawn in Section 4.

2. Methodology

The Adaptive Parameter VAR-KF Model (APVAR-KF)

Consider the vector autoregressive (VAR) model for a stationary n-dimensional state vector at time instant k, $x(k) \in \mathbb{R}^n$. The VAR model of order p, denoted by VAR(p), has the form [10]

$$x(k) = c + B_1 x(k-1) + B_2 x(k-2) + \cdots + B_p x(k-p) + \eta(k) \tag{1}$$

where $c \in \mathbb{R}^n$ is an intercept vector, B_i for $i = 1, 2, \ldots, p$ is an $n \times n$ matrix of autoregressive coefficients and η is an n-dimensional error vector.

Specifically, we assume the VAR model of order one, VAR(1), which can be expressed as

$$x(k) = c + Bx(k-1) + \eta(k) \tag{2}$$

Equation (2) is treated as a state-space dynamical system in the KF method. This equation also signifies the validity of the linearity assumption of the KF through the VAR process. Let $y(k)$ be the q-dimensional observation vector, which is related to the model state by the following equation:

$$y(k) = Hx(k) + \mu(k) \tag{3}$$

where $H \in \mathbb{R}^{q \times n}$ is an observation operator and $\mu \in \mathbb{R}^q$ is an observational error vector. To introduce time-variation parameters into the state Equation (2), we assume that the parameter transition equations follow a random walk process; therefore, for $i, j = 1, 2, \ldots, n$,

$$c_i(k) = c_i(k-1) + \delta_i(k-1), \quad b_{ij}(k) = b_{ij}(k-1) + \varepsilon_{ij}(k-1) \tag{4}$$

where c_i and b_{ij} for $i, j = 1, 2, \ldots, n$ are coefficient components of matrices c and B, respectively, and δ_i and ε_{ij} represent random noises, which are assumed to have the same distribution as η_i.

By treating the parameters as additional state variables, they are concatenated to the model state vector in order to form a single vector $z(k) = [x(k), c(k), \beta(k)]$, where $c(k) = [c_1(k), c_2(k), \ldots, c_n(k)]$ and $\beta(k) = [b_{11}(k), b_{12}(k), \ldots, b_{nn}(k)]$. The modified state propagation equation becomes

$$z(k) = \tilde{B}(k)z(k-1) + \tilde{\eta}(k) \tag{5}$$

where $\tilde{\eta}(k)$ is the zero-mean white noise with covariance matrix Q. The model coefficient matrix is formulated as

$$\tilde{B}(k) = \begin{bmatrix} B(k-1) & I_n & 0 \\ 0 & I_n & 0 \\ 0 & 0 & I_{n^2} \end{bmatrix}$$

where I_j denotes the $j \times j$ identity matrix and 0 is a zero matrix of appropriate size. The elements of $B(k-1)$ are the parameter estimates from the previous time step, where the elements of $B(1)$ are computed by the least square method.

The observation equation is subsequently modified as

$$y(k) = \tilde{H}z(k) + \mu(k) \tag{6}$$

where the observation operator $\tilde{H} = [H \ 0] \in \mathbb{R}^{n \times (n^2 + 2n)}$ and the observation noise term $\mu(k) \in \mathbb{R}^n$ is assumed to be an independent and identically distributed observational Gaussian noise with associated error covariance matrix R. Since a volatility persistence is usually detected in financial and macroeconomic time series, we therefore incorporated a volatility feature through the observational covariance matrix R, which was modeled by the generalized autoregressive conditional heteroscedasticity (GARCH) process [21]. In particular, the multivariate BEKK [22] representation was selected to parametrize the GARCH model as the matrix R is guaranteed to be positive definite with unrestricted parameterizations. The BEKK(1,1) specification is written as

$$R(k) = D'D + A'\mu(k-1)\mu'(k-1)A + M'R(k-1)M \tag{7}$$

where D is restricted to be a lower triangular matrix representing constant components and A denotes an ARCH coefficient matrix that describes the effects of both own and cross fluctuations. The coefficient matrix M characterizes the GARCH effects reflecting the degree of its own and cross volatility persistence. To estimate the elements of these parameter matrices, we made use of the quasi-maximum likelihood [23] estimation, in which the likelihood function is given by

$$L(\theta) = \sum_{k=1}^{T} \left(-\frac{n}{2}\ln(2\pi) - \frac{1}{2}\left(\ln|R(k;\theta)| + \mu'(k)R^{-1}(k;\theta)\mu(k) \right) \right) \tag{8}$$

where T is the number of observations and θ denotes an unknown parameter vector.

Similar to the KF process, the APVAR-KF method comprises two steps: the forecast (prediction) and analysis (update) steps. In the forecast step, the aggregated state vector $z(k)$ is propagated through the governing Equation (5). The resulting estimates are subsequently integrated with observation information in the analysis step to produce the optimal estimates. Superscripts f and a stand for forecast and analysis estimates, respectively, and we assumed the initial state estimate, $z^f(1)$, to be a Gaussian vector of zero mean with corresponding error covariance matrix $P^f(1)$. A description of how the error covariance matrices $P^f(1)$ and Q in the KF step are attained is given in Section 3.

The Forecast Step

Given that the analysis mean $z^a(k-1)$ and its corresponding analysis covariance matrix $P^a(k-1)$ are available, the forecast state $z^f(k)$ can be obtained through

$$z^f(k) = \tilde{B}(k)z^a(k-1) \tag{9}$$

and the forecast covariance matrix

$$P^f(k) = \tilde{B}(k)P^a(k-1)\tilde{B}'(k) + Q. \tag{10}$$

The Analysis Step

The analysis state $z^a(k)$ and analysis covariance $P^a(k)$ are expressed as

$$z^a(k) = z^f(k) + G(y(k) - \tilde{H}z^f(k)), \tag{11}$$

$$P^a(k) = (I - G\tilde{H})P^f(k), \tag{12}$$

where the Kalman gain matrix, G, determines the weight attributed to recent measurements, and is given by

$$G = P^f(k)\tilde{H}'(\tilde{H}P^f(k)\tilde{H}' + R(k))^{-1}. \tag{13}$$

3. Data and Simulation Results

To evaluate the efficiency of our proposed method, the monthly historical data used in this study include the stock market index, real effective exchange rate (REER) and consumer price index (CPI) of Thailand and Indonesia spanning from January 1997 to May 2021. The stock exchange of Thailand (SET) index data were collected from the Stock Exchange of Thailand website [24] and the REER and CPI data were acquired from the Bank of Thailand website [25], whereas the Jakarta stock exchange (JKSE) composite index was obtained from the investing.com database [26] and its REER and CPI data were taken from the Federal Reserve Economic Data (FRED) statistics [27]. The dataset is divided into two groups: the data from January 1997 to March 2021 were utilized for the training phase and data from April 2021 to May 2021 were treated as the testing phase. These raw data were transformed into monthly returns by taking the first logarithm difference. A z-score normalization [28] was subsequently applied to these return time series in order to adjust the range variation to comparable scales. The normalized returns were constructed by extracting the average from attribute values and dividing by the corresponding standard deviation.

3.1. Granger Causality Analysis

This section demonstrates an assessment of the interactions between different pairs of time series using the bivariate Granger causality test [29]. This analysis helps us to determine whether lagged values of one variable are linearly informative in forecasting another variable. Given two stationary variables $x_1(k)$ and $x_2(k)$ at time instant k, the bivariate Granger causality test follows a pair of regression equations:

$$x_1(k) = \sum_{j=1}^{J} a_j x_1(k-j) + \sum_{j=1}^{J} b_j x_2(k-j) + u_1(k) \tag{14}$$

$$x_2(k) = \sum_{j=1}^{J} c_j x_1(k-j) + \sum_{j=1}^{J} d_j x_2(k-j) + u_2(k) \tag{15}$$

where u_1 and u_2 are random disturbances and J is the maximum lag order. From the equations above, a unidirectional causality from the variable x_2 to variable x_1 is indicated if $\sum_{j=1}^{k} b_j$ in Equation (14) is significantly different to zero by F-statistics whereas $\sum_{j=1}^{k} c_j$ in the Equation (15) is not significant.

Table 1 presents the results of the Granger causality test for the direction of causality (F-statistics and p-value in parenthesis) among the normalized returns of the SET index, REER and CPI. The results show that the CPI does Granger-cause the SET index and REER at a 1% level of significance. Although the null hypothesis, which states that REER does not Granger-cause the CPI and SET index, is accepted, the null hypothesis in the opposite direction is rejected with a significance level of 1%. In the case of Indonesia, Table 2 reveals a two-way directional relationship between the CPI and JKSE index, and also between CPI and REER at a 5% level of significance. In addition, there is a unidirection causality running from the JKSE index to REER. With regard to the causality direction, the sufficient condition for the cointegration between two variables is that the Granger causality must exist in at least one direction [30]. Since our results indicate unidirectional causality between each pair of variables, it therefore suggests that all factors can be included in the model.

Table 3 presents some descriptive statistics of the monthly normalized return series. All normalized return series for both Thailand and Indonesia are highly leptokurtic and skewed with respect to the normal distribution, as indicated by the kurtosis and skewness measures. These results can be further confirmed by the Jarque–Bera test in which the normality hypothesis is rejected at a 1% significant level for all three variables. Similarly, the

ARCH test rejects the null hypothesis of homoscedasticity at a 1% level of significance for all variables except the JKSE index with a 5% level of significance. This suggests the validation of the GARCH model in capturing the volatility interaction among variables, resulting in a plausible assumption of the observational covariance matrix in Equation (7). Since the VAR approach requires the data input to be stationary prior to the model implementation to avoid spurious regressions, the presence of unit roots was examined by a standard augmented Dickey–Fuller (ADF) test [31,32]. The ADF test well rejects the null hypothesis, with a statistical significance of 1% for every variable, which provides strong evidence of stationarity in the normalized return series for both countries.

Table 1. Pairwise Granger causality test of the normalized return series of Thailand.

Dependent Variable	F-Statistics Test		
	SET Index (Prob. Values)	**REER (Prob. Values)**	**CPI (Prob. Values)**
SET index		21.2505	1.6713
		(0.0000) ***	(0.1971)
REER	1.0583		0.2262
	(0.3045)		(0.6347)
CPI	10.6046	15.7289	
	(0.0013) ***	(0.0000) ***	

Notes: *** denotes significance at the 1%.

Table 2. Pairwise Granger causality test of the normalized return series of Indonesia.

Dependent Variable	F-Statistics Test		
	JKSE Index (Prob. Values)	**REER (Prob. Values)**	**CPI (Prob. Values)**
JKSE index		21.0083	4.3218
		(0.0000) ***	(0.0385) **
REER	0.5960		93.2368
	(0.4408)		(0.0000) ***
CPI	6.5980	5.8258	
	(0.0107) **	(0.0164) **	

Notes: ** and *** denote significance at the 5% and 1%, respectively.

Table 3. Descriptive statistics of normalized return series.

Variable	Stock Index	REER	CPI
Thailand			
Skewness	−0.4037	−1.7799	−0.8021
Kurtosis	6.1646	25.6633	11.7886
Maximum	3.5321	5.5006	4.6552
Minimum	−4.5314	−7.9811	−5.8710
Jarque–Bera	128.8900 ***	6359.4434 ***	964.3984 ***
ARCH test	8.1017 ***	37.2224 ***	29.8895 ***
ADF	−11.2860 ***	−11.1884 ***	−9.4615 ***
Indonesia			
Skewness	−1.2042	−3.5107	4.7035
Kurtosis	8.6444	39.2024	31.6584
Maximum	3.1898	3.6852	8.6086
Minimum	−5.0722	−9.8602	−1.3426
Jarque–Bera	455.0480 ***	16432.3347 ***	10993.3566 ***
ARCH test	5.7372 **	20.9877 ***	47.8145 ***
ADF	−12.2780 ***	−12.8950 ***	−7.2100 ***

Notes: ** and *** denote significance at the 5% and 1%, respectively.

3.2. Results

A prior requirement for a Kalman-filter-based recursive algorithm is the specification of an initial state vector, as well as its error covariance matrix. At the initial time instant $k = 1$, we used the actual initial data during our sample period along with the coefficients estimated from the ordinary least square method to be the elements of the initial state vector, $z^f(1)$. The corresponding error covariance matrix $P^f(1)$ is assumed to be equal to the process noise covariance matrix Q, which is often assigned to be arbitrarily constant. We estimated the matrix Q through an average of sample covariances of the state prediction errors. The reference state vector at time instant k, $z^{\text{ref}}(k)$, corresponds to a collection of the actual data and parameters evaluated from the VAR(1) model, and this gives $z^{\text{ref}}(k) = [x^{\text{ref}}(k), c^{\text{ref}}(k), \beta^{\text{ref}}(k)]$. The noisy state-parameter vector $z(k)$ was sampled from a Gaussian distribution with mean equal to $z^{\text{ref}}(k)$, and the standard deviation was set to 25% of the reference values. The matrix Q was thus constructed using the following estimation:

$$Q = \frac{1}{m} \sum_{k=1}^{m} (z^{\text{ref}}(k) - \tilde{B}(k)z(k-1)(z^{\text{ref}}(k) - \tilde{B}(k)z(k-1))^T \tag{16}$$

where m is the number of time instants. There are three sample periods used to approximate the matrix Q, ranging from the first 12 months up to 60 months: January 1997–December 1997, January 1997–December 1999 and January 1997–December 2001. The resulting matrix Q applied to the APVAR-KF method was calculated on a statistical basis through the use of Monte Carlo simulations; that is, the matrix Q was determined by taking an average of over 50 experiments for each time instant. The results presented for the APVAR-KF method were obtained from the best-tuned values of the matrices $P^f(1)$ and Q, which relied on the optimal achievable values of MAPE in the training period.

To demonstrate the performance of the APVAR-KF method in estimation and prediction, the classical vector autoregressive model of order one, VAR(1), was taken as a benchmark scheme. Meanwhile, an augmentation between the VAR model and KF with fixed model coefficients in Equation (2), the VAR-KF method, was additionally computed to illustrate the effects of a two-step procedure with and without time-variant model parameters upon the forecasting accuracy. The mean absolute percentage error (MAPE) and root mean square error (RMSE) were used as the performance evaluation indicators. They are formulated as follows:

$$\text{MAPE} = \frac{1}{T} \sum_{k=1}^{T} \left(\left| \frac{\tilde{z}(k) - \hat{z}(k)}{\tilde{z}(k)} \right| \times 100 \right)$$

and

$$\text{RMSE} = \sqrt{\frac{1}{T} \sum_{k=1}^{T} (\tilde{z}(k) - \hat{z}(k))^2}$$

where T is the total number of simulations, $\tilde{z}(k)$ represents the actual measured data and $\hat{z}(k)$ denotes the estimated value.

Table 4 displays the estimation efficiency during the training period through the MAPE and RMSE statistics. According to MAPE and RMSE measures, both hybrid models have a superior estimation performance to the single model with lower MAPE and RMSE values for all variables of both countries. In the case of Thailand, the average MAPE values of VAR(1), VAR-KF and APVAR-KF models are 2.3460%, 2.0556% and 1.4089%, respectively. The VAR-KF approach reduces MAPE and RMSE values by over 10% compared to the benchmark model, whereas those of APVAR-KF by up to 40%. The same estimation pattern can be seen for Indonesia, where the overall improvement when using hybrid models is above 40%. These findings suggest that, by augmenting the Kalman filter in the VAR model, a significant improvement in the estimation accuracy is attained. When comparing among hybrid models, the APVAR-KF model exhibits better MAPE and RMSE values for all variables. The APVAR-KF model improves the quality of the overall estimation by approximately 30% for Thailand and approximately 6% for Indonesia regarding the MAPE values.

Table 4. Mean absolute percentage errors and root mean square errors during the training phase (January 1997–March 2021).

Variable	MAPE			RMSE		
	VAR(1)	VAR-KF	APVAR-KF	VAR(1)	VAR-KF	APVAR-KF
Thailand						
SET index	5.5705	5.0223	3.4655	54.8660	50.5360	34.8130
REER	1.1547	0.8920	0.5892	2.0094	1.5422	1.1419
CPI	0.3129	0.2525	0.1721	0.4243	0.3529	0.2415
Average error	2.3460	2.0556	1.4089	31.6991	29.1913	20.1106
Indonesia						
JKSE index	5.2524	2.8055	2.7349	158.7900	70.9540	69.4390
REER	2.7593	1.6239	1.4992	3.3597	2.3697	2.2680
CPI	0.5095	0.3356	0.2279	0.4702	0.2827	0.1906
Average error	2.8404	1.5883	1.4873	91.6984	40.9885	40.1122

To assess the predictability using initial states acquired from three models, the state estimates in March 2021 were treated as the initial state vector for the underlying dynamical Equation (2) to forecast the state values of April 2021 and May 2021 (the testing phase). There are two different scenarios with respect to the model coefficients. The coefficients remain unchanged from the training phase for the VAR(1) and VAR-KF approaches, whereas those that relied on the APVAR-KF method are based on the parameter estimates in March 2021.

Table 5 demonstrates the forecasting performance in April 2021 of three models in terms of MAPE and RMSE criteria. The hybrid models in comparison with the VAR(1) model for Thailand yield a higher forecasting accuracy for all factors, with the average MAPE being 0.8303% and 0.6213%. These are, respectively, equivalent to a 18.8695% and 39.2900% improvement for the VAR-KF and APVAR-KF models, with the SET index being best improved. Similarly, both VAR-KF and APVAR-KF models achieve a better performance than the benchmark method for Indonesia, with a considerable improvement in the REER variable. Most errors attained from the APVAR-KF model are less than those of the VAR-KF method, except the REER variable of Indonesia, where the errors of using time-variant parameters are slightly greater than using fixed parameters. This indicates that the first time step prediction can predominantly be improved by exploiting the adjustable model parameters.

Table 5. Mean absolute percentage errors and root mean square errors of April 2021.

Variable	MAPE			RMSE		
	VAR(1)	VAR-KF	APVAR-KF	VAR(1)	VAR-KF	APVAR-KF
Thailand						
SET index	1.0765	0.6936	0.1845	17.0420	10.9800	2.9212
REER	0.6606	0.6199	0.5341	0.7087	0.6650	0.5730
CPI	1.3333	1.1776	1.1454	1.3397	1.1832	1.1509
Average error	1.0235	0.8303	0.6213	9.8780	6.3876	1.8427
Indonesia						
JKSE index	0.9919	0.5629	0.4873	59.4720	33.7510	29.2170
REER	0.9884	0.0666	0.0844	0.8703	0.0586	0.0743
CPI	0.3198	0.3065	0.2595	0.3773	0.3616	0.3062
Average error	0.7667	0.3120	0.2771	34.3405	19.4873	16.8694

Table 6 reports the prediction efficiency of May 2021 forecasts. The VAR-KF and APVAR-KF models continue to outperform the traditional VAR(1) method for Thailand, with lower errors for all variables. By comparing among different hybrid algorithms, the APVAR-KF model provides better results for the REER and CPI, with a lower average MAPE of 0.8725%. Nevertheless, a different result arises for Indonesia, where the REER forecasts of the two-step methods are worse than the VAR(1) model despite the fact that the JKSE index and CPI errors achieved by the APVAR-KF technique are lowest among all of the individual algorithms. The predicted values for June 2021 are not shown in this report, considering that the error trends are similar to those in May 2021. The APVAR-KF model remains providing superior predictions for all variables of Thailand and for the main stock market price index of Indonesia.

Table 6. Mean absolute percentage errors and root mean square errors of May 2021.

Variable	MAPE			RMSE		
	VAR(1)	**VAR-KF**	**APVAR-KF**	**VAR(1)**	**VAR-KF**	**APVAR-KF**
Thailand						
SET index	0.8830	0.2855	0.3552	14.0720	4.5503	5.6597
REER	2.4384	2.3650	2.2208	2.5718	2.4943	2.3423
CPI	0.2959	0.0811	0.0415	0.2946	0.0807	0.0413
Average error	1.2058	0.9105	0.8725	8.2608	2.9963	3.5365
Indonesia						
JKSE index	2.8246	2.3965	2.2698	167.9900	142.5300	135.0000
REER	0.0547	1.2434	1.4632	0.0486	1.1049	1.3002
CPI	0.5235	0.5752	0.5049	0.6197	0.6809	0.5977
Average error	1.1343	1.4050	1.4126	96.9897	82.2931	77.9467

Figures 1 and 2 depict a visual comparison between the normalized return estimates and actual data of all three variables of Thailand and Indonesia from July 2018 to May 2021. The plots of actual data and their corresponding estimates over the whole study period can be seen in Figures A1 and A2. The discrepancy between the estimated values derived from all approaches and actual data appears to be minor over the tranquil period. In the course of the COVID-19 outbreak, when drastic changes in economic and financial situations took place, the APVAR-KF method performs best in capturing these abrupt changes in all variables, followed by the VAR-KF and VAR(1) models. These results may reflect that a variation in parameters allows the model to better track the actual data, especially during times of high uncertainty. This may be due to that fact that the coefficients of a model system are sequentially updated using recent observations, causing the underlying model to be able to forecast abruptly changing trends. For Indonesia, it appears that the forecasting results derived from the hybrid models exhibit similar increasing trends to the REER actual data, with relatively lower slopes during the testing phase, whereas the opposite trend direction pattern is found in the VAR(1) method. Although the results in Table 6 indicate a better REER forecasting ability when using the VAR(1) model, a further examination of how the trend direction changes can be of importance.

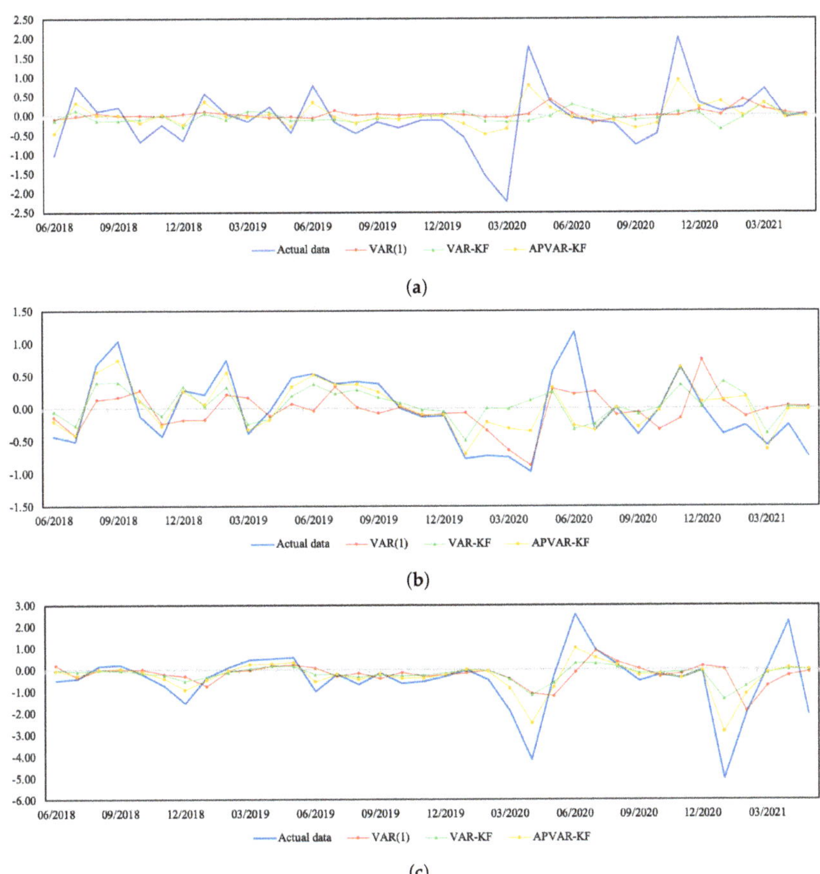

Figure 1. A comparison between the actual data and the estimated values from three methods for Thailand during July 2018–May 2021. (**a**) Normalized SET index return; (**b**) normalized real effective exchange rate return; (**c**) normalized CPI return.

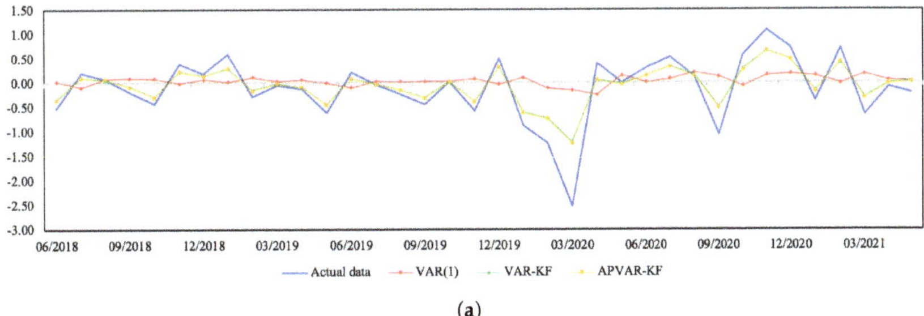

(**a**)

Figure 2. *Cont.*

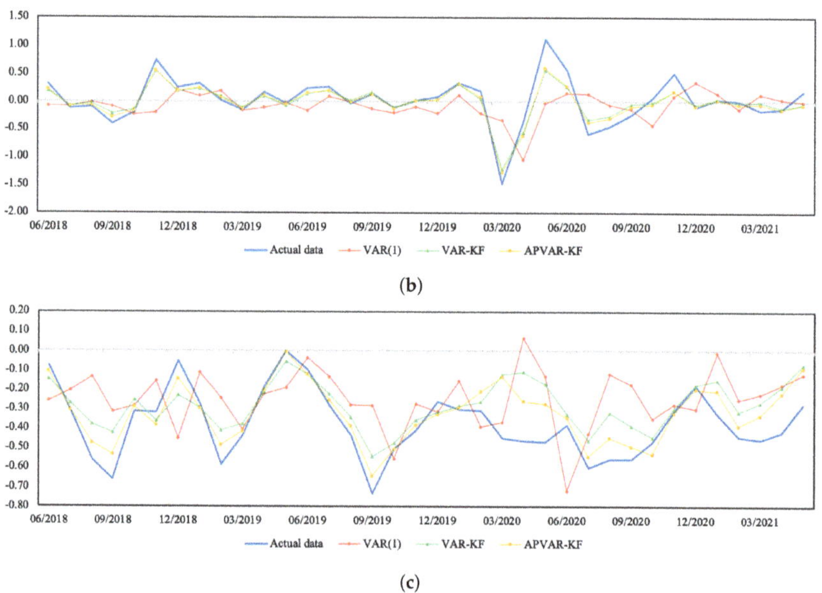

(b)

(c)

Figure 2. A comparison between the actual data and the estimated values from three methods for Indonesia during July 2018–May 2021. (**a**) Normalized JKSE index return; (**b**) normalized real effective exchange rate return; (**c**) normalized CPI return.

4. Conclusions and Discussion

Forecasting economic and financial time series can have substantial implications for the implementation of monetary policies and regulations and for an individual investor's investment decision. This paper is designed to model and forecast the complex interactions between the economic factors and financial market by introducing the hybrid APVAR-KF model for joint state parameter estimation. The method combines the Kalman filter with the VAR model, in which, the observational error covariance matrix is implemented using the multivariate BEKK-GARCH representation.

In addition to providing the best estimation performance, the APVAR-KF technique tends to offer satisfactory short-term predictions and future trend patterns. As presented in Figures A1 and A2, the coefficient of determination (R-square) values between the observed data and the estimates of the APVAR-KF model range from 0.6109 to 0.8713, or 61.09% to 87.13%, which suggests that our proposed model has the ability to capture the dynamics of economic and financial time series.

In this regard, the benefits of the APVAR-KF model in estimation and prediction may be attributed to two reasons. The first reason is that this is a two-step process in which the Kalman filter provides a mechanism that can extract discriminative information from the training data. Another reason is that adaptive parameters can enhance the hybrid performance, creating a plausible model structure that adjusts to a change in state characteristics over time. This is considerably beneficial, especially when an unexpected fluctuation caused by economic instability occurs. Despite the favorable results of this study, the assumption of lag one in the VAR step can be a limitation of the method. The VAR model specification with higher lag orders and an inclusion of more macro factors, can be of particular interest. However, this is an apparent tradeoff problem between a more elaborate model and a heavy computational burden due to a large dimension of the state space. Ensemble-based filters that allow for the error covariance matrices to be computed without a moment closure assumption can potentially provide computational feasibility and efficiency. Due to the sensitivity of the process noise covariance Q to the prediction performance of the APVAR-KF approach, another challenge concerns the selection criteria of the matrix Q. In this work, we used the sample covariance calculated from a particular time period to represent the process noise statistics. Instead, other techniques, including the adaptive Q algorithm, covariance inflation and some rigorous optimization approaches, can be adopted, especially under some complex and dynamic circumstances.

Author Contributions: N.P. and N.C. designed the research, implemented the numerical experiments and analyzed the results. N.C. wrote the manuscript and N.P. and N.C. approved the final manuscript. All authors have read and agreed to the published version of the manuscript.

Funding: This research was funded by the Research Fund for DPST Graduate with First Placement [Grant no. 029/2559], the Institute for the Promotion of Teaching Science and Technology (IPST), Thailand.

Acknowledgments: The authors would like to thank Thanasak Mouktonglang and Sompop Moonchai for their helpful insight and suggestions. This research was supported by Chiang Mai University and the Centre of Excellence in Mathematics, CHE, Thailand. Nawinda Chutsagulprom was supported by the Research Fund for DPST Graduate with First Placement [Grant no. 029/2559], the Institute for the Promotion of Teaching Science and Technology (IPST), Thailand, under the mentoring of Thanasak Mouktonglang.

Conflicts of Interest: The authors declare no conflict of interest.

Abbreviations

The following abbreviations are used in this manuscript:

VAR	Vector Autoregressive Model
KF	Kalman Filter
GARCH	Generalized Autoregressive Conditional Heteroscedasticity
SET	Stock Exchange of Thailand
JKSE	Jakarta Stock Exchange Composite Index
REER	Real Effective Exchange Rate
CPI	Consumer Price Index
MAPE	Mean Absolute Percentage Error
RMSE	Root Mean Square Error

Appendix A

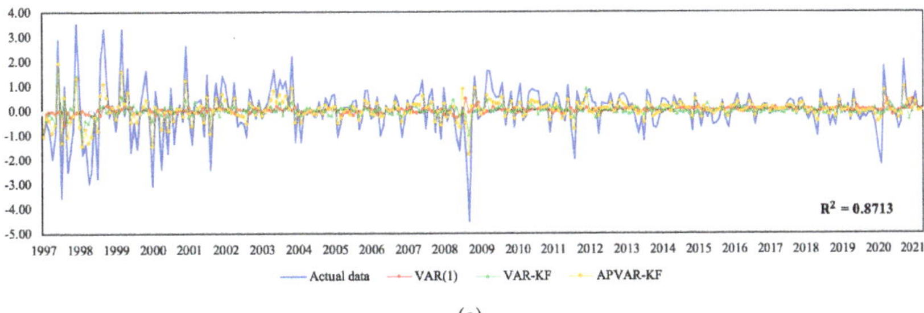

(a)

Figure A1. *Cont.*

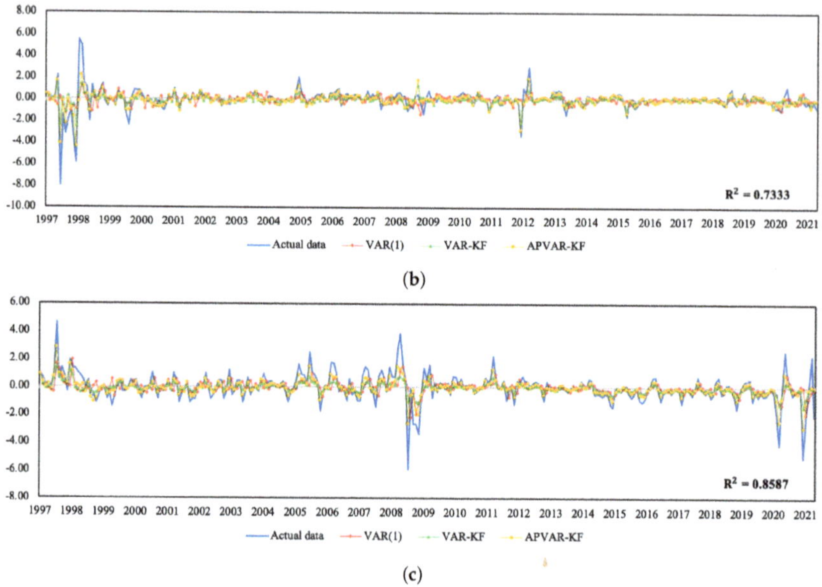

Figure A1. Plots of the actual data and the estimated values from three methods for Thailand. (**a**) Normalized SET index return; (**b**) normalized real effective exchange rate return; (**c**) normalized CPI return.

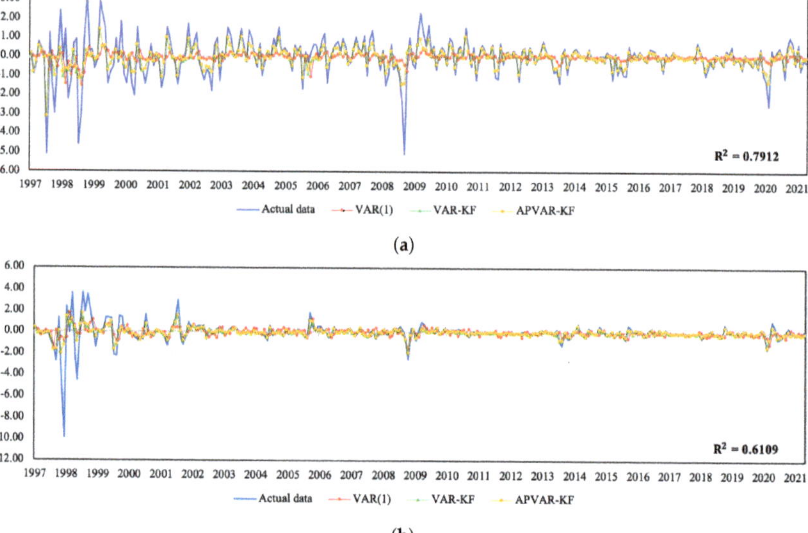

Figure A2. *Cont.*

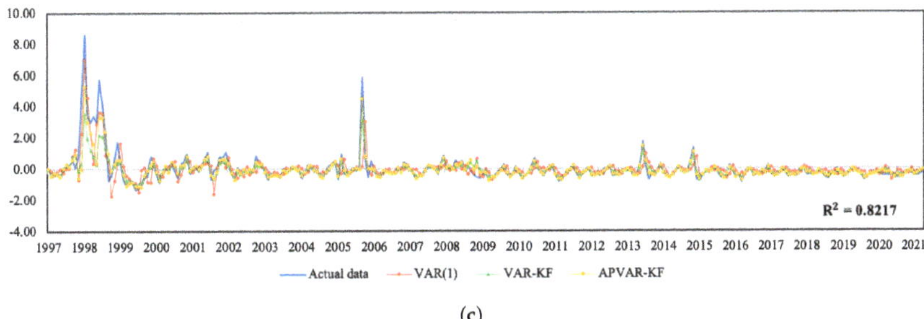

(c)

Figure A2. Plots of the actual data and the estimated values from three methods for Indonesia. (**a**) Normalized SET index return; (**b**) normalized real effective exchange rate return; (**c**) normalized CPI return.

References

1. Ariyo, A.A.; Adewumi, A.O.; Ayo, C.K. Stock price prediction using the ARIMA model. In Proceedings of the 2014 UKSim-AMSS 16th International Conference on Computer Modelling and Simulation, Cambridge, UK, 26–28 March 2014; pp. 106–112. [CrossRef]
2. Junior, P.R.; Salomon, F.L.R.; de Oliveira Pamplona, E. ARIMA: An applied time series forecasting model for the Bovespa stock index. *Appl. Math.* **2014**, *5*, 3383. [CrossRef]
3. Moghaddam, A.H.; Moghaddam, M.H.; Esfandyari, M. Stock market index prediction using artificial neural network. *J. Econ. Financ. Adm. Sci.* **2016**, *21*, 89–93. [CrossRef]
4. Qiu, M.; Song, Y.; Akagi, F. Application of artificial neural network for the prediction of stock market returns: The case of the Japanese stock market. *Chaos Solit. Fract.* **2016**, *85*, 1–7. [CrossRef]
5. Shen, S.; Jiang, H.; Zhang, T. *Stock Market Forecasting Using Machine Learning Algorithms*; Department of Electrical Engineering, Stanford University: Stanford, CA, USA, 2012; pp. 1–5.
6. Ren, R.; Wu, D.D.; Liu, T. Forecasting stock market movement direction using sentiment analysis and support vector machine. *IEEE Syst. J.* **2018**, *13*, 760–770. [CrossRef]
7. Yang, E.; Kim, S.H.; Kim, M.H.; Ryu, D. Macroeconomic shocks and stock market returns: The case of Korea. *Appl. Econ.* **2018**, *50*, 757–773. [CrossRef]
8. Antonakakis, N.; André, C.; Gupta, R. Dynamic spillovers in the United States: Stock market, housing, uncertainty, and the macroeconomy. *South. Econ. J.* **2016**, *83*, 609–624. [CrossRef]
9. Vrugt, E.B. US and Japanese macroeconomic news and stock market volatility in Asia-Pacific. *Pac.-Basin Financ. J.* **2009**, *17*, 611–627. [CrossRef]
10. Sims, C.A. Macroeconomics and reality. *Econom. J. Econom. Soc.* **1980**, 1–48. [CrossRef]
11. Suharsono, A.; Aziza, A.; Pramesti, W. Comparison of vector autoregressive (VAR) and vector error correction models (VECM) for index of ASEAN stock price. *AIP Publ.* **2017**, *1913*, 020032. [CrossRef]
12. Öğünç, F.; Akdoğan, K.; Başer, S.; Chadwick, M.G.; Ertuğ, D.; Hülagü, T.; Kösem, S.; Özmen, M.U.; Tekatlı, N. Short-term inflation forecasting models for Turkey and a forecast combination analysis. *Econ. Model.* **2013**, *33*, 312–325. [CrossRef]
13. Primiceri, G.E. Time varying structural vector autoregressions and monetary policy. *Rev. Econ. Stud.* **2005**, *72*, 821–852. [CrossRef]
14. D'Agostino, A.; Gambetti, L.; Giannone, D. Macroeconomic forecasting and structural change. *J. Appl. Econom.* **2013**, *28*, 82–101. [CrossRef]
15. Bekiros, S.; Gupta, R.; Paccagnini, A. Oil price forecastability and economic uncertainty. *Econ. Lett.* **2015**, *132*, 125–128. [CrossRef]
16. Kumar, M. A time-varying parameter vector autoregression model for forecasting emerging market exchange rates. *Int. J. Econ. Sci. Appl. Res.* **2010**, *3*, 21–39.
17. Korobilis, D. VAR forecasting using Bayesian variable selection. *J. Appl. Econom.* **2013**, *28*, 204–230. [CrossRef]
18. Kalman, R.E. A new approach to linear filtering and prediction problems. *J. Basic Eng.* **1960**, *82*, 35–45. [CrossRef]
19. Bekiros, S. Forecasting with a state space time-varying parameter VAR model: Evidence from the Euro area. *Econ. Model.* **2014**, *38*, 619–626. [CrossRef]
20. Koop, G.; Korobilis, D. Large time-varying parameter VARs. *J. Econom.* **2013**, *177*, 185–198. [CrossRef]
21. Bollerslev, T. Generalized autoregressive conditional heteroskedasticity. *J. Econom.* **1986**, *31*, 307–327. [CrossRef]
22. Engle, R.F.; Kroner, K.F. Multivariate simultaneous generalized ARCH. *Econom. Theory* **1995**, *11*, 122–150. [CrossRef]
23. Bollerslev, T.; Wooldridge, J.M. Quasi-maximum likelihood estimation and inference in dynamic models with time-varying covariances. *Econom. Rev.* **1992**, *11*, 143–172. [CrossRef]
24. Stock Exchange of Thailand Website. 2021. Available online: https://classic.set.or.th/en/market/market_statistics.html (accessed on 5 October 2021).
25. Bank of Thailand Website. 2021. Available online: https://bot.or.th (accessed on 5 October 2021).

26. Investing Website. 2022. Available online: https://www.investing.com/indices/idx-composite-historical-data (accessed on 5 October 2021).
27. Federal Reserve Economic Data website. 2022. Available online: https://fred.stlouisfed.org/ (accessed on 5 October 2021).
28. Kang, S.H.; Yoon, S.M. Long memory features in the high frequency data of the Korean stock market. *Phys. Stat. Mech. Its Appl.* **2008**, *387*, 5189–5196. [CrossRef]
29. Granger, C.W. Investigating causal relations by econometric models and cross-spectral methods. *Econom. J. Econom. Soc.* **1969**, 424–438. [CrossRef]
30. Engle, R.F.; Granger, C.W. Co-integration and error correction: Representation, estimation, and testing. *Econom. J. Econom. Soc.* **1987**, *55*, 251–276. [CrossRef]
31. Dickey, D.A.; Fuller, W.A. Distribution of the estimators for autoregressive time series with a unit root. *J. Am. Stat. Assoc.* **1979**, *74*, 427–431. [CrossRef]
32. Dickey, D.A.; Fuller, W.A. Likelihood ratio statistics for autoregressive time series with a unit root. *Econom. J. Econom. Soc.* **1981**, *49*, 1057–1072. [CrossRef]

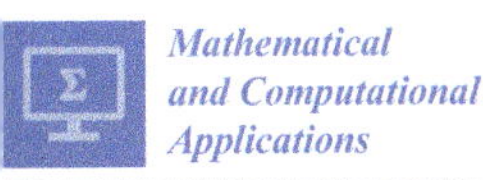

Mathematical and Computational Applications

Article

The Arctan Power Distribution: Properties, Quantile and Modal Regressions with Applications to Biomedical Data

Suleman Nasiru [1], Abdul Ghaniyyu Abubakari [1] and Christophe Chesneau [2,*]

[1] Department of Statistics and Actuarial Science, School of Mathematical Sciences, C. K. Tedam University of Technology and Applied Sciences, Kassena-Nankana Navrongo-Kologo Road, Navrongo P.O. Box 24, Upper East, Ghana
[2] Department of Mathematics, LMNO, CNRS-Université de Caen, Campus II, Science 3, 14032 Caen, France
* Correspondence: christophe.chesneau@unicaen.fr

Abstract: The usefulness of (probability) distributions in the field of biomedical science cannot be underestimated. Hence, several distributions have been used in this field to perform statistical analyses and make inferences. In this study, we develop the arctan power (AP) distribution and illustrate its application using biomedical data. The distribution is flexible in the sense that its probability density function exhibits characteristics such as left-skewedness, right-skewedness, and J and reversed-J shapes. The characteristic of the corresponding hazard rate function also suggests that the distribution is capable of modeling data with monotonic and non-monotonic failure rates. A bivariate extension of the AP distribution is also created to model the interdependence of two random variables or pairs of data. The application reveals that the AP distribution provides a better fit to the biomedical data than other existing distributions. The parameters of the distribution can also be fairly accurately estimated using a Bayesian approach, which is also elaborated. To end the study, the quantile and modal regression models based on the AP distribution provided better fits to the biomedical data than other existing regression models.

Keywords: quantile regression; modal regression; biomedical; unit distribution; skewed data

Citation: Nasiru, S.; Abubakari, A.G.; Chesneau, C. The Arctan Power Distribution: Properties, Quantile and Modal Regressions with Applications to Biomedical Data. *Math. Comput. Appl.* **2023**, *28*, 25. https://doi.org/10.3390/mca28010025

Academic Editor: Sandra Ferreira

Received: 22 December 2022
Revised: 9 February 2023
Accepted: 10 February 2023
Published: 14 February 2023

1. Introduction

Parametric statistical techniques have been used in biomedical studies to conduct analyses and draw conclusions. These parametric analyses, however, are constrained by some assumptions about (probability) distributions. Thus, the task of selecting an appropriate distribution for such analyses is incredibly essential. In addition, it is nontrivial, as the use of an incorrect distribution will result in misleading inferences. Knowing which distribution to use in biomedical modeling has become increasingly important as it is used to develop new parametric regression models for modeling the relationship between endogenous variables and a set of exogenous variables. These new regression models often provide a good fit with minimal loss of information compared to the existing ones. This has triggered new interest in developing regression models using extended or modified forms of existing distributions.

Among the distributions used for developing the regression models, those that are defined on the unit interval have received much attention due to the small loss of information they offer in modeling data on this interval. Some of these distributions include the unit folded normal distribution [1], bounded truncated Cauchy power exponential distribution [2], unit exponentiated Fréchet distribution [3], log XLindley (LXL) distribution [4], unit Chen distribution [5], unit Burr XII distribution (UBXII) [6], unit generalized half-normal distribution [7], unit Burr III (UBIII) distribution [8], unit Lindley distribution [9], unit Gompertz distribution [10], unit improved second degree Lindley (UISDL) distribution [11], unit Weibull distribution [12], and exponentiated Topp–Leone distribution [13].

Despite the existence of these distributions, it is worth noting that the behavior of humans or organisms is nondeterministic, and a single distribution cannot be selected in all situations to describe or model these traits. Therefore, we develop a new distribution called the arctan power (AP) distribution for modeling data on the unit interval based on the following motivations:

1. Develop a flexible unit distribution that is able to model data that are left-skewed, right-skewed, symmetric, J, and reversed-J shapes.
2. Develop a unit distribution capable of modeling data with increasing, bathtub, and modified upside-down bathtub hazard rate functions (HRFs).
3. Develop quantile regression for modeling response variables that are skewed or contain extreme values.
4. Develop modal regression for modeling response variables that are asymmetric or heavy-tailed.

The article is organized into eight sections. Section 2 describes the development of the AP distribution. Section 3 presents their statistical properties. Section 4 shows the construction of a possible bivariate extension of the AP distribution. Nine frequentist approaches to estimating the involved parameters are proposed in Section 5. The frequentist and Bayesian univariate applications of the distribution are given in Section 6. Section 7 is devoted to the quantile and modal regressions based on the AP distribution and their applications. The conclusion of the study is presented in Section 8.

2. Development of AP Distribution

Suppose that a random variable, X, follows the arctan uniform (AU) distribution. Then, according to [14], the cumulative distribution function (CDF) and probability density function (PDF) of X are, respectively, given by

$$F_X(x; \alpha) = \frac{\arctan(\alpha x)}{\arctan(\alpha)}, \alpha > 0, x \in (0,1) \tag{1}$$

and

$$f_X(x; \alpha) = \frac{\alpha}{\arctan(\alpha)(1 + \alpha^2 x^2)}, x \in (0,1). \tag{2}$$

The proposed AP distribution is obtained using the power transformation $Y = X^{1/\beta}, \beta > 0$. The motivations for introducing the power parameter, β, are to improve the tail properties of the new distribution, making it capable of handling both monotonic and non-monotonic HRFs. Other researchers have used the power transformation approach to modify existing continuous distributions. See, for instance, [15–17]. Hence, using standard mathematical developments, the CDF of Y is obtained as

$$\begin{aligned} F_Y(y; \alpha, \beta) &= F_X(y^\beta; \alpha) \\ &= \frac{\arctan(\alpha y^\beta)}{\arctan(\alpha)}, \alpha > 0, \beta > 0, y \in (0,1). \end{aligned} \tag{3}$$

The PDF and HRF are, respectively, given by

$$f_Y(y; \alpha, \beta) = \frac{\alpha \beta y^{\beta-1}}{\arctan(\alpha)(1 + \alpha^2 y^{2\beta})}, y \in (0,1) \tag{4}$$

and

$$h_Y(y; \alpha, \beta) = \frac{\alpha \beta y^{\beta-1}}{(\arctan(\alpha) - \arctan(\alpha y^\beta))(1 + \alpha^2 y^{2\beta})}, y \in (0,1). \tag{5}$$

Basically, when $\alpha \to 0^+$, the PDF of the AP distribution reduces to the one of the power distribution. As $\alpha \to 0^+$ and $\beta = 1$, the PDF of the AP distribution reduces to the one of the standard uniform distribution. Furthermore, when $\beta = 1$, the PDF of the AP distribution reduces to the one of the AU distribution.

The expanded form of the PDF is often useful when deriving the statistical properties of the distribution. Thus, using the arctangent function expansion indicated as follows: $\arctan(z) = \sum_{k=0}^{\infty} \frac{(-1)^k z^{2k+1}}{2k+1}, |z| < 1$ (see [18]) and $\alpha \in (0,1)$, the CDF of Y can be expressed as

$$F_Y(y; \alpha, \beta) = \sum_{k=0}^{\infty} \frac{(-1)^k \alpha^{2k+1} y^{(2k+1)\beta}}{(2k+1)\arctan(\alpha)}, y \in (0,1). \tag{6}$$

Differentiating the expanded form of the CDF in Equation (6), the corresponding PDF is given by

$$f_Y(y; \alpha, \beta) = \sum_{k=0}^{\infty} \frac{(-1)^k \beta \alpha^{2k+1} y^{(2k+1)\beta-1}}{\arctan(\alpha)}, y \in (0,1). \tag{7}$$

The PDF and HRF plots are shown in Figure 1 for some given parameter values. In it, the PDF exhibits left-skewed, right-skewed, J, and reversed-J shapes. This makes the AP distribution superior to the AU distribution, which exhibits only J shapes. On this side, the HRF displays increasing, bathtub, and modified upside-down bathtub shapes.

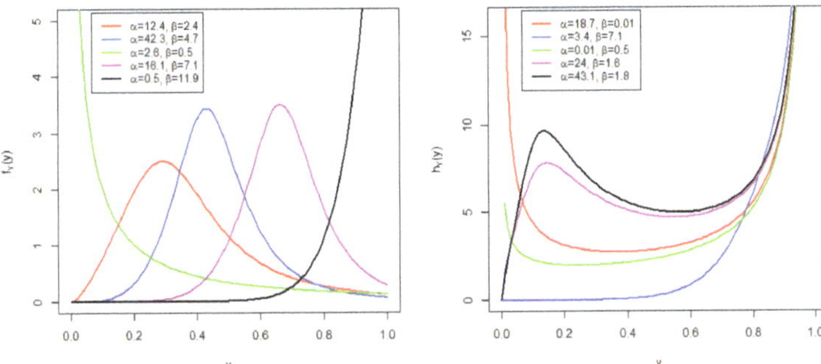

Figure 1. PDF (**left**) and HRF (**right**) plots.

3. Some Statistical Properties

In this section, some statistical properties of the AP distribution are presented.

3.1. Mode

The mode of a distribution is a useful measure of central tendency. It can be used as it for data measured on the nominal, ordinal, interval, or ratio scale. The AP distribution has a unique mode when $\beta > 1$, and it is expressed in the result below.

Proposition 1. *The mode of the AP distribution is given by*

$$\text{mode} = \left(\frac{\beta-1}{\alpha^2(\beta+1)} \right)^{\frac{1}{2\beta}}, \beta > 1. \tag{8}$$

Proof. To establish this expression, it is essential to locate the critical point(s) of the PDF. A critical point of the PDF is a point of the PDF, or equivalently, the logarithm of the PDF, where its derivative is zero or infinity. Taking the logarithm of the PDF and differentiating, we have

$$\frac{d \log f_Y(y; \alpha, \beta)}{dy} = \frac{\beta - 1 - \alpha^2(\beta+1)y^{2\beta}}{y(1 + \alpha^2 y^{2\beta})}.$$

Equating the derivative to zero and simplifying yields the mode. This completes the proof. □

3.2. Quantile Function

The quantile function can be used to generate random observations from the AP distribution and to compute shape-related metrics like skewness and kurtosis.

Proposition 2. *The quantile function of the AP distribution is given by*

$$Q(u; \alpha, \beta) = \left[\frac{\tan(u \arctan(\alpha))}{\alpha} \right]^{\frac{1}{\beta}}, u \in (0, 1). \tag{9}$$

Proof. The quantile function is the solution $Q(u; \alpha, \beta)$ of the following nonlinear equation: $F_Y(Q(u; \alpha, \beta); \alpha, \beta) = u$ for all $u \in (0, 1)$. After some simplifications, letting $y = Q(u; \alpha, \beta)$ in the CDF and equating the CDF to $u \in (0, 1)$ yields the quantile function. This completes the proof. □

It is important to note that the quantile function of the AP distribution is uniquely determined with simple trigonometric and power functions.

The median $Q(0.5; \alpha, \beta)$, first quartile $Q(0.25; \alpha, \beta)$, and upper quartile $Q(0.75; \alpha, \beta)$ are obtained, respectively, by substituting 0.5, 0.25, and 0.75 into the quantile function. The Bowley's (BS) measure of skewness and the Moors' (MK) measure of kurtosis can then be calculated using the quantiles. They are, respectively, given by

$$\mathrm{BS} = \frac{Q(0.75; \alpha, \beta) + Q(0.25; \alpha, \beta) - 2Q(0.5; \alpha, \beta)}{Q(0.75; \alpha, \beta) - Q(0.25; \alpha, \beta)},$$

and

$$\mathrm{MK} = \frac{Q(0.375; \alpha, \beta) - Q(0.125; \alpha, \beta) + Q(0.875; \alpha, \beta) - Q(0.625; \alpha, \beta)}{Q(0.75; \alpha, \beta) - Q(0.25; \alpha, \beta)}.$$

The plots of the Bowley's coefficient of skewness and Moor's coefficient of kurtosis are displayed in Figure 2. Both the skewness and kurtosis are affected by changes in the values of the parameters. From this figure, we can observe that the AP distribution can be left-skewed or right-skewed.

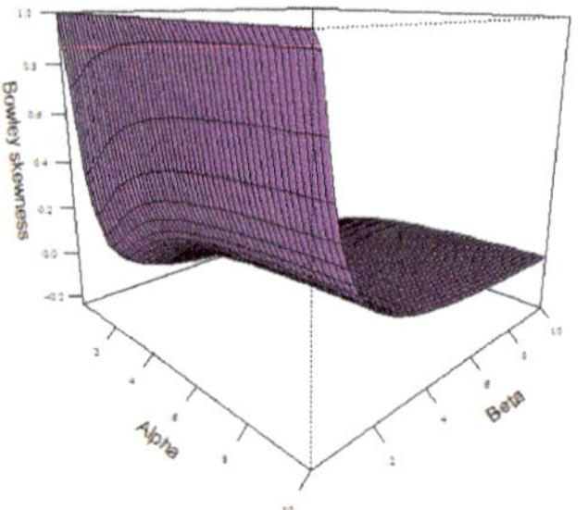
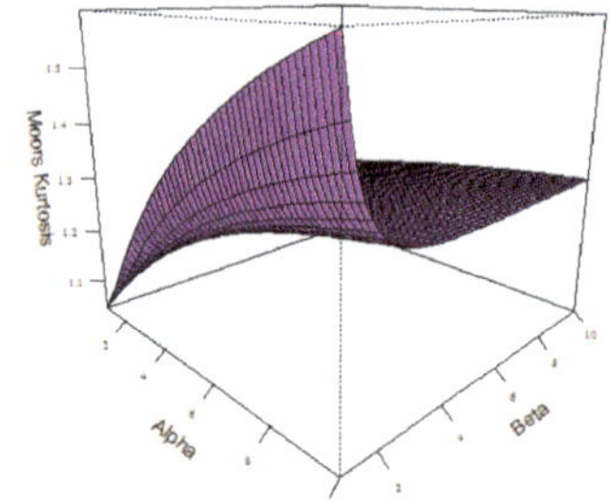

Figure 2. Skewness (**left**) and Kurtosis (**right**) plots.

3.3. Moments and Generating Function

The moments are useful for estimating measures of central tendency, dispersion, and shapes. The generating functions can be used to estimate the moments, if they exist in the mathematical sense.

Proposition 3. *For $\alpha \in (0, 1)$, the rth raw moment of an AP random variable Y is given by*

$$\mu'_r = \sum_{k=0}^{\infty} \frac{(-1)^k \beta \alpha^{2k+1}}{(r + (2k+1)\beta)\arctan(\alpha)}, r = 1, 2, \dots \tag{10}$$

Proof. The rth raw moment by definition is given by $\mu'_r = E(Y^r) = \int_0^1 y^r f_Y(y; \alpha, \beta)dy$. Thus, we obtain

$$\mu'_r = \sum_{k=0}^{\infty} \frac{(-1)^k \beta \alpha^{2k+1}}{\arctan(\alpha)} \int_0^1 y^{r+(2k+1)\beta-1} dy.$$

After some algebraic simplifications, the raw moment of the AP random variable is obtained. This completes the proof. $\square$

The incomplete moment is very useful when computing measures of inequalities, such as the Lorenz and Bonferroni curves.

Proposition 4. *For $\alpha \in (0,1)$, the rth incomplete moment of an AP random variable Y is given by*

$$\vartheta_r(y) = \sum_{k=0}^{\infty} \frac{(-1)^k \beta \alpha^{2k+1} y^{r+(2k+1)\beta}}{(r+(2k+1)\beta)\arctan(\alpha)}, r = 1, 2, \ldots \tag{11}$$

Proof. By definition, $\vartheta_r(y) = E(Y^r \mathbf{1}\{Y < y\}) = \int_0^y z^r f_Y(z; \alpha, \beta)dz$. Hence, substituting the expanded PDF into the definition and simplifying it completes the proof. $\square$

The Lorenz and Bonferroni curves are obtained, respectively, as

$$L_F(y) = \frac{1}{\mu} \int_0^y z f_Y(z; \alpha, \beta)dz$$

and

$$B_F(y) = \frac{1}{\mu F_Y(y; \alpha, \beta)} \int_0^y z f_Y(z; \alpha, \beta)dz,$$

where $\mu = \mu'_1$ is the mean.

Figure 3 displays the plots of the Lorenz and Bonferroni curves of the AP distribution for some selected parameter values. For the Lorenz curve, when $L_F(y) = y$, the minimal point of inequality is obtained. When $B_F(y) = y$, the so-called equidistributional line for the Bonferroni curve is obtained.

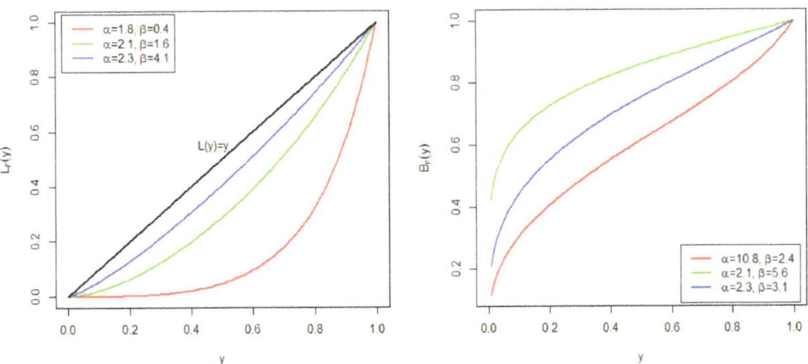

Figure 3. Plots of Lorenz curve (**left**) and Bonferroni curve (**right**).

When non-central moments of a random variable exist, they can be found using the moment-generating function (MGF).

Proposition 5. *For $\alpha \in (0,1)$, the MGF of an AP random variable Y is given by*

$$M_Y(t) = \sum_{r=0}^{\infty} \sum_{k=0}^{\infty} \frac{(-1)^k t^r \beta \alpha^{2k+1}}{r!(r + (2k+1)\beta)\arctan(\alpha)}. \tag{12}$$

Proof. Using the definition $M_Y(t) = E(e^{tY}) = \int_0^1 e^{ty} f_Y(y; \alpha, \beta)\, dy$ and applying the Taylor series expansion, we get

$$M_Y(t) = \sum_{r=0}^{\infty} \frac{t^r}{r!} \mu'_r$$

Hence, substituting the rth non-central moment completes the proof. $\square$

3.4. Order Statistics

Order statistics are very useful in extreme value analysis. They can be used to determine the behavior of the minimum and maximum value. Consider the order statistics $Y_{1:n} \leq Y_{2:n} \leq \ldots \leq Y_{n:n}$ from the AP distribution. Then, the PDF of $Y_{k:n}, k = 1, 2, \ldots, n$ is

$$f_{k:n}(y; \alpha, \beta) = C_{k:n}[F_Y(y; \alpha, \beta)]^{k-1}[1 - F_Y(y; \alpha, \beta)]^{n-k} f_Y(y; \alpha, \beta),$$

where the factor constant is given by

$$C_{k:n} = \frac{n!}{(k-1)!(n-k)!}.$$

Using the standard binomial expansion, we can express this PDF as

$$f_{k:n}(y; \alpha, \beta) = C_{k:n} \sum_{j=0}^{n-k} (-1)^j \binom{n-k}{j} [F_Y(y; \alpha, \beta)]^{k+j-1} f_Y(y; \alpha, \beta).$$

Hence, we obtain

$$f_{k:n}(y; \alpha, \beta) = \frac{\alpha \beta y^{\beta-1} C_{k:n}}{\arctan(\alpha)(1 + \alpha^2 y^{2\beta})} \sum_{j=0}^{n-k} (-1)^j \binom{n-k}{j} \left[\frac{\arctan(\alpha y^\beta)}{\arctan(\alpha)} \right]^{k+j-1}. \tag{13}$$

The minimum $(Y_{1:n})$ and maximum $(Y_{n:n})$ order statistics can serve to investigate the minimum and maximum failure time of a system, respectively. The PDF of $Y_{1:n}$ is given by

$$\begin{aligned} f_{1:n}(y; \alpha, \beta) &= n f_Y(y; \alpha, \beta)[1 - F_Y(y; \alpha, \beta)]^{n-1} \\ &= \frac{n\alpha\beta y^{\beta-1}(\arctan(\alpha) - \arctan(\alpha y^\beta))^{n-1}}{(1 + \alpha^2 y^{2\beta})(\arctan(\alpha))^n} \end{aligned}$$

and the PDF of $Y_{n:n}$ is

$$\begin{aligned} f_{n:n}(y; \alpha, \beta) &= n f_Y(y; \alpha, \beta)[F_Y(y; \alpha, \beta)]^{n-1} \\ &= \frac{n\alpha\beta y^{\beta-1}(\arctan(\alpha y^\beta))^{n-1}}{(1 + \alpha^2 y^{2\beta})(\arctan(\alpha))^n}. \end{aligned}$$

The minimum and maximum (min-max) plot of the order statistics can be used to describe whether the distribution is symmetrical or skewed. The min-max plots depend on $E(Y_{1:n})$ and $E(Y_{n:n})$. The min-max plots for some chosen parameter values for the AP distribution are shown in Figure 4. This figure reveals that the AP distribution can be right-skewed, left-skewed, or symmetric.

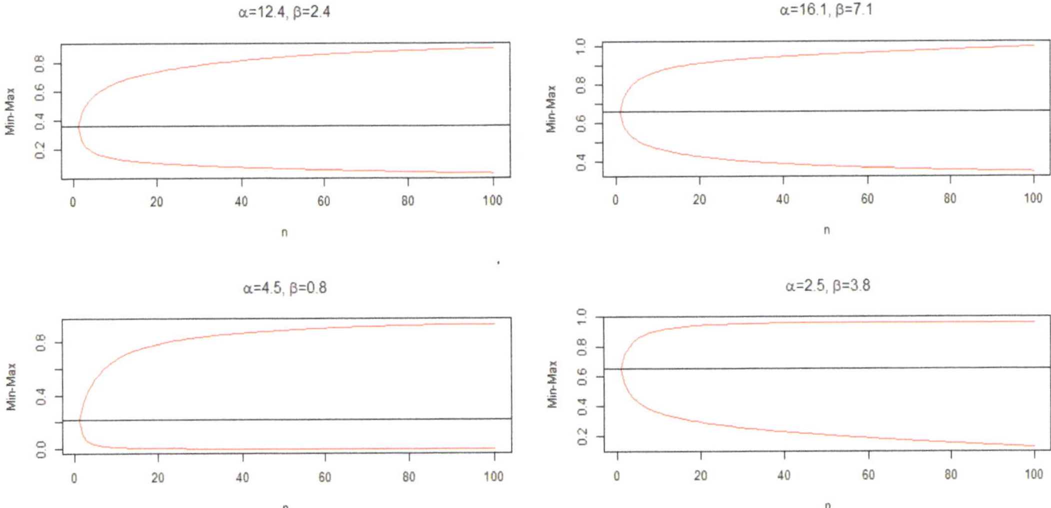

Figure 4. Min-max plots for the AP distribution.

4. Bivariate AP Distribution

The development of bivariate distributions is very useful in the context of investigating the joint relationship between two random variables. For example, one may be interested in studying the relationship between the human development index and literacy rate of a country, the maternal mortality rate and literacy rate, or rainfall and temperature, among others. There are different methods of developing bivariate distributions. One way to do this is to use copula functions (see [19]). However, in this study, we follow the approach used by [20,21]. Let (X, Y) be a bivariate continuous random vector. The CDF of the bivariate AP (BAP) distribution with parameters $\alpha, \beta, \rho_1, \rho_2, \rho_3$, where $\alpha > 0, \beta > 0$, $-1 < \rho_1 + \rho_3 < 1, -1 < \rho_2 + \rho_3 < 1, x \in (0,1)$ and $y \in (0,1)$, is given by

$$F_{XY}(x,y;\varsigma) = \frac{\arctan(\alpha x^\beta)\arctan(\alpha y^\beta)(\arctan(\alpha))^{-2}}{\left[1 + (\rho_1 + \rho_3)\left(\frac{\arctan(\alpha)-\arctan(\alpha x^\beta)}{\arctan(\alpha)}\right) + (\rho_2 + \rho_3)\left(\frac{\arctan(\alpha)-\arctan(\alpha y^\beta)}{\arctan(\alpha)}\right)\right]^{-1}}, \tag{14}$$

where $\varsigma = (\alpha, \beta, \rho_1, \rho_2, \rho_3)$. The plots of the CDF of the BAP distribution for the given parameter values are shown in Figure 5:

(a) $\alpha = 8.5, \beta = 2.5, \rho_1 = 0.4, \rho_2 = 0.1, \rho_3 = 0.2,$
(b) $\alpha = 4.5, \beta = 8.2, \rho_1 = -0.3, \rho_2 = 0.4, \rho_3 = -0.2$ and
(c) $\alpha = 3.4, \beta = 6.2, \rho_1 = 0.3, \rho_2 = 0.4, \rho_3 = -0.6.$

These plots reveal different concave and convex shapes for the chosen parameter values. The PDF of the BAP distribution is given by

$$f_{XY}(x,y;\varsigma) = \frac{(\alpha\beta)^2(xy)^{\beta-1}(\arctan(\alpha))^{-2}\left[1 + (\alpha x^\beta)^2 + (\alpha y^\beta)^2 + \alpha^4(xy)^{2\beta}\right]^{-1}}{\left[1 + (\rho_1 + \rho_3)\left(\frac{\arctan(\alpha)-\arctan(\alpha x^\beta)}{\arctan(\alpha)}\right) + (\rho_2 + \rho_3)\left(\frac{\arctan(\alpha)-\arctan(\alpha y^\beta)}{\arctan(\alpha)}\right)\right]^{-1}}, \tag{15}$$

The PDF plots of the BAP distribution for the following selected parameter values are displayed in Figure 6:

(a) $\alpha = 8.5, \beta = 2.5, \rho_1 = 0.4, \rho_2 = 0.1, \rho_3 = 0.2,$
(b) $\alpha = 4.5, \beta = 8.2, \rho_1 = -0.3, \rho_2 = 0.4, \rho_3 = -0.2$ and
(c) $\alpha = 3.4, \beta = 2.5, \rho_1 = 0.3, \rho_2 = 0.4, \rho_3 = -0.6.$

These plots display left-skewed, right-skewed, and approximate symmetrical shapes.

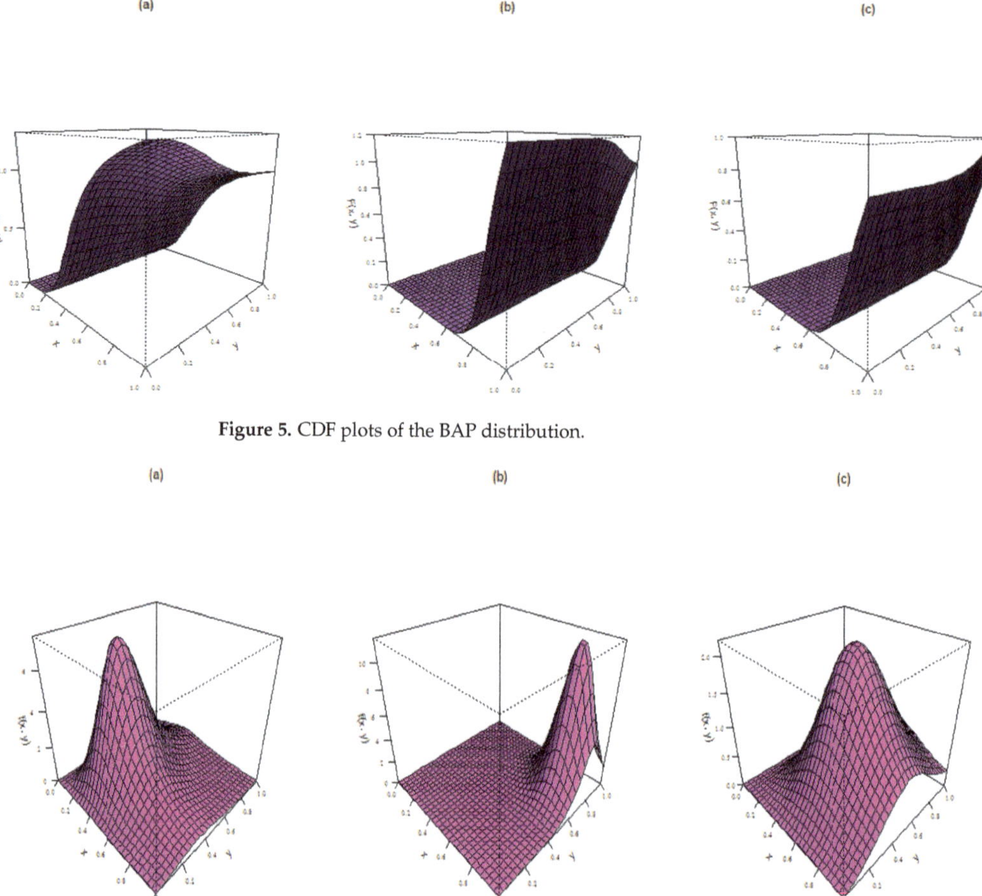

Figure 5. CDF plots of the BAP distribution.

Figure 6. PDF plots of the BAP distribution.

5. Estimation Methods and Simulations

This section presents nine frequentist estimation procedures for estimating the parameters of the AP distribution. These are the maximum likelihood (ML) estimation, ordinary least squares (OLS), weighted least squares (WLS), Cramér–von Mises (CVM) estimation, Anderson–Darling (AD) estimation, percentile estimation (PE), and product spacing estimations.

5.1. Maximum Likelihood Estimation

Let $y_1, y_2, \ldots, y_n$ be independent and identically random observations of sample size n from the AP distribution. Suppose that $\xi = (\alpha, \beta)'$ is the vector of parameters; then, the total log-likelihood function is

$$\ell(\xi) = n\log(\alpha\beta) - n\log(\arctan(\alpha)) + (\beta - 1)\sum_{i=1}^{n}\log(y_i) - \sum_{i=1}^{n}\log(1 + \alpha^2 y_i^{2\beta}). \tag{16}$$

The total likelihood function can be maximized directly with respect to the parameters α and β to obtain the ML estimates of the parameters. Alternatively, these estimates can be obtained by equating the score functions to zero and solving the resulting system of equations simultaneously. The score functions, obtained by differentiating Equation (16) with respect to the parameters, are given by

$$\frac{\partial \ell(\xi)}{\partial \alpha} = \frac{n}{\alpha} - \frac{n}{(1+\alpha^2)\arctan(\alpha)} - \sum_{i=1}^{n} \frac{2\alpha y_i^{2\beta}}{1+\alpha^2 y_i^{2\beta}} \tag{17}$$

and

$$\frac{\partial \ell(\xi)}{\partial \beta} = \frac{n}{\beta} + \sum_{i=1}^{n} \log(y_i) - \sum_{i=1}^{n} \frac{2\alpha^2 \log(y_i) y_i^{2\beta}}{1+\alpha^2 y_i^{2\beta}}. \tag{18}$$

The score functions do not have a closed form, thus, the resulting system of equations are solved numerically to obtain the estimates $\hat{\alpha}$ and $\hat{\beta}$.

5.2. Ordinary and Weighted Least Squares Estimation

Consider an ordered random sample $y_{(1)}, y_{(2)}, \ldots, y_{(n)}$ of size n from the AP distribution; then, the OLS estimates, $\hat{\alpha}_{OLS}$ and $\hat{\beta}_{OLS}$, of the parameters are obtained by minimizing the function

$$OLS = \sum_{i=1}^{n} \left(\frac{\arctan(\alpha y_{(i)}^{\beta})}{\arctan(\alpha)} - \frac{i}{n+1} \right)^2, \tag{19}$$

with respect to the parameters α and β. The OLS estimates can also be obtained by numerically solving the nonlinear equations

$$\sum_{i=1}^{n} \left(\frac{\arctan(\alpha y_{(i)}^{\beta})}{\arctan(\alpha)} - \frac{i}{n+1} \right) \pi_s(y_{(i)}; \alpha, \beta) = 0, \; s = 1, 2, \tag{20}$$

where

$$\pi_1(y; \alpha, \beta) = \frac{2y_{(i)}^{\beta}}{\arctan(\alpha)(1+\alpha^2 y_{(i)}^{2\beta})} - \frac{2\arctan(\alpha y_{(i)}^{\beta})}{(\arctan(\alpha))^2(1+\alpha^2)} \tag{21}$$

and

$$\pi_2(y; \alpha, \beta) = \frac{2y_{(i)}^{\beta}}{\arctan(\alpha)(1+\alpha^2 y_{(i)}^{2\beta})}. \tag{22}$$

The WLS estimates, $\hat{\alpha}_{WLS}$ and $\hat{\beta}_{WLS}$, of the parameters are obtained by minimizing the function

$$\sum_{i=1}^{n} \frac{(n+1)^2(n+2)}{i(n-i+1)} \left(\frac{\arctan(\alpha y_{(i)}^{\beta})}{\arctan(\alpha)} - \frac{i}{n+1} \right)^2, \tag{23}$$

with respect to the parameters α and β. Alternatively, the WLS estimates are obtained by numerically solving the nonlinear equations

$$\sum_{i=1}^{n} \frac{(n+1)^2(n+2)}{i(n-i+1)} \left(\frac{\arctan(\alpha y_{(i)}^{\beta})}{\arctan(\alpha)} - \frac{i}{n+1} \right) \pi_s(y_{(i)}; \alpha, \beta) = 0, \; s = 1, 2, \tag{24}$$

where $\pi_s(y; \alpha, \beta)$, $s = 1, 2$ are defined in Equations (21) and (22).

5.3. Cramér–Von Mises Estimation

Given that $y_{(1)}, y_{(2)}, \ldots, y_{(n)}$ are the ordered observations of size n from the AP distribution, the CVM estimates, $\hat{\alpha}_{CVM}$ and $\hat{\beta}_{CVM}$, of the parameters are obtained by minimizing the function

$$CVM = \frac{1}{12n} + \sum_{i=1}^{n} \left(\frac{\arctan(\alpha y_{(i)}^{\beta})}{\arctan(\alpha)} - \frac{2i-1}{2n} \right)^2, \tag{25}$$

with respect to the parameters α and β. The CVM estimates can also be obtained by solving the nonlinear equation

$$\sum_{i=1}^{n} \left(\frac{\arctan(\alpha y_{(i)}^{\beta})}{\arctan(\alpha)} - \frac{2i-1}{2n} \right) \pi_s(y_{(i)}; \alpha, \beta) = 0, \ s = 1, 2, \tag{26}$$

where $\pi_s(y; \alpha, \beta)$, $s = 1, 2$ are given in Equations (21) and (22).

5.4. Anderson–Darling Estimation

Let $y_{(1)}, y_{(2)}, \ldots, y_{(n)}$ be ordered observations of size n from the AP distribution. The AD estimates, $\hat{\alpha}_{AD}$ and $\hat{\beta}_{AD}$, of the parameters of the AP distribution are obtained by minimizing the function

$$AD = -n - \frac{1}{n} \sum_{i=1}^{n} (2i-1) \left[\log \left(\frac{\arctan(\alpha y_{(i)}^{\beta})}{\arctan(\alpha)} \right) - \log \left(\frac{\arctan(\alpha) - \arctan(\alpha y_{(i)}^{\beta})}{\arctan(\alpha)} \right) \right], \tag{27}$$

with respect to the parameters α and β.

5.5. Percentile Estimation

Let $y_{(1)}, y_{(2)}, \ldots, y_{(n)}$ be ordered observations of size n from the AP distribution, and $u_i = i/(n+1)$. The percentile estimates, $\hat{\alpha}_{PE}$ and $\hat{\beta}_{PE}$, of the parameters of the AP distribution are obtained by minimizing the function

$$PE = \sum_{i=1}^{n} \left[y_{(i)} - \left(\frac{\tan(u_i \arctan(\alpha))}{\alpha} \right)^{1/\beta} \right]^2, \tag{28}$$

with respect to the parameters α and β.

5.6. Product Spacing Estimations

In this subsection, the maximum product spacing (*MPS*) and minimum spacing distance (MSD) estimation methods are discussed. The *MPS* estimation method is based on the Kullback–Leibler information measure. Let us consider the uniform spacing

$$\begin{aligned} D_i &= F_Y(y_{(i)}; \alpha, \beta) - F_Y(y_{(i-1)}; \alpha, \beta) \\ &= \frac{\arctan(\alpha y_{(i)}^{\beta})}{\arctan(\alpha)} - \frac{\arctan(\alpha y_{(i-1)}^{\beta})}{\arctan(\alpha)}, \end{aligned}$$

where $F_Y(y_{(0)}; \alpha, \beta) = 0$, $F_Y(y_{(n+1)}; \alpha, \beta) = 1$ and $D_0(\alpha, \beta) + D_1(\alpha, \beta) + \ldots + D_{n+1}(\alpha, \beta) = 1$. The *MPS* estimates, $\hat{\alpha}_{MPS}$ and $\hat{\beta}_{MPS}$, of the parameters are obtained by directly maximizing the logarithm of the geometric mean of the spacing given by

$$MPS = \frac{1}{n+1} \sum_{i=1}^{n+1} \log D_i(\alpha, \beta), \tag{29}$$

with respect to the parameters α and β.

The *MSD* estimates, $\hat{\alpha}_{MSD}$ and $\hat{\beta}_{MSD}$, of the parameters of the AP distribution are obtained my minimizing the function

$$MSD = \sum_{i=1}^{n} \Delta\left(D_i(\alpha, \beta), \frac{1}{n+1}\right), \tag{30}$$

where $\Delta(a, b)$ represents an appropriate distance. Several choices of $\Delta(a, b)$ exist. However, in this study, we employ the absolute $|a - b|$ and absolute-logarithm $|\log(a) - \log(b)|$ distances. Hence, the minimum spacing absolute distance ($MSAD$) and minimum spacing absolute-logarithm (MSALD) estimates of the parameters are obtained by minimizing the functions

$$MSAD = \sum_{i=1}^{n} \left| D_i(\alpha, \beta) - \frac{1}{n+1} \right| \tag{31}$$

and

$$MSAD = \sum_{i=1}^{n} \left| \log(D_i(\alpha, \beta)) - \log\left(\frac{1}{n+1}\right) \right|, \tag{32}$$

where $D_i(\alpha, \beta) \neq \frac{1}{n+1}$ and $\log(D_i(\alpha, \beta)) \neq \log(\frac{1}{n+1})$.

5.7. Monte Carlo Simulation

In this section, we conduct Monte Carlo simulation studies to investigate how the various estimation techniques perform with regards to estimating the parameter of the AP distribution. The exercise is carried out with two sets of parameter values, which are $\alpha = 0.8, \beta = 0.4$ and $\alpha = 4.5, \beta = 6.2$. The simulation experiments are repeated 5000 times using the sample sizes $n = 25, 50, 100, 250$ and 350. The average estimates (AE), average absolute bias (AB), and root mean square error (RMSE) of the parameters are estimated and reported in Tables 1 and 2. We observe that as the sample size increases, the AE of the parameters approaches the true parameter values. Furthermore, the ABs and RMSEs of the parameters decrease as the sample size increases for all the estimation methods used. Thus, the various estimation methods produce consistent estimates for the parameters of the AP distribution. However, none of the estimation methods proves to be superior to the others.

Table 1. AE, AB, and RMSE for $\alpha = 0.8$ and $\beta = 0.4$.

Parameter	n	ML	MPS	OLS	WLS	AD	CVM	PE	MADS	MALDS
					AE					
	25	0.7609	1.1013	0.4303	0.5079	0.5634	0.6210	0.8969	0.1730	0.5673
	50	0.8989	1.1131	0.6865	0.7679	0.7794	0.8387	0.9400	0.1718	0.5865
α	100	0.5186	0.6330	0.5285	0.5408	0.5316	0.6020	0.7364	0.3013	0.4153
	250	0.7563	0.8212	0.6438	0.6947	0.6821	0.6737	0.6598	0.4516	0.5850
	350	0.8082	0.8765	0.7720	0.8039	0.7933	0.7969	0.6947	0.5602	0.7547
	25	0.4217	0.4674	0.3992	0.4005	0.4065	0.4237	0.4895	0.3323	0.4021
	50	0.4294	0.4580	0.4086	0.4158	0.4174	0.4258	0.4584	0.2995	0.3967
β	100	0.3903	0.4039	0.3947	0.3944	0.3926	0.4016	0.4371	0.3582	0.3858
	250	0.4035	0.4115	0.3938	0.3975	0.3966	0.3974	0.4061	0.3673	0.3940
	350	0.3949	0.4026	0.3907	0.3944	0.3931	0.3936	0.3904	0.3719	0.3899
					AB					
	25	0.5584	0.6872	0.6047	0.5382	0.5453	0.6459	0.7676	0.6845	0.6637
	50	0.5308	0.6270	0.5159	0.5405	0.4941	0.5491	0.9510	0.6712	0.6083
α	100	0.6628	0.6447	0.7083	0.6909	0.6867	0.6793	0.8618	0.5800	0.6805
	250	0.2803	0.2719	0.3670	0.3164	0.3256	0.3616	0.4728	0.5443	0.4994
	350	0.2584	0.2666	0.2306	0.2376	0.2389	0.2336	0.4586	0.4518	0.3332

Table 1. *Cont.*

Parameter	n	ML	MPS	OLS	WLS	AD	CVM	PE	MADS	MALDS
β	25	0.0701	0.1000	0.0807	0.0724	0.0686	0.0844	0.1327	0.2182	0.1001
	50	0.0442	0.0643	0.0495	0.0435	0.0428	0.0580	0.1059	0.1275	0.0445
	100	0.0504	0.0530	0.0493	0.0493	0.0490	0.0500	0.0657	0.0640	0.0480
	250	0.0270	0.0286	0.0352	0.0306	0.0314	0.0358	0.0534	0.0557	0.0356
	350	0.0226	0.0222	0.0243	0.0176	0.0192	0.0243	0.0520	0.0428	0.0268
RMSE										
α	25	0.6832	0.8824	0.6642	0.6196	0.6373	0.7498	0.9374	0.7249	0.7684
	50	0.6291	0.7570	0.6603	0.6831	0.5963	0.7164	1.4860	0.7176	0.6671
	100	0.7322	0.7492	0.7848	0.7744	0.7611	0.7921	0.9537	0.6576	0.7420
	250	0.3359	0.3366	0.4614	0.3988	0.4108	0.4615	0.5893	0.6260	0.5625
	350	0.3129	0.3093	0.3154	0.3086	0.3098	0.3142	0.5602	0.5355	0.4107
β	25	0.0910	0.1217	0.1029	0.0918	0.0880	0.1174	0.1684	0.2464	0.1214
	50	0.0542	0.0782	0.0607	0.0559	0.0493	0.0712	0.1646	0.1592	0.0603
	100	0.0612	0.0655	0.0627	0.0618	0.0606	0.0652	0.0875	0.0981	0.0604
	250	0.0337	0.0362	0.0402	0.0364	0.0374	0.0411	0.0679	0.0696	0.0446
	350	0.0259	0.0259	0.0293	0.0242	0.0249	0.0289	0.0619	0.0560	0.0337

Table 2. AE, AB, and RMSE for $\alpha = 4.5$ and $\beta = 6.2$.

Parameter	n	ML	MPS	OLS	WLS	AD	CVM	PE	MADS	MALDS
AE										
α	25	7.0765	10.3643	5.9141	5.8055	6.6186	7.5983	4.8574	1.2794	8.3329
	50	5.0499	5.9801	4.8062	4.7651	4.7680	5.3690	4.1797	3.3758	5.4587
	100	4.3862	4.8383	4.1504	4.2629	4.2891	4.3589	3.9500	3.6863	4.3552
	250	4.3660	4.5560	4.2758	4.3155	4.3307	4.3597	4.1551	3.9716	4.4893
	350	4.3334	4.4767	4.2076	4.2748	4.2766	4.2668	4.2163	4.1250	4.3294
β	25	6.4914	7.3170	5.9496	5.9510	6.2163	6.5382	5.5927	3.3139	5.9368
	50	6.1885	6.6336	5.9530	6.0059	6.0516	6.2226	5.7082	4.6987	6.1925
	100	6.2534	6.5278	6.0770	6.1657	6.1849	6.2094	5.9914	5.5851	6.2811
	250	6.1297	6.2481	6.0714	6.1025	6.1135	6.1240	6.0026	5.7696	6.1201
	350	6.0608	6.1514	5.9857	6.0232	6.0258	6.0232	5.9824	5.8618	6.0932
AB										
α	25	3.4127	5.9293	3.3920	3.1570	3.4268	4.2622	2.7449	3.2862	5.8446
	50	1.8288	2.1741	2.1320	1.9167	1.7383	2.2757	1.7767	2.4817	2.5227
	100	1.0012	0.9566	1.0738	1.0249	1.0781	1.0474	1.0521	1.5290	1.2026
	250	0.8031	0.8054	0.8103	0.7709	0.7570	0.7912	0.8309	1.2029	1.0822
	350	0.6395	0.6136	0.6138	0.6133	0.6086	0.6041	0.6972	0.8890	0.5945
β	25	1.2038	1.4981	1.3240	1.2379	1.1823	1.3926	1.2174	2.9029	1.2698
	50	0.9340	0.9660	1.0599	0.9933	0.9327	1.0433	1.0666	2.1079	1.2164
	100	0.5449	0.5436	0.5723	0.5544	0.5715	0.5383	0.5769	0.9975	0.6254
	250	0.4017	0.4156	0.4049	0.4016	0.3959	0.4026	0.4575	0.7574	0.6456
	350	0.3707	0.3538	0.3835	0.3678	0.3652	0.3723	0.4190	0.5258	0.3588

Table 2. *Cont.*

Parameter	n	ML	MPS	OLS	WLS	AD	CVM	PE	MADS	MALDS
					RMSE					
α	25	9.0289	16.6588	7.7515	7.0825	8.9366	10.7903	5.1325	3.5862	19.9363
	50	3.1101	4.1306	3.7004	2.9429	2.7048	4.3787	2.2720	3.1047	4.0033
	100	1.2746	1.4424	1.3415	1.3020	1.3645	1.3743	1.1958	2.0602	1.7619
	250	1.0203	1.0631	1.0172	1.0052	0.9906	1.0217	1.0439	1.6323	1.3097
	350	0.7575	0.7559	0.7539	0.7476	0.7376	0.7427	0.8050	1.2130	0.7278
β	25	1.5369	2.0307	1.6441	1.5388	1.5357	1.7984	1.4325	3.3678	1.7998
	50	1.2005	1.3372	1.3614	1.2314	1.1733	1.3964	1.2318	2.6988	1.5320
	100	0.6942	0.7728	0.7270	0.6891	0.7131	0.7296	0.6689	1.5722	0.8371
	250	0.5388	0.5432	0.5306	0.5343	0.5215	0.5232	0.5916	0.9666	0.7900
	350	0.4264	0.4122	0.4673	0.4368	0.4343	0.4534	0.4743	0.6624	0.4570

6. Empirical Application

In this section, we present frequentist and Bayesian applications of the AP distribution using biomedical data.

6.1. Frequentist Application

In this subsection, the univariate application of the AP distribution is illustrated using the ML estimation approach. The illustration is done using data on the recovery rates for viable CD34+ cells of 239 patients who agreed to an autologous peripheral blood stem cell (PBSC) transplant after myeloablative doses of chemotherapy between the years 2003 and 2008 at the Edmonton Hematopoietic Stem Cell Lab in the Cross Cancer Institute-Alberta Health Services. The data can be found in the simplexreg package developed by [22]. Ref. [6] recently fitted the unit Burr XII (UBXII) distribution to improve the recovery rates for viable CD34+ cells. The AP distribution is fitted to the recovery rates in this study, and its performance is compared to the AU distribution [14], unit power Weibull (UPW) distribution [23], log-XLindley (LXL) distribution [4], unit Lindley (UL) distribution [9], unit improved second degree Lindley (UISDL) distribution [11], bounded Marshall–Olkin extended exponential (BMOEE) distribution [24], unit Burr III (UBIII) distribution [8], unit Gompertz (UG) distribution [10], unit Weibull (UW) distribution [12], exponentiated Topp–Leone (ETL) distribution [13], Kumaraswamy distribution [25], and beta distribution. The performances of the distributions are compared using the log-likelihood (ℓ), Akaike information criterion (AIC), AIC difference (DAIC), Bayesian information criterion (BIC), Anderson–Darling (AD) test, Cramér–von Mises (CVM) test, and Kolmogorov–Smirnov (KS) test. The distribution with the highest value of ℓ and lowest values of AIC, BIC, AD, CVM, and KS is considered to be the best. The DAIC is computed as $\text{DAIC}_i = \text{AIC}_i - \text{AIC}_{\min}, i = 1, 2, \ldots, S$, where S is the number of distributions under comparison. The best distribution satisfies $\text{DAIC} = 0$. If $\text{DAIC} > 2$, then the difference in performance between the two models is significant. Before fitting the models to the recovery rate for viable CD34+ cells, we explore their characteristics. From the kernel density, boxplot, and violin plots shown in Figure 7, we observe that the recovery rate for viable CD34+ cells is left-skewed. Hence, a distribution capable of modeling left-skewed data is required, which is the case for the AP distribution.

Table 3 presents the ML estimates of the parameters with their respective standard errors in brackets. The AP distribution appears to be the best model since it has the highest log-likelihood values and the smallest values for the AIC, BIC, AD, CVM, and KS. The p-values of the AD, CVM, and KS tests are given in parentheses. The p-values also indicate that the AP distribution is the best. Furthermore, looking at the DAIC values, the AP distribution significantly performs better than the other fitted distributions. Comparing the goodness-of-fit statistics of the AP and AU distributions, it can be concluded that the induction of the new parameter has greatly improved the performance of the AP distribution, making it superior to the AU distribution.

Table 3. Parameter estimates, standard errors, goodness-of-fit tests.

Model	Parameter	ℓ	AIC	DAIC	BIC	AD	CVM	K-S
AP	$\alpha = 5.0250(0.9841)$ $\beta = 8.1856(0.6324)$	194.5900	-385.1756	0.0000	-378.2227	0.3670 (0.8806)	0.0461 (0.8999)	0.0430 (0.7694)
AU	$\alpha = 2.5208 \times 10^{-14}(0.0828)$	0.0000	2.0000	387.1756	5.4765	131.0700 (<0.0001)	28.2090 (<0.0001)	0.5572 (<0.0001)
Beta	$\alpha = 8.6671(0.8063)$ $\beta = 2.2859(0.1962)$	191.8700	-379.7345	5.4411	-372.7816	0.8732 (0.4310)	0.1402 (0.4213)	0.0650 (0.2647)
Kumaraswamy	$\alpha = 6.6942(0.4546)$ $\beta = 2.4355(0.2411)$	190.7600	-377.5820	7.5936	-370.5751	1.1438 (0.2899)	0.1916 (0.2845)	0.0723 (0.1646)
UBIII	$\alpha = 6.4356(0.5341)$ $\beta = 1.5532(0.0695)$	192.5000	-381.0031	4.1725	-374.0501	0.7758 (0.4987)	0.1191 (0.4996)	0.0535 (0.4997)
BMOEE	$\alpha = 7.6885(1.7248)$ $\beta = 9.6771(0.7554)$	192.4200	-380.8355	4.3401	-373.8825	0.6848 (0.5715)	0.0866 (0.6551)	0.0489 (0.6182)
UG	$\alpha = 1.0457(0.2360)$ $\beta = 2.3734(0.3237)$	177.0300	-350.0612	35.1144	-343.1082	4.9419 (0.0031)	0.7829 (0.0080)	0.1106 (0.0058)
UW	$\alpha = 8.0560(0.8314)$ $\beta = 1.6182(0.0791)$	192.0200	-380.0314	5.1442	-373.0785	0.8636 (0.4373)	0.1328 (0.4467)	0.0557 (0.4486)
ETL	$\alpha = 14.9326(1.3241)$ $\beta = 0.8641(0.0718)$	192.6800	-381.3601	3.8155	-374.4072	0.6705 (0.5838)	0.0996 (0.5873)	0.0520 (0.5370)
UBXII	$\alpha = 10.0760(1.0039)$ $\beta = 1.7321(0.0787)$	193.5000	-383.0054	2.1702	-376.0525	0.5806 (0.6664)	0.0887 (0.6437)	0.0522 (0.5321)
UISDL	$\alpha = 0.3571(0.0134)$	54.2900	-106.5865	278.5891	-103.1101	34.4330 (<0.0001)	20.1010 (<0.0001)	0.2851 (<0.0001)
UL	$\alpha = 0.2424(0.0112)$	97.6400	-193.2741	191.9015	-189.7976	20.1010 (<0.0001)	4.0961 (<0.0001)	0.2365 (<0.0001)
LXL	$\alpha = 4.2040(0.2569)$	154.6800	-307.3564	77.8192	-303.8799	15.7970 (<0.0001)	3.0033 (<0.0001)	0.2010 (<0.0001)
UPW	$\alpha = 500.0000(8.1076 \times 10^{-6})$ $\beta = 2.4183(9.9309 \times 10^{-2})$ $\lambda = 0.0372(3.5461 \times 10^{-3})$	168.2600	-330.5111	54.6645	-320.0817	5.3084 (0.0021)	0.8375 (0.0059)	0.1152 (0.0035)

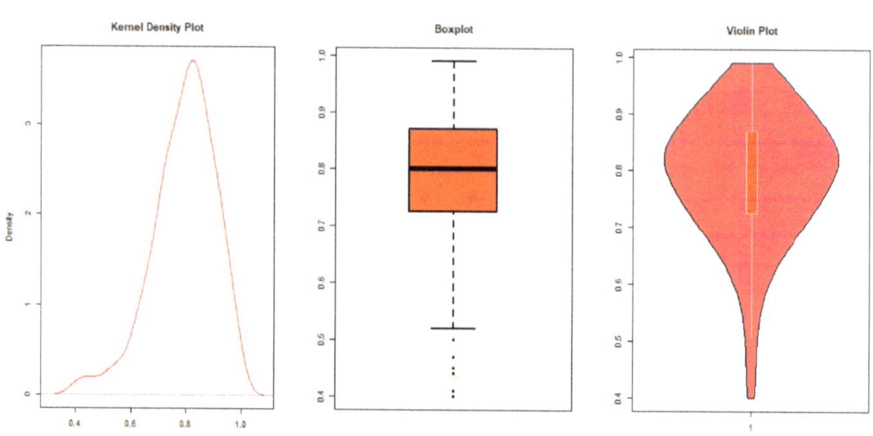

Figure 7. Kernel density, boxplot, and violin plots.

Figure 8 displays the histogram of the data and the estimated PDF of the AP distribution on the one hand and the empirical CDF and the estimated CDF of the AP distribution on the other hand, using the estimates of the parameter. This figure suggests that the AP distribution provides good fit to the data.

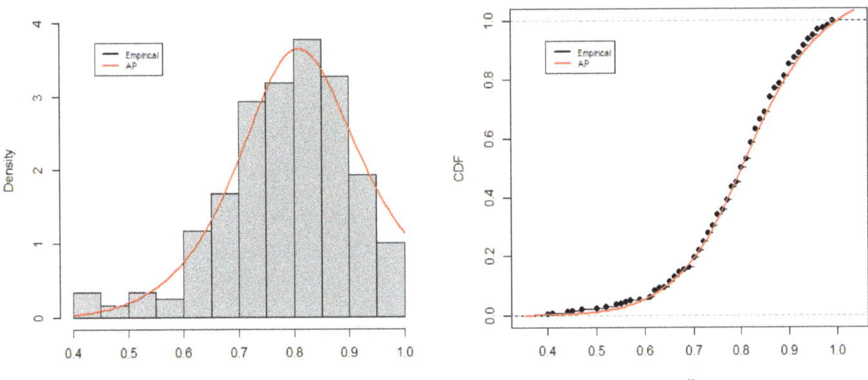

Figure 8. Histogram and estimated PDF (**left**), and empirical CDF and estimated CDF (**right**).

Figure 9 displays the probability-probability (P-P) plots of the fitted distributions. This figure suggests that the AP distribution provides a good fit to the data as its expected and observed probabilities cluster along the diagonal line.

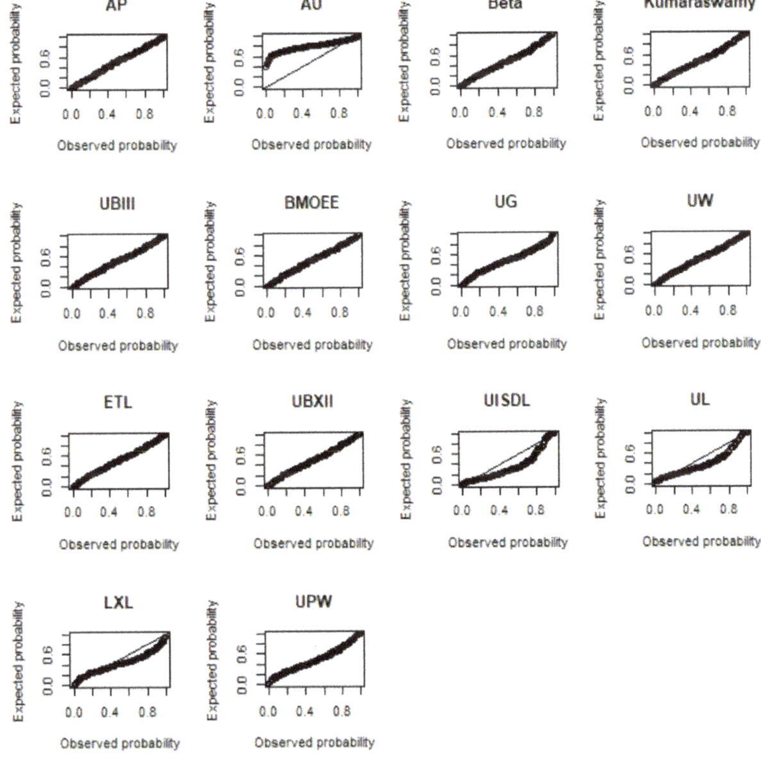

Figure 9. P-P plots of the fitted distributions.

The profile log-likelihood plots of the estimated parameters of the AP distribution are shown in Figure 10. These plots suggest that the ML estimates of the parameters are unique and denote the true maxima.

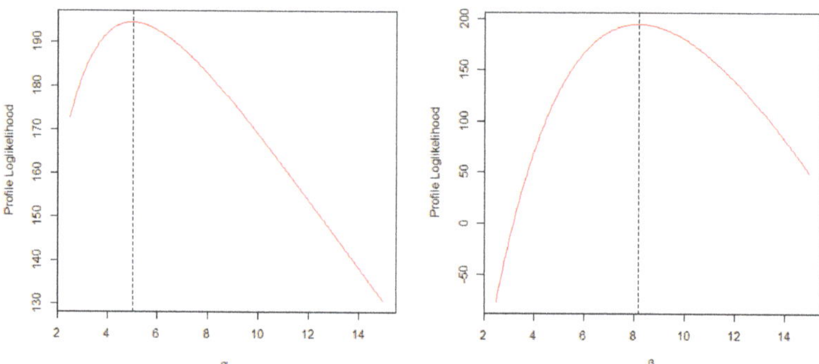

Figure 10. Profile log-likelihood plots of the estimated parameters of the AP distribution.

6.2. Bayesian Application

In this subsection, we demonstrate how to use the Bayesian approach to estimate the parameters of the AP distribution. To proceed, we need to first establish the prior distributions for the parameters, as it is very essential in Bayesian estimation. In this study, we use the non-informative gamma distribution as the prior distribution. Numerous studies have recommended the use of this approach (see [26,27]). Thus, the prior distributions of the parameters are

$$\pi(\alpha) \sim Gamma(a_1, b_1) = \frac{b_1^{a_1}}{\Gamma(a_1)} \alpha^{a_1 - 1} e^{-b_1 \alpha}, a_1 > 0, b_1 > 0, \alpha > 0$$

and

$$\pi(\beta) \sim Gamma(a_2, b_2) = \frac{b_2^{a_2}}{\Gamma(a_2)} \beta^{a_2 - 1} e^{-b_2 \beta}, a_2 > 0, b_2 > 0, \beta > 0$$

The joint PDF of the prior distributions of the parameters is given by

$$\pi(\alpha, \beta) = \pi(\alpha)\pi(\beta).$$

The joint posterior PDF is therefore given by

$$P(\alpha, \beta | y) \propto \prod_{i=1}^{n} f_Y(y_i; \alpha, \beta) \times \pi(\alpha, \beta),$$

where $\prod_{i=1}^{n} f_Y(y_i; \alpha, \beta)$ is the likelihood function of the AP distribution. The joint posterior PDF is not analytically tractable; hence, we employ the Markov Chain Monte Carlo (MCMC) approach to obtain samples from which features of the marginal distributions can be inferred. The following hyperparameter values $a_1 = a_2 = b_1 = b_2 = 0.001$ are considered for the analysis. The analysis is performed using the R2jags package in R (see [28]) and the data described in Section 6.1. We use three parallel chains, each with 40,000 iterations and a burn-in of 5000. Hence, posterior sample of size 7000 and thinning interval 5 is used in the analysis. Table 4 presents the mean estimate, Monte Carlo standard error (SE), posterior standard deviation (SD), and other numerical summaries of the posterior distribution. From the results, the MCMC algorithm has converged because the potential reduction scale factor ($\hat{R}$) is approximately 1 and the effective sample size (neff) is greater than 400. The estimated deviance information criterion (DIC) is -385.2000. It can be observed that the Bayesian estimates and ML estimates of the parameters are quite close.

Table 4. Posterior summaries of the parameters of the AP distribution.

Parameter	Estimate	SE	SD	2.50%	50%	97.50%	$\hat{R}$	Neff
α	5.0600	0.0107	1.0150	3.3760	4.9540	7.3560	1.0010	5500
β	8.1600	0.0066	0.6300	6.9640	8.1490	9.4110	1.0010	6200

We investigate the convergence of the chains visually using the trace, ergodic mean, and autocorrelation plots. The trace plots shown in Figure 11 suggest a stationary pattern and thus convergence of the chains.

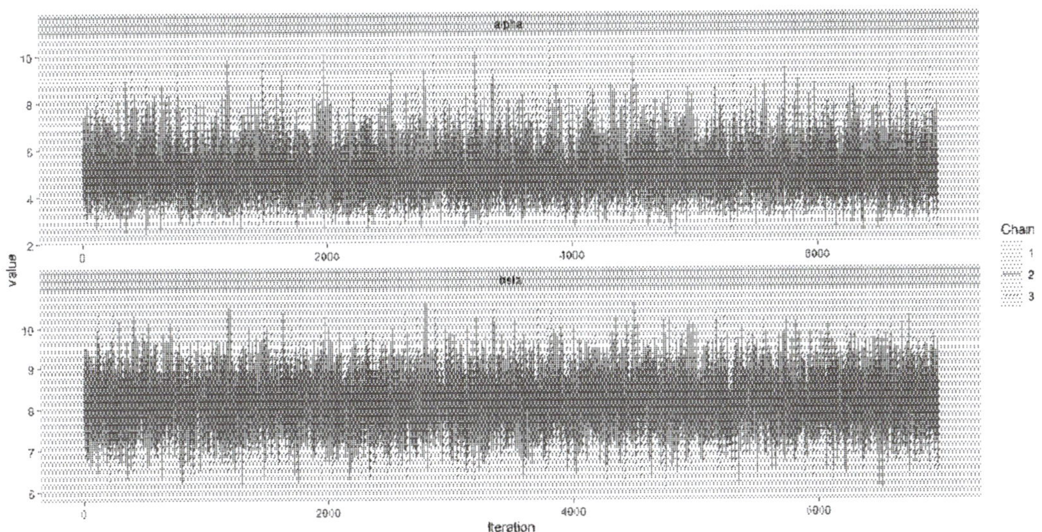

Figure 11. The AP distribution posterior parameters trace plots.

The ergodic mean plots (Figure 12) of the parameters clearly show that the chains have converged after 3000 iterations.

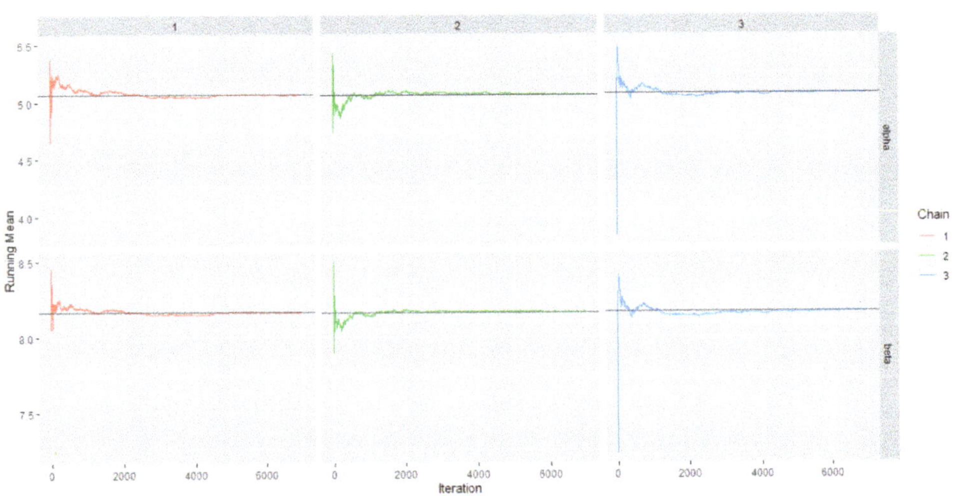

Figure 12. The AP distribution posterior parameters ergodic mean plots.

The rapid decay of the autocorrelation plots, as shown in Figure 13, suggests good mixing of the chains and the convergence of the MCMC algorithm.

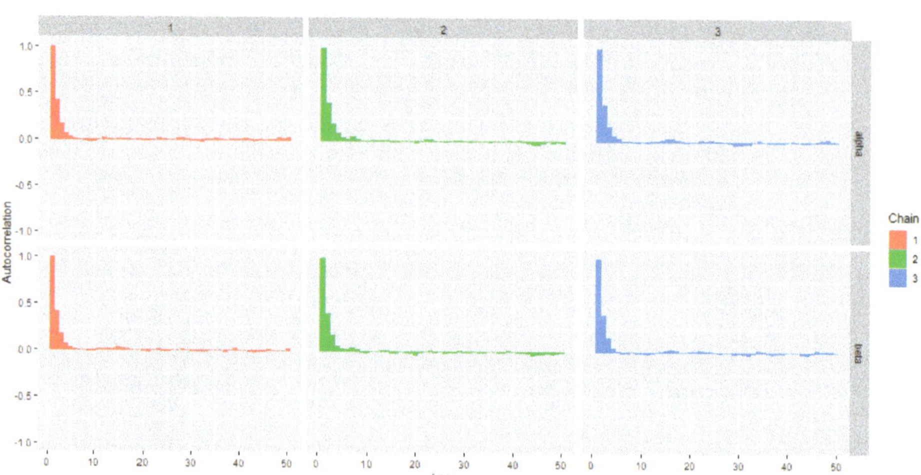

Figure 13. The AP distribution posterior parameters autocorrelation plots.

7. Regression Models

In this section, the quantile and modal regression models are developed for investigating the relationship between a dependent variable and a set of independent variable (s).

7.1. Quantile Regression Model

When investigating the influence of covariates on a skewed, bounded response variable, the beta regression model cannot produce reliable results since it models the conditional mean of the response variable. This is because the mean is not an appropriate measure of central tendency when the data are skewed. Thus, a regression model that is not influenced by outliers is required. The quantile regression is appropriate when dealing with skewed response variables. In this subsection, the AP quantile regression model is developed. To this aim, we re-parameterize the PDF of the AP distribution in terms of its quantile function. Let $\eta = Q(u; \alpha, \beta), \eta \in (0, 1)$, making β the subject in the quantile function, and we have $\beta = (\log(\eta))^{-1} \log(\alpha^{-1} \tan(u \arctan(\alpha)))$. Hence, the re-parametrized PDF in terms of the quantile function is given by

$$f_Y(y; \alpha, \eta) = \frac{\alpha(\log(\eta))^{-1} \lambda y^{(\log(\eta))^{-1}\lambda - 1}}{\arctan(\alpha)(1 + \alpha^2 y^{2(\log(\eta))^{-1}\lambda})}, \tag{33}$$

where $\lambda = \log(\alpha^{-1} \tan(u \arctan(\alpha)))$ and η is the quantile parameter. Suppose that $y_1, y_2, ..., y_n$ are random observations from the AP distribution and z_i is non-random covariates. The AP quantile regression model is thus given by

$$\eta_i = g^{-1}(z_i^{\mathrm{T}} \delta)$$

where $\delta = (\delta_0, \delta_1, \delta_2, \ldots, \delta_p)^{\mathrm{T}}$ is the vector of coefficients of the covariates to be estimated, $z_i^{\mathrm{T}} = (1, z_{i1}, z_{i2}, \ldots, z_{ip})$ is the known i^{th} vector of independent variables, and $g(\cdot)$ is an appropriate link function that relates the independent variables to the conditional quantile of the dependent variable. When $u = 0.5$, the median regression is obtained. Although different link functions exist for modeling bounded response variables, in this study, the logit link function is used due to the easy interpretation of the parameters. Hence, we have

$$\log\left(\frac{\eta_i}{1-\eta_i}\right) = \delta_0 + \delta_1 z_{i1} + \delta_2 z_{i2} + \ldots + \delta_p z_{ip}$$

The log-likelihood for estimating the parameters of the regression model is

$$\ell = n\log(\alpha) - n\log(\arctan(\alpha)) + n\log(\lambda) - \sum_{i=1}^{n}\log(\log(\eta_i)) + \sum_{i=1}^{n}\left((\log(\eta_i))^{-1}\lambda - 1\right)\log(y_i)$$
$$- \sum_{i=1}^{n}\log\left(1 + \alpha^2 y_i^{2(\log(\eta_i))^{-1}\lambda}\right). \tag{34}$$

Maximizing the log-likelihood function in Equation (34) with respect to the involved parameters gives the estimates of the parameters of the model. For more information on the development of parametric quantile regressions, we refer the readers to [2,3,6].

7.2. Modal Regression

When the response variable is heavy-tailed or asymmetric, modal regression is known to give a better fit than the conditional mean or median regression [29]. It is also established that the prediction intervals from modal regression possess a higher coverage probability than the mean-based prediction interval (see [29,30]). This subsection presents the modal-based regression using the AP distribution. Suppose that the transformation $(\alpha, \beta) \to (\eta, \varphi)$ is one-to-one, where $\eta \in (0,1)$ is the mode and $\varphi > 1$ is a precision/shape parameter. Then the PDF of the AP distribution can be re-parameterized in terms of the mode (see [29]). Let $\beta = \varphi$, then $\alpha = \eta^{-\varphi}(\varphi+1)^{-1/2}(\varphi-1)^{1/2}$ and the PDF of the AP distribution in terms of mode is given by

$$f_Y(y;\eta,\varphi) = \frac{\eta^{-\varphi}\varphi(\varphi+1)^{-1/2}(\varphi-1)^{1/2}y^{\varphi-1}}{\arctan(\eta^{-\varphi}(\varphi+1)^{-1/2}(\varphi-1)^{1/2})(1+\eta^{-2\varphi}(\varphi+1)^{-1}(\varphi-1)y^{2\varphi})}. \tag{35}$$

The modal regression is given by

$$\eta_i = h^{-1}(z_i^{\mathrm{T}}\delta)$$

where $\delta = (\delta_0, \delta_1, \delta_2, \ldots, \delta_p)^{\mathrm{T}}$ is the vector of unknown parameters to be estimated, $z_i^{\mathrm{T}} = (1, z_{i1}, z_{i2}, \ldots, z_{ip})$ are the known i^{th} vector of covariates and $h(\cdot)$ is an appropriate link function that links the covariates to the conditional mode of the response variable. The logit link function is adopted since the mode of the AP distribution lies on $(0, 1)$. Thus, we have

$$\log\left(\frac{\eta_i}{1-\eta_i}\right) = \delta_0 + \delta_1 z_{i1} + \delta_2 z_{i2} + \ldots + \delta_p z_{ip}$$

The log-likelihood for estimating the parameters of the model is given by

$$\ell = n\log(\varphi(\varphi+1)^{-1/2}(\varphi-1)^{1/2}) - \varphi\sum_{i=1}^{n}\log(\eta_i) + (\varphi-1)\sum_{i=1}^{n}\log(y_i) - \sum_{i=1}^{n}\log(\arctan(\eta_i^{-\varphi}(\varphi+1)^{-1/2}(\varphi-1)^{1/2})) -$$
$$\sum_{i=1}^{n}\log\left(1 + \eta_i^{-2\varphi}(\varphi+1)^{-1}(\varphi-1)y_i^{2\varphi}\right). \tag{36}$$

The estimates of the parameters of the modal regression are obtained by maximizing Equation (36) with respect to the involved parameters.

7.3. Residual Analysis

Investigating how well a model fits a given data set is very important. Hence, the adequacy of the model is often examined using the residuals from the fitted model. The Cox–Snell and randomized quantile residuals are used to assess the performance of the regression models in this study.

Thus, the Cox–Snell residuals (see [31]) are used to assess the adequacy of the regression models. The Cox–Snell residuals are defined as

$$e_i = -\log(1 - F_Y(y_i; \hat{\delta})), i = 1, 2, ..., n$$

where $\hat{\delta}$ is the vector of the estimated parameters of the regression models. The Cox–Snell residuals are expected to be standard exponentially distributed if the models provide good fit to the data.

Assessing the randomized quantile residuals of the model is another alternative for examining the adequacy of the regression model. The randomized quantile residual is given by

$$e_i = \Phi^{-1}(F_Y(y_i; \hat{\delta})), i = 1, 2, ..., n,$$

where $\Phi^{-1}(\cdot)$ is the quantile of the standard normal distribution. If the regression model provides good fit to the data, the randomized quantile residuals are expected to follow the standard normal distribution (see [32]).

7.4. Monte Carlo Simulation for Regression Models

In this section, Monte Carlo simulation experiments are carried out to assess how the ML estimates perform with regards to estimating the parameters of the AP quantile and modal regressions. The simulations for the quantile regression are carried out using the conditional median. The conditional median in this case is the median of the response variable given the values of the covariates. The experiment is replicated 5000 times for each sample size $n = 50, 150, 250, 350, 450$, and 550. For the first scenario, the following parameter combinations are used for the quantile and modal regressions, respectively: $(\delta_0, \delta_1, \delta_2, \alpha) = (0.8, 0.3, 0.6, 1.5)$ and $(\delta_0, \delta_1, \delta_2, \varphi) = (0.8, 0.3, 0.6, 1.5)$. In the second scenario, the parameter following combinations are used, respectively, for the quantile and modal regressions: $(\delta_0, \delta_1, \delta_2, \alpha) = (0.1, 0.4, 0.8, 1.3)$ and $(\delta_0, \delta_1, \delta_2, \varphi) = (0.1, 0.4, 0.8, 1.3)$. The following regression structure with two covariates is employed during the simulation for both regression models:

$$\log\left(\frac{\eta_i}{1 - \eta_i}\right) = \delta_0 + \delta_1 z_{i1} + \delta_2 z_{i2}, i = 1, 2, ..., n.$$

The covariate, z_{i1}, is generated from a standard normal distribution and z_{i2} is from a t distribution with four degrees of freedom. The covariates are held fixed during the simulation process. The observations for the response variable are generated using the inversion method for both the quantile and modal regressions. The performance of the estimation method is assessed using the average estimate (AE), absolute bias (AB), and root mean square error (RMSE). The results in Tables 5 and 6 reveal that the AEs approach the true parameter values as the sample size increases. Furthermore, the ABs and RMSEs decrease as the sample size increases. Hence, the estimates of the parameters for both models are consistent based on the ML technique.

Table 5. Simulation results for the first scenario.

Parameter	n	AP Quantile Regression			Parameter	n	AP Modal Regression		
		AE	AB	RMSE			AE	AB	RMSE
δ_0	50	0.7659	0.2028	0.2533	δ_0	50	0.6495	0.5931	0.6372
	150	0.7870	0.1286	0.1586		150	0.7551	0.5240	0.5771
	250	0.7837	0.1041	0.1304		250	0.7015	0.4583	0.5226
	350	0.7953	0.0896	0.1104		350	0.7526	0.4226	0.4880
	450	0.7990	0.0868	0.1071		450	0.7674	0.3745	0.4419
	550	0.7990	0.0681	0.0844		550	0.7668	0.3499	0.4195

Table 5. *Cont.*

Parameter	n	AP Quantile Regression			Parameter	n	AP Modal Regression		
		AE	AB	RMSE			AE	AB	RMSE
δ_1	50	0.4010	0.3256	0.3983	δ_1	50	0.7202	0.6676	0.7959
	150	0.3266	0.1974	0.2407		150	0.6208	0.5630	0.7027
	250	0.3308	0.1737	0.2122		250	0.6470	0.5746	0.7074
	350	0.3119	0.1443	0.1742		350	0.5695	0.5176	0.6518
	450	0.3012	0.1403	0.1711		450	0.5439	0.4813	0.6098
	550	0.2951	0.1044	0.1309		550	0.4965	0.4450	0.5669
δ_2	50	0.6015	0.0893	0.1157	δ_2	50	0.5921	0.3502	0.4263
	150	0.6045	0.0480	0.0614		150	0.6143	0.2171	0.2787
	250	0.6057	0.0381	0.0469		250	0.6090	0.1694	0.2232
	350	0.6006	0.0325	0.0410		350	0.6183	0.1563	0.2020
	450	0.6001	0.0291	0.0371		450	0.6174	0.1259	0.1659
	550	0.6017	0.0272	0.0336		550	0.6187	0.1193	0.1569
α	50	1.8184	0.7279	0.8795	φ	50	1.6644	0.2465	0.2948
	150	1.6469	0.4111	0.5266		150	1.5793	0.1477	0.1879
	250	1.5957	0.3058	0.3971		250	1.5376	0.1026	0.1333
	350	1.5689	0.2526	0.3190		350	1.5289	0.0840	0.1100
	450	1.5586	0.2227	0.2891		450	1.5216	0.0721	0.0931
	550	1.5412	0.2047	0.2602		550	1.5085	0.0693	0.0870

7.5. Application of Regression Models

The use of quantile and modal regressions is demonstrated in this subsection. The application of the quantile regression is illustrated via the conditional median regression by setting $u = 0.5$. The application of the models is illustrated by regressing the recovery rates for viable CD34+ cells of 239 patients described in Section 6 on the following covariates: gender (z_{i1}, 0 for female and 1 for male), chemotherapy (z_{i2}, 0 for receiving chemotherapy on a one-day protocol and 1 for a three-day protocol), and adjusted patient's age (z_{i3}, that is the current age minus 40). Ref. [6] fitted the UBXII median regression with the following results: AIC $= -384.2649$ and BIC $= -366.8826$. The authors showed that the UBXII median regression performs better than the Kumaraswamy median regression with the following results: AIC $= -375.6599$ and BIC $= -358.2775$, and beta mean regression with the following results: AIC $= -381.7912$ and BIC $= -364.4089$. The exploratory analysis in Section 6.1 suggests that the response variable is left-skewed or contains some extreme values. This is an indication that robust regression models are required for modeling the data, and thus our choice of using the median and modal regressions is appropriate. We adopt the following regression structure:

$$\log\left(\frac{\eta_i}{1 - \eta_i}\right) = \delta_0 + \delta_1 z_{i1} + \delta_2 z_{i2} + \delta_3 z_{i3}, i = 1, 2, ..., 239$$

to model the data. Table 7 displays the estimates of the model parameters, standard errors, p-values, and information criteria. From the information criteria, the AP regressions (median and modal) perform better than the UBXII median, Kumaraswamy median, and beta mean regressions. Since DAIC > 2, the AP regressions perform significantly better than the compared regressions. Comparing the AP median regression with the modal regression, it can be said that the AP median regression performs better than the modal regression. From Table 7, it can be seen that the parameter δ_1 is not statistically significant at 5% level of significance. Hence, the variable gender has no significant effect on the recovery rate. The parameters δ_2 and δ_3 are statistically significant at the 5% level of significance. This implies that the recovery rate of older patients is higher than that of younger ones. Furthermore, the recovery rate of patients who receive chemotherapy on a three-day protocol is higher than that of those who receive chemotherapy on a one-day protocol.

Table 6. Simulation results for the second scenario.

Parameter	n	AP Quantile Regression			Parameter	n	AP Modal Regression		
		AE	AB	RMSE			AE	AB	RMSE
δ_0	50	0.1667	0.1496	0.1906	δ_0	50	0.3746	0.3802	0.6027
	150	0.1484	0.1207	0.1502		150	0.3336	0.3336	0.5220
	250	0.1136	0.0907	0.1097		250	0.2376	0.2422	0.3747
	350	0.1171	0.0845	0.1021		350	0.2302	0.2282	0.3570
	450	0.1164	0.0842	0.1028		450	0.2165	0.2085	0.3172
	550	0.1122	0.0714	0.0856		550	0.1841	0.1748	0.2572
δ_1	50	0.4049	0.3025	0.3523	δ_1	50	0.5759	0.5815	0.6773
	150	0.3681	0.1882	0.2312		150	0.4728	0.4831	0.5746
	250	0.4042	0.1654	0.2011		250	0.4892	0.4385	0.5127
	350	0.3862	0.1498	0.1808		350	0.4187	0.3793	0.4540
	450	0.3912	0.1453	0.1771		450	0.4457	0.3684	0.4586
	550	0.3730	0.1047	0.1324		550	0.3974	0.3408	0.4147
δ_2	50	0.7935	0.1038	0.1363	δ_2	50	0.8970	0.3344	0.4124
	150	0.8057	0.0546	0.0699		150	0.8773	0.2046	0.2720
	250	0.8013	0.0426	0.0519		250	0.8651	0.1441	0.2004
	350	0.8008	0.0364	0.0457		350	0.8471	0.1296	0.1734
	450	0.7987	0.0327	0.0414		450	0.8440	0.1052	0.1468
	550	0.8050	0.0326	0.0394		550	0.8339	0.1025	0.1397
α	50	1.2087	0.3183	0.4361	φ	50	1.4403	0.2164	0.2713
	150	1.2667	0.2281	0.2932		150	1.3604	0.1242	0.1589
	250	1.2719	0.1967	0.2448		250	1.3258	0.0870	0.1127
	350	1.2930	0.1702	0.2034		350	1.3211	0.0700	0.0911
	450	1.2871	0.1632	0.1971		450	1.3153	0.0609	0.0785
	550	1.2919	0.1546	0.1845		550	1.3063	0.0588	0.0739

Table 7. Estimates, standard errors, and information criteria for the regression models.

AP Quantile Regression				AP Modal Regression			
Parameter	Estimate	Standard Error	p-Value	Parameter	Estimate	Standard Error	p-Value
δ_0	1.0119	0.1226	<0.0001	δ_0	0.8903	0.1715	<0.0001
δ_1	0.0533	0.0912	0.5585	δ_1	0.0921	0.1235	0.4560
δ_2	0.2392	0.0940	0.0110	δ_2	0.3153	0.1559	0.0432
δ_3	0.0169	0.0049	0.0006	δ_3	0.0253	0.0082	0.0020
α	5.6100	1.1128	<0.0001	φ	8.4244	0.6471	<0.0001
	$\ell = 201.1400$				$\ell = 199.7300$		
	AIC $= -392.2835$				AIC $= -389.4540$		
	BIC $= -374.9012$				BIC $= -372.0717$		

The adequacy of the fitted regression models is assessed by examining the residuals of the fitted models. The P-P plots and half-normal plots with simulated envelopes of the randomized quantile residuals in Figure 14 indicate that the models are adequate.

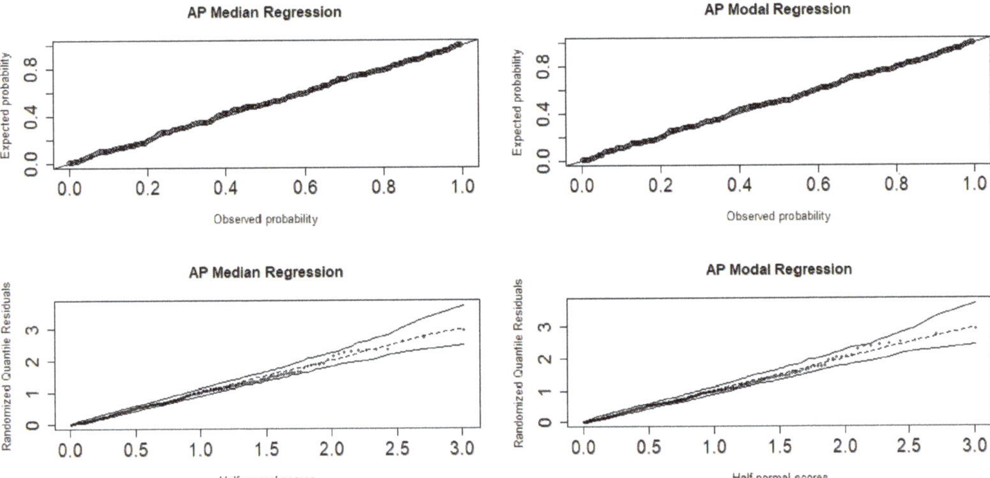

Figure 14. P-P (**top**) and half-normal (**bottom**) plots of the randomized quantile residuals.

The P-P and quantile-quantile (Q-Q) plots with simulated envelopes of the Cox–Snell residuals shown in Figure 15 again affirm that the fitted models are adequate.

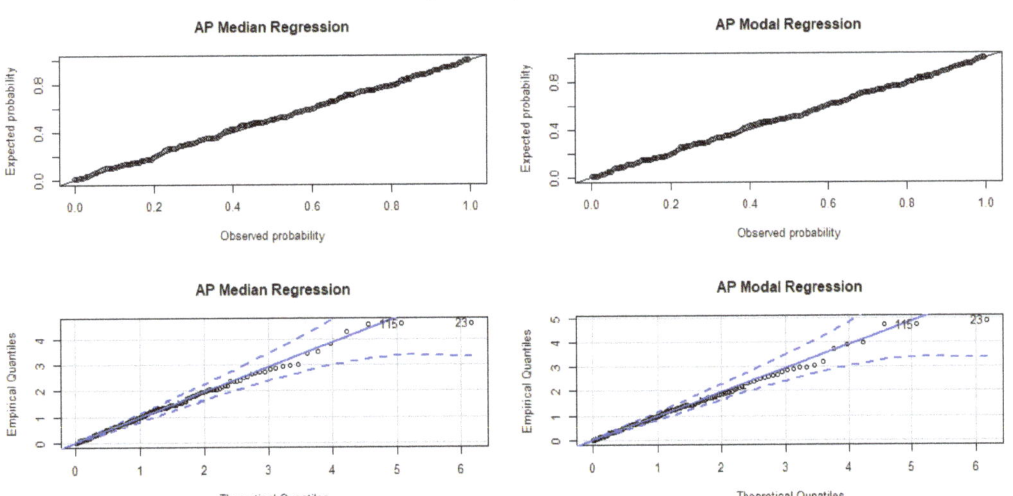

Figure 15. P-P (**top**) and Q-Q (**bottom**) plots of the Cox–Snell residuals.

8. Conclusions

In this study, the AP distribution and its associated quantile and modal regressions were developed. The PDF of the AP distribution exhibits flexible shapes such as left-skewed, right-skewed, J, and reversed-J shapes. This makes the distribution a suitable candidate for fitting data with such characteristics. The corresponding HRF also suggests that the distribution is capable of fitting data with monotonic and non-monotonic failure rates. We explored the performance of nine frequentist estimation procedures for estimating the parameters of the distribution using Monte Carlo simulations, and the results revealed that most of the procedures are consistent with regards to estimating the parameters. A biomedical application of the distribution showed that the model provides a good fit to the data. A Bayesian illustration of how to apply the distribution showed that the approach is able to estimate the parameters of the distribution very well. The applications of the elaborated quantile and modal regressions demonstrated that the new regression models outperformed some existing regression models. The future perspective of this work is to demonstrate the Bayesian applications of the quantile and modal regressions.

Author Contributions: Conceptualization, S.N., A.G.A. and C.C.; Data curation, S.N., A.G.A. and C.C.; Methodology, S.N., A.G.A. and C.C.; Supervision, S.N. and C.C.; Validation, S.N. and C.C.; Visualization, S.N. and A.G.A.; Writing, S.N. and A.G.A.; Review and editing, S.N. and C.C. All authors have read and agreed to the published version of the manuscript.

Funding: This research received no external funding.

Data Availability Statement: The data used in this study can be found in the simplexreg package of the R software developed by [22].

Acknowledgments: We express our sincere gratitude to the editor and reviewers whose constructive criticism improved the content of the article.

Conflicts of Interest: The authors declare no conflict of interest.

References

1. Korkmaz, M.Ç.; Chesneau, C.; Korkmaz, Z.S. The unit folded normal distribution: A new unit probability distribution with the estimation procedures, quantile regression modeling and educational attainment applications. *J. Reliab. Stat. Stud.* **2022**, *15*, 261–298. [CrossRef]
2. Nasiru, S.; Abubakari, A.G.; Chesneau, C. New lifetime distribution for modeling data on the unit interval: Properties, application and quantile regression. *Math. Comput. Appl.* **2022**, *27*, 105. [CrossRef]
3. Abubakari, A.G.; Luguterah, A.; Nasiru, S. Unit exponentiated Fréchet distribution: Actuarial measures, quantile regression and applications. *J. Indian Soc. Probab. Stat.* **2022**, *23*, 387–424. [CrossRef]
4. Eliwa, M.S.; Ahsan-ul-Haq, M.; Al-Bossly, A.; El-Morshedy, M. Properties and estimation techniques with application to model data from SC16 and P3 algorithms. *Math. Probl. Eng.* **2022**, *2022*, 9289721. [CrossRef]
5. Korkmaz, M.Ç.; Emrah, A.; Chesneau, C.; Yousof, H.M. On the unit-Chen distribution with associated quantile regression and applications. *Math. Slovaca* **2022**, *72*, 765–786. [CrossRef]
6. Korkmaz, M.Ç.; Chesneau, C. On the unit Burr XII distribution with the quantile regression modeling and applications. *Comput. Appl. Math.* **2021**, *40*, 29. [CrossRef]
7. Korkmaz, M.Ç. The unit generalized half normal distribution: A new bounded distribution with inference and application. *UPB Sci. Bull. Ser. A* **2020**, *82*, 133–140.
8. Modi, K.; Gill, V. Unit Burr-III distribution with application. *J. Stat. Manag. Syst.* **2019**, *23*, 579–592. [CrossRef]
9. Mazucheli, J.; Menezes, A.F.B.; Chakraborty, S. On the one parameter unit-Lindley distribution and its associated regression model for proportion data. *J. Appl. Stat.* **2019**, *46*, 700–714. [CrossRef]
10. Mazucheli, J.; Menezes, A.F.; Dey, S. Unit-Gompertz distribution with applications. *Statistica* **2019**, *79*, 25–43.
11. Altun, E.; Cordeiro, G.M. The unit-improved second-degree Lindley distribution: Inference and regression modeling. *Comput. Stat.* **2019**, *35*, 259–279. [CrossRef]
12. Mazucheli, J.; Menezes, A.F.; Ghitany, M.E. The unit Weibull distribution and associated inference. *J. Appl. Probab. Stat.* **2018**, *13*, 1–22.
13. Pourdarvish, A.; Mirmostafaee, S.M.T.K.; Naderi, K. The exponentiated Topp-Leone distribution: Properties and application. *J. Appl. Environ. Biol. Sci.* **2015**, *5*, 251–256.
14. Kharazmi, O.; Alizadeh, M.; Contreras-Reyes, J.E.; Haghbin, H. Arctan-based family of distributions: Properties, survival regression, Bayesian analysis and applications. *Axioms* **2022**, *11*, 399. [CrossRef]
15. Al-Mofleh, H.; Afify, A.Z.; Ibrahim, N.A. A new extended two-parameter distribution: Properties, estimation methods and, applications in medicine and geology. *Mathematics* **2020**, *8*, 1578. [CrossRef]
16. Iqbal, Z.; Tahir, M.M.; Riaz, N.; Ali, S.A.; Ahmad, M. Generalized inverted Kumaraswamy distribution: Properties and application. *Open J. Stat.* **2017**, *7*, 645–662. [CrossRef]
17. Iqbal, Z.; Hasnain, S.A.; Salman, M.; Ahmad, M.; Hamedani, G.G. Generalized exponentiated moment exponential distribution. *Pak. J. Stat.* **2014**, *30*, 537–554.
18. Gradshteyn, I.S.; Ryzhik, I.M. *Tables of Integrals, Series and Products*, 7th ed.; Elsevier/Academic Press: Amsterdam, The Netherlands, 2007.
19. Sklar, A. Random variables, joint distribution functions and copulas. *Kybernetika* **1973**, *9*, 449–460.

20. Elhassanein, A. On statistical properties of a new bivariate modified Lindley distribution with an application to financial data. *Complexity* **2022**, *2022*, 2328831. [CrossRef]
21. Ganji, M.; Bevrani, H.; Hami, N. A new method for generating continuous bivariate families. *J. Iran. Stat. Soc.* **2018**, *17*, 109–129. [CrossRef]
22. Zhang, P.; Qiu, Z.; Shi, C. Simplexreg: An R package for regression analysis of proportional data using the simplex distribution. *J. Stat. Softw.* **2016**, *71*, 1–21. [CrossRef]
23. Bantan, R.A.R.; Shafiq, S.; Tahir, M.H.; Elhassanein, A.; Jamal, F.; Almutiry, W.; Elgarhy, M. Statistical analysis of COVID-19 data: Using a new univariate and bivariate statistical model. *J. Funct. Spaces* **2022**, *2022*, 2851352. [CrossRef]
24. Ghosh, I.; Dey, S.; Kumar, D. Bounded M-O extended exponential distribution with applications. *Stoch. Qual. Control.* **2019**, *34*, 35–51. [CrossRef]
25. Kumaraswamy, P. A Generalized probability density function for double-bounded random processes. *J. Hydrol.* **1980**, *46*, 79–88. [CrossRef]
26. Muse, A.H.; Chesneau, C.; Ngesa, O.; Mwalili, S. Flexible parametric accelerated hazard model: Simulation and application to censored lifetime data with crossing survival curves. *Math. Comput. Appl.* **2022**, *27*, 104. [CrossRef]
27. Khan, S.A. Exponentiated Weibull regression for time-to-event data. *Lifetime Data Anal.* **2018**, *24*, 328–354. [CrossRef]
28. Su, Y.S.; Yajima, M. R2jags: A Package for Running Jags from R. 2012. Available online: https://CRAN.R-project.org/package=R2jags (accessed on 21 December 2022).
29. Menezes, A.F.B.; Mazucheli, J.; Chakraborty, S. A collection of parametric modal regression models for bounded data. *J. Biopharm. Stat.* **2021**, *31*, 490–506. [CrossRef]
30. Yao, W.; Li, L. A new regression model. *Scand. J. Stat.* **2014**, *41*, 656–671. [CrossRef]
31. Cox, D.R.; Snell, E.J. A general definition of residuals. *J. R. Stat. Soc. Ser. B* **1968**, *30*, 248–275. [CrossRef]
32. Dunn, P.K.; Smyth, G.K. Randomized quantile residuals. *J. Comput. Graph. Stat.* **1996**, *5*, 236–244.

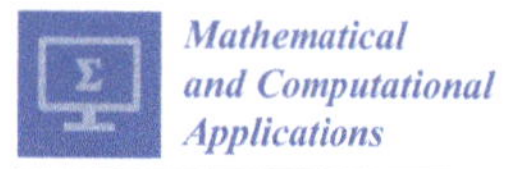
Mathematical and Computational Applications

Article

Prediction Interval for Compound Conway–Maxwell–Poisson Regression Model with Application to Vehicle Insurance Claim Data

Jahnavi Merupula [1], V. S. Vaidyanathan [1] and Christophe Chesneau [2,*]

[1] Department of Statistics, Pondicherry University, Puducherry 605014, India
[2] Department of Mathematics, LMNO, University of Caen, 14032 Caen, France
* Correspondence: christophe.chesneau@unicaen.fr

Abstract: Regression models in which the response variable has a compound distribution have applications in actuarial science. For example, the aggregate claim amount in a vehicle insurance portfolio can be modeled using a compound Poisson distribution. In this paper, we propose a regression model, wherein the response variable is assumed to have a compound Conway–Maxwell–Poisson (CMP) distribution. This distribution is a parsimonious two-parameter Poisson distribution that accounts for both over- and under-dispersed count data, making it more suitable for application in various fields. A two-part methodology in the framework of a generalized linear model is proposed to estimate the parameters. Additionally, a method to obtain the prediction interval of the response variable is developed. The workings of the proposed methodology are illustrated through simulated data. An application of the compound CMP regression model to real-life vehicle insurance claims data is presented.

Keywords: aggregate claims distribution; compound CMP regression model; generalized linear models; prediction intervals

Citation: Merupula, J.; Vaidyanathan, V.S.; Chesneau, C. Prediction Interval for Compound Conway–Maxwell–Poisson Regression Model with Application to Vehicle Insurance Claim Data. *Math. Comput. Appl.* **2023**, *28*, 39. https://doi.org/10.3390/mca28020039

Academic Editor: Sandra Ferreira

Received: 16 January 2023
Revised: 15 February 2023
Accepted: 27 February 2023
Published: 9 March 2023

1. Introduction

Compound regression models have applications in various research fields, including economics and finance. In economic consumer theory, for example, compound Poisson regression models are often used to examine the factors that account for the expenditures incurred by tourists during their stay at a location. The factors may include length of stay, type of holiday accommodations, age, occupation, socio-economic status of the tourist, etc. See Gómez-Déniz and Pérez-Rodríguez [1]. In actuarial risk theory, the aggregate claim amount incurred by the insurance company against the claims made by the policyholders is modeled using compound models. See Klugman et al. [2] and Bahnemann [3] for a detailed discussion on compound models, their distributional properties and applications in insurance claim modeling. Jørgensen and Paes De Souza [4] applied the compound Poisson regression model to determine the impact on the conditional mean of the aggregate claim amount caused by factors such as age and model of the vehicle, exposure, deductibles, etc., in the context of car insurance. In this paper, we propose a compound regression model using a two-parameter Poisson distribution. On this topic, some mathematical backgrounds are presented below in order to fix the notations. Let

$$S = \sum_{j=1}^{N} Y_j, \tag{1}$$

denote the random sum, where the distributions of the random variables N and $Y_1, Y_2, \ldots, Y_N$ are assumed to be discrete and continuous, respectively. Moreover, (Y_j)s are assumed to be independent and identically distributed. Therefore, in the sequel, we refer to Y_js as Y.

Further, N and Y in general are assumed to be independent. The above-mentioned S is a compound random variable. Suppose Y_j represents the claim amounts on an insurance portfolio, N denotes the number of claims made, then S represents the aggregate claim amount. When N has a Poisson distribution, the distribution of S is known as the compound Poisson distribution. Though the Poisson distribution is often used in constructing compound distributions, it is not suitable for modeling over- or under-dispersed count data. As an alternative to the Poisson distribution, one can use a generalized Poisson distribution (Consul and Jain [5]) to model count data that are either over- or under-dispersed. Recently, Shmueli et al. [6] studied a two-parameter Poisson distribution developed by Conway and Maxwell [7] known as the Conway–Maxwell–Poisson (CMP) distribution. This is a two-parameter flexible generalization of the Poisson distribution that can model both over- and under-dispersed data and has the feature to include the Poisson, geometric and Bernoulli distributions as special cases. A detailed discussion on the properties of this distribution and its applications can be found in Sellers et al. [8]. Also, Sellers and Premeaux [9] contains a detailed review on CMP regression models. In the context of compound distributions, assuming the CMP and binomial distributions for N and Y in Equation (1), a discrete compound CMP-binomial distribution is developed by Saavithri et al. [10].

Considering the Poisson distribution as the counting distribution, compound Poisson regression models are available in the literature. See Frees et al. [11], Andersen and Bonat [12], and Delong et al. [13]. However, its applicability is limited to data with equi-dispersed counts. To allow for flexibility in the compound regression models in terms of accommodating dispersed counts, a counting distribution that can model both over- and under-dispersed data should be considered. This serves as motivation to use the CMP distribution as the counting distribution to build a compound regression model.

The goal of this work is to create a regression model for S using a CMP distribution for N. The present work is novel because of the distribution used for N and its convolution with the distribution of Y. The problem of obtaining prediction intervals for the response variable S is also addressed. The parameters of the compound regression model are estimated using the generalized linear model (GLM) approach in two cases. In the first case, we assume that data on S are available but not on N and Y. We assume data on both N and Y are available in the latter case. For this case, a two-part likelihood-based estimation procedure is developed within the framework of the GLM. A methodology to obtain the prediction interval (PI) for the response variable of the proposed compound regression model is developed.

The rest of the paper is organized as follows: The compound CMP regression model is given in Section 2. In Section 3, the estimation of the parameters of the proposed regression model using the GLM approach is discussed. Section 4 deals with the suggested methodology for obtaining the prediction intervals for the compound CMP regression model. A numerical illustration of the estimation procedure using simulated data and an application to real-life vehicle insurance claims data is presented in Section 5. The conclusion of the paper is given in Section 6.

2. Compound CMP Regression Model

The probability mass function (pmf) of the random variable N having the CMP distribution is given by

$$P(N = n) = \frac{\lambda^n}{(n!)^\nu Z(\lambda, \nu)}, \qquad n = 0, 1, 2, \ldots, \ \lambda > 0, \ \nu \geq 0, \tag{2}$$

where $Z(\lambda, \nu) = \sum_{j=0}^{\infty} \lambda^j / (j!)^\nu$ is the normalizing constant. Some important remarks on this distribution are given below. The parameters λ and ν are the location and dispersion parameters, respectively. This pmf is not defined for $\lambda \geq 1$ and $\nu = 0$. The mean and variance of N are given by $E(N) = \lambda \dfrac{\partial \ln Z(\lambda, \nu)}{\partial \lambda}$ and $V(N) = \lambda \dfrac{\partial E(N)}{\partial \lambda}$, respectively. When

$v = 1$, the CMP distribution reduces to the Poisson distribution. For $v > 1$, the distribution is under-dispersed, and for $v < 1$, it is over-dispersed.

Since the location parameter λ of the CMP distribution does not represent its mean, a mean reparameterized form of the distribution is used in building the compound regression model. The pmf of N under the mean-reparametrization is given by

$$P(N = n) = \left(\mu_1 + \frac{e^\phi - 1}{2e^\phi} \right)^{ne^\phi} \frac{(n!)^{-e^\phi}}{Z(\mu_1, \phi)}, \quad n = 0, 1, 2, \ldots, \mu_1 > 0, \phi \in \mathbb{R}, \tag{3}$$

where $Z(\mu_1, \phi) = \sum_{j=0}^{\infty} \left(\mu_1 + \frac{e^\phi - 1}{2e^\phi} \right)^{je^\phi} \frac{1}{(j!)^{e^\phi}}$ is the normalizing constant. When $\phi = 0$, the distribution reduces to the Poisson distribution. For $\phi > 0$, the distribution is under-dispersed, and for $\phi < 0$, it is over-dispersed. See Ribeiro Jr et al. [14]. Here, $\mu_1 \approx \lambda^{1/v} - \frac{v - 1}{2v}$ corresponds to the mean of the distribution and $\phi = \ln(v)$. This approximation works reasonably well for $v \leq 1$ or $\lambda > 10^v$. The mean and variance of N are $E(N) = \mu_1$ and $V(N) = \mu_1 e^{-\phi}$, respectively.

Convolutions can be used to obtain the probability density function (pdf) of the random sum S defined in Equation (1). In Equation (1), $N = 0$ implies $S = 0$. Let p_0 denote the probability mass at $S = 0$. Since S is not continuous at zero, the pdf of S is represented as a generalized pdf in terms of Dirac delta function as

$$f(s) = p_0 \delta(s) + \sum_{i=1}^{\infty} g_Y^{*i}(s) P(N = i), \quad s \geq 0, \tag{4}$$

where $\delta(s)$ is the Dirac delta function such that $\int_0^\infty \delta(s)ds = 1$. Here, $P(N = i)$ denotes the pmf of the CMP distribution defined in Equation (3), and $g_Y^{*i}(.)$ denotes the pdf of the i-fold convolution of Y, whose distribution is assumed to be continuous with support in $\mathbb{R}^+$. Note that $p_0 = P(N = 0) = Z(\mu_1, \phi)^{-1}$. In this paper, the distribution of Y is considered to be a mean reparameterized gamma distribution. Based on Jorgensen [15] (Chapter 3), the pdf of Y is given by

$$g_Y(y; \mu_2, \psi) = \frac{1}{\Gamma(\psi)} \left(\frac{\psi}{\mu_2} \right)^\psi y^{\psi-1} \exp\left(-\frac{\psi y}{\mu_2} \right), \quad y > 0, \mu_2 > 0, \psi > 0, \tag{5}$$

where μ_2 denotes the mean of Y, ψ denotes the dispersion parameter and $\Gamma(.)$ denotes the gamma function. This form is taken for mathematical convenience and to accommodate asymmetry in the distribution of Y. For example, in the context of insurance claim modeling, the individual claim amounts are always positive and often right-skewed. Since the gamma distribution is closed under convolution, we obtain

$$g_Y^{*i}(y) = \frac{1}{\Gamma(\psi)} \left(\frac{\psi}{i\mu_2} \right)^\psi y^{\psi-1} \exp\left(-\frac{\psi y}{i\mu_2} \right), \quad y > 0, \mu_2 > 0, \psi > 0. \tag{6}$$

Using Equations (3) and (6) in Equation (4), we obtain

$$f(s) = p_0 \delta(s) + \frac{s^{\psi-1} \psi^\psi}{Z(\mu_1, \phi) \mu_2^\psi \Gamma(\psi)} \sum_{i=1}^{\infty} \left(\mu_1 + \frac{e^\phi - 1}{2e^\phi} \right)^{ie^\phi} \frac{(i!)^{-e^\phi}}{i^\psi} \exp\left(\frac{-\psi s}{i\mu_2} \right), \quad s \geq 0. \tag{7}$$

The pdf of S defined in Equation (7) is called the compound CMP gamma pdf. For the random sum defined in Equation (1), we have

$$\begin{cases} E(S) = E(N)E(Y), \\ V(S) = E(N)V(Y) + V(N)[E(Y)]^2. \end{cases} \tag{8}$$

See, for instance, Bahnemann [3] (Chapter 4). Using Equation (8), the mean and variance of the compound CMP gamma distribution given in Equation (7) are obtained as

$$\begin{cases} E(S) = \mu_1\mu_2, \\ V(S) = \mu_2^2\mu_1[\psi^{-1} + e^{-\phi}]. \end{cases} \tag{9}$$

To build a compound regression model for S, let $X = (\vec{1}, \vec{X}_1, \vec{X}_2, \ldots, \vec{X}_p)$ denote the design matrix where $\vec{X}_i, i = 1, 2, \ldots, p$ are the column vectors corresponding to the covariates $X_i, i = 1, 2, \ldots, p$ and $\vec{1}$ is the vector of $1's$. Following the GLM procedure given in De Jong et al. [16] (Chapter 5), the model is built by regressing S on X using the log-link function. This is because the log-link function guarantees that the expected value of the response variable is positive. Let μ denote the expected value of S. Then, the compound CMP gamma regression model is given by

$$\mu = \exp(X\delta), \tag{10}$$

where $\delta = (\delta_0, \delta_1, \ldots, \delta_p)'$ is a $(p+1) \times 1$ vector of regression parameters. In the context of modeling vehicle insurance claims data, S may denote the aggregate claim amount, and the covariates may denote the driver's age, vehicle type, and so on. In the sequel, the method of estimating the regression parameters using the likelihood approach is discussed.

3. Parameter Estimation

Consider a sample $\vec{s} = (s_1, s_2, \ldots, s_r)'$ of r observations on S. Let $D(> 0)$ positive values in $\vec{s}$ and $r - D$ zeros exist. Note that D can be assimilated to be random and $D \sim Binomial(r, 1 - p_0)$, where $p_0 = Z(\mu_1, \phi)^{-1}$. Therefore, the likelihood function L based on $\vec{s}$ and $D = d$ is

$$\begin{aligned} L &= \binom{r}{d} p_0^{r-d}(1 - p_0)^d \prod_{k=1}^{d} f(s_k^+) \\ &= \binom{r}{d} \left(\frac{1}{Z(\mu_1, \phi)}\right)^{r-d} \left(1 - \frac{1}{Z(\mu_1, \phi)}\right)^d \prod_{k=1}^{d} f(s_k^+), \end{aligned} \tag{11}$$

where $f(s_k^+) = \dfrac{s_k^{\psi-1}\psi^\psi}{(Z(\mu_1, \phi) - 1)\mu_2^\psi\Gamma(\psi)} \sum_{i=1}^{\infty}\left(\mu_1 + \dfrac{e^\phi - 1}{2e^\phi}\right)^{ie^\phi} \dfrac{(i!)^{-e^\phi}}{i^\psi} \exp\left(\dfrac{-\psi s_k}{i\mu_2}\right).$

Thus, the log-likelihood function l based on $\vec{s}$ and $D = d$ is obtained as

$$\begin{aligned} l(\mu_1, \mu_2, \phi, \psi; \vec{s}) = {} & \ln\left(\binom{r}{d}\right) - r\ln(Z(\mu_1, \phi)) + (\psi - 1)\sum_{k=1}^{d}\ln(s_k) - \sum_{k=1}^{d}\psi\ln(\mu_2) + d\psi\ln(\psi) \\ & - d\ln(\Gamma(\psi)) + \sum_{k=1}^{d}\ln\left[\sum_{i=1}^{\infty}\left(\mu_1 + \frac{e^\phi - 1}{2e^\phi}\right)^{ie^\phi}\frac{(i!)^{-e^\phi}}{i^\psi}\exp\left(\frac{-\psi s_k}{i\mu_2}\right)\right]. \end{aligned} \tag{12}$$

Since $E(N) = \mu_1$ and $E(Y) = \mu_2$, from Equation (9), we obtain $\mu = \mu_1\mu_2$. Let the elements of the design matrix X be $x_{kl}, l = 0, 1, \ldots, p; k = 1, 2, \ldots, d$ with the kth row given by $x_k = (1, x_{k1}, x_{k2}, \ldots, x_{kp})$. Replacing μ_2 with $\dfrac{\mu}{\mu_1}$ and μ with $\exp(X\delta)$ in Equation (12), the log-likelihood function based on $\vec{s}$ and $D = d$ becomes

$$\begin{aligned} l(\delta, \mu_1, \phi, \psi; \vec{s}) = {} & \ln\left(\binom{r}{d}\right) - r\ln(Z(\mu_1, \phi)) + (\psi - 1)\sum_{k=1}^{d}\ln(s_k) - \sum_{k=1}^{d}\psi\ln\left(\frac{e^{\sum_{l=0}^{p}x_{kl}\delta_l}}{\mu_1}\right) \\ & + d\psi\ln(\psi) - d\ln(\Gamma(\psi)) + \sum_{k=1}^{d}\ln\left[\sum_{i=1}^{\infty}\left\{\left(\mu_1 + \frac{e^\phi - 1}{2e^\phi}\right)^{ie^\phi}\frac{(i!)^{-e^\phi}}{i^\psi}\exp\left(\frac{-\psi s_k\mu_1}{ie^{\sum_{l=0}^{p}x_{kl}\delta_l}}\right)\right\}\right]. \end{aligned} \tag{13}$$

The maximum likelihood (ML) estimates of the parameters in Equation (13) can be obtained by solving the $(p+4)$ log-likelihood equations simultaneously. However, these equations are non-linear, and therefore closed-form solutions cannot be obtained. Hence, iterative algorithms based on numerical methods can be used to solve the equations to get the estimates for the parameters. Let $\hat{\delta}$ denote the ML estimate of δ. By the asymptotic property of the ML estimators, for large r, the following distribution approximation holds:

$$\Sigma_\delta^{1/2}(\hat{\delta} - \delta) \sim \mathcal{N}_{p+1}(\mathbf{0}, I),$$

where δ and Σ_δ denote the mean vector and the covariance matrix of $\hat{\delta}$, respectively. Using Equation (10), an estimate of the expected value of S given the covariates X can be obtained as $\hat{\mu} = \exp\left(X\hat{\delta}\right)$.

Assume that data on S are unavailable, but data on N and Y are. This can happen in such situations as, for example, when modeling the aggregate claim amount when one has data on the claim frequency (N) and the individual claim amounts (Y). Using N and Y, we can compute the value of S and then build the regression model using the method described above. However, it is computationally more challenging to compute the estimates due to the presence of an infinite sum in the log-likelihood function. To reduce the computational difficulty, we can use N and Y to build two separate regression models to obtain $\hat{\mu}$. Towards this, a two-part GLM methodology is proposed to estimate μ assuming N and Y to be (1) independent and (2) dependent.

3.1. Independent Compound Regression Model

Using Equation (9), we have $\mu = \mu_1\mu_2$. The proposed two-part GLM method is implemented by building two separate regression models, namely, the CMP regression model and the gamma regression model, for the means of N and Y, respectively. Given the data on N, Y and X, the estimated mean of S is computed as $\hat{\mu} = \hat{\mu}_1\hat{\mu}_2$. Here, $\hat{\mu}_1$ and $\hat{\mu}_2$ are obtained by regressing N and Y separately on X. Using the log-link function, we have $\mu_1 = E(N) = e^{X\alpha}, \mu_2 = E(Y) = e^{X\beta}$, where $\alpha = (\alpha_0, \alpha_1, \ldots, \alpha_p)'$ and $\beta = (\beta_0, \beta_1, \ldots, \beta_p)'$ denote the set of regression parameters.

Let $\vec{n} = (n_1, \ldots, n_m)'$ denote m observations on N. For each $n_k > 0$, let there be n_k observations on Y denoted by $y_{kj}, j = 1, 2, \ldots, n_k, k = 1, 2, \ldots, m$. Let $\vec{y} = (\bar{y}_1, \bar{y}_2, \ldots, \bar{y}_m)'$ where $\bar{y}_k = \begin{cases} \sum_{j=1}^{n_k} y_{kj}/n_k & \text{if } n_k > 0 \\ 0 & \text{if } n_k = 0. \end{cases}$

Let the design matrix X be of order $m \times (p+1)$ with elements $x_{kl}, k = 1, 2, \ldots, m; l = 0, 1, \ldots, p$. Since the distribution of Y has positive support, zeros in $\vec{y}$, if any, are not to be considered. The corresponding sample observation in $\vec{y}$ and the observed covariate matrix X are not included when building the gamma regression model. Let q denote the number of observations for which $\bar{y}_k = 0, k = 1, 2, \ldots, m$ and let $t = m - q$. Following Garrido et al. [17], the distribution of $Y \sim gamma(\mu_2, \psi)$ is equivalent to $\bar{Y}|N \sim gamma\left(\mu_2, \dfrac{\psi}{N}\right)$ for independently identically distributed $Y_1, \ldots, Y_N$. Using the pmf of N given in Equation (3) with $\mu_1 = e^{X\alpha}$, the corresponding log-likelihood function is given by

$$l(\alpha, \phi; \vec{n}) = \sum_{k=1}^{m} e^\phi \left[n_k \ln\left(e^{\sum_{l=0}^{p} x_{kl}\alpha_l} + \frac{e^\phi - 1}{2e^\phi} \right) - \ln(n_k!) \right] - \sum_{k=1}^{m} \ln\left(Z(e^{\sum_{l=0}^{p} x_{kl}\alpha_l}, \phi) \right). \quad (14)$$

The ML estimates for the $(p+1)$ regression parameters are obtained by simultaneously solving the corresponding log-likelihood equations. Let $\hat{\alpha} = (\hat{\alpha}_0, \hat{\alpha}_1, \ldots, \hat{\alpha}_p)'$ denote the ML estimate of α. Then the ML estimate of μ_1 is obtained as $\hat{\mu}_1 = e^{X\hat{\alpha}}$. In similar lines, the

ML estimate of β, namely, $\hat{\beta} = (\hat{\beta}_0, \hat{\beta}_1, \ldots, \hat{\beta}_p)'$, is obtained using the likelihood function corresponding to the conditional pdf of $\bar{Y}$ given $N = n$. The conditional pdf is given by

$$f(\bar{y}|n; \mu_2, \psi) = \frac{1}{\Gamma(\psi/n)} \left(\frac{\psi/n}{\mu_2}\right)^{\psi/n} \bar{y}^{(\psi/n)-1} \exp\left(-\frac{\psi\bar{y}}{n\mu_2}\right), \quad \bar{y} > 0. \tag{15}$$

Taking $\mu_2 = e^{X\beta}$ in Equation (15), the log-likelihood function is obtained as

$$l(\beta, \psi; \vec{y}) = -t \ln\left(\Gamma\left(\frac{\psi}{n}\right)\right) + \frac{t\psi}{n} \ln\left(\frac{\psi}{n}\right) + \sum_{k=1}^{t} \left[\left(\frac{\psi}{n} - 1\right) \ln(\bar{y}_k) - \frac{\psi\bar{y}_k}{ne^{\sum_{l=0}^{p} x_{kl}\beta_l}} - \frac{\psi}{n} \sum_{l=0}^{p} x_{kl}\beta_l\right]. \tag{16}$$

The likelihood equations for α and β are, respectively, given by

$$\sum_{k=1}^{m} x_{kl}\left(n_k - e^{\sum_{l=0}^{p} x_{kl}\alpha_l}\right) = 0 \tag{17}$$

and

$$\sum_{k=1}^{t} \frac{x_{kl}n_k}{e^{\sum_{l=0}^{p} x_{kl}\beta_l}} \left(\bar{y}_k - e^{\sum_{l=0}^{p} x_{kl}\beta_l}\right) = 0, \quad l = 0, 1, \ldots, p. \tag{18}$$

Since Equations (17) and (18) are non-linear, iterative procedures can be used to solve them. As an alternate, one can use the in-built functions `cmp()` and `glm(., family='gamma')` available in R to obtain $\hat{\alpha}$ and $\hat{\beta}$. Using $\hat{\alpha}$ and $\hat{\beta}$, the ML estimate of the expected value of S, namely, $\hat{\mu} = \hat{\mu}_1\hat{\mu}_2$, can be computed. By the asymptotic property of the ML estimators, we have

$$\Sigma_{\alpha}^{1/2}(\hat{\alpha} - \alpha) \sim \mathcal{N}_{p+1}(\mathbf{0}, I)$$

and

$$\Sigma_{\beta}^{1/2}(\hat{\beta} - \beta) \sim \mathcal{N}_{p+1}(\mathbf{0}, I).$$

Here, α and Σ_{α} denote the mean vector and covariance matrix of $\hat{\alpha}$, respectively. Similarly, β and Σ_{β} denote the mean vector and covariance matrix of $\hat{\beta}$, respectively. The standard errors of $\hat{\alpha}$ and $\hat{\beta}$ are the square root of the diagonal elements of the corresponding covariance matrices. Since $\hat{\alpha}$ and $\hat{\beta}$ do not have closed-form expressions, their standard errors can be obtained using the sample Hessian matrix. The sample Hessian matrices of $\hat{\alpha}$ and $\hat{\beta}$, namely, $H_{\hat{\alpha}}$ and $H_{\hat{\beta}}$, are given by $H_{\hat{\alpha}} = e^{\hat{\phi}}e^{X\hat{\alpha}}XX'$ and $H_{\hat{\beta}} = \hat{\psi}XX'$, respectively. Since the expressions of the standard errors of the parameters α and β contain the dispersion parameters ϕ and ψ, respectively, they may be estimated using the following formulas:

$$\hat{\phi} = \ln\left\{(m - (p+1)) \sum_{k=1}^{m} \frac{\hat{\mu}_{1k}}{(n_k - \hat{\mu}_{1k})^2}\right\} \tag{19}$$

and

$$\hat{\psi} = \frac{1}{(t - (p+1))} \sum_{k=1}^{t} \left(\frac{\bar{y}_k - \hat{\mu}_{2k}}{\hat{\mu}_{2k}}\right)^2, \tag{20}$$

where $\hat{\mu}_{1k}$ and $\hat{\mu}_{2k}$ are the estimated values of μ_1 and μ_2, respectively, corresponding to the kth observation.

3.2. Dependent Compound Regression Model

Although independence between N and Y is commonly assumed in compound regression models, it is rarely observed in practice. For instance, in the framework of modeling the aggregate claim amounts, it is typical to observe that the claim amounts depend on the claim frequency as well. See, for example, the work of Garrido et al. [17]. As a result, N is included as a covariate in the regression model of $\bar{Y}$. Let θ represent the regression parameter associated with N. Since S denotes a random sum, it can be written as $S = N\bar{Y}$. The GLM of S through the log-link function is given by Garrido et al. [17] as

$$\mu = e^{X\beta} M'_N(\theta),$$

where $M'_N(\theta)$ represents the derivative of the moment generating function of N with respect to θ. Taking N as CMP, $M'_N(\theta)$ is obtained as

$$M'_N(\theta) = \sum_{n=0}^{\infty} n e^{\theta n} \left(\mu_1 + \frac{e^{\phi} - 1}{2e^{\phi}} \right)^{ne^{\phi}} \frac{(n!)^{-e^{\phi}}}{Z(\mu_1, \phi)}.$$

Note that if $\theta = 0$, i.e., when N is independent of $\bar{Y}$, $M'_N(\theta) = E(N)$, and thus the dependent compound regression model will coincide with the independent compound regression model. The pdf of S under dependent case is given by

$$f_S(s) = f_{\bar{Y}|N}(\bar{y}|n) f_N(n),$$

where $f_{\bar{Y}|N}(\bar{y}|n)$ is indicated in Equation (15) with $\mu_2 = \mu_\theta$ and $\psi = \psi_\theta$. The corresponding log-likelihood function is

$$l(\alpha, \beta, \phi, \psi, \theta) = l(\alpha, \phi; \vec{n}) + l(\beta, \psi, \theta; \vec{\bar{y}}|\vec{n}),$$

where $l(\alpha, \phi; \vec{n})$ corresponds to Equation (14). Let the ML estimates of α, β and θ be denoted as $\tilde{\alpha}, \tilde{\beta}$ and $\tilde{\theta}$, where $\tilde{\alpha}$ is obtained using Equation (17). The function $l(\beta, \psi, \theta; \vec{\bar{y}}|\vec{n})$ corresponds to Equation (16) with μ_2 replaced with μ_θ. To obtain the estimates of β and θ, the GLM of $E(\bar{Y}|N, X)$ is used with the log-link function and is defined by $\mu_\theta = e^{X\beta + \theta N}$. The corresponding likelihood equations of the regression parameters are

$$\sum_{k=1}^{t} \frac{n_k x_{kl}}{e^{\sum_{l=0}^{p} x_{kl}\beta_l + \theta n_k}} \left(\bar{y}_k - e^{\sum_{l=0}^{p} x_{kl}\beta_l + \theta n_k} \right) = 0 \tag{21}$$

and

$$\sum_{k=1}^{t} \frac{n_k^2}{e^{\sum_{l=0}^{p} x_{kl}\beta_l + \theta n_k}} \left(\bar{y}_k - e^{\sum_{l=0}^{p} x_{kl}\beta_l + \theta n_k} \right) = 0, \quad l = 0, 1, \ldots, p. \tag{22}$$

The dispersion parameter ψ_θ can be estimated using

$$\hat{\psi}_\theta = \frac{1}{(t - (p+1))} \sum_{k=1}^{t} \left(\frac{\bar{y}_k - \hat{\mu}_{\theta k}}{\hat{\mu}_{\theta k}} \right)^2,$$

where $\hat{\mu}_{\theta k}$ is the estimated value of μ_θ corresponding to the kth observation. In addition, $\tilde{\beta}$ and $\tilde{\theta}$ can be obtained by solving Equations (21) and (22) through iterative algorithms. Thus, the estimate of μ is given by $\tilde{\mu} = e^{X\tilde{\beta}} M'_N(\tilde{\theta})$. Denote $\beta_\theta = \begin{bmatrix} \beta \\ \theta \end{bmatrix}_{(p+2) \times 1}$ and its ML

estimate as $\tilde{\beta}_\theta = \begin{bmatrix} \tilde{\beta} \\ \tilde{\theta} \end{bmatrix}_{(p+2) \times 1}$. By the asymptotic property of the ML estimators, we have

$$\Sigma_{\beta_\theta}^{1/2}(\tilde{\beta}_\theta - \beta_\theta) \sim \mathcal{N}_{p+2}(\mathbf{0}, I).$$

Here, β_θ and Σ_{β_θ} denote the mean vector and covariance matrix of $\tilde{\beta}_\theta$, respectively. The standard error of $\tilde{\beta}_\theta$ corresponds to the square root of the diagonal elements of the sample Hessian matrix, which is given by $H_{\tilde{\beta}_\theta} = \hat{\psi}_\theta X^{*\prime} A X^*$, where X^* is a matrix of order $t \times (p+2)$ that denotes the design matrix which includes $\vec{n}$. A is a $t \times t$ diagonal matrix with positive elements of $\vec{n}$. Note that $H_{\tilde{\alpha}} = H_{\hat{\alpha}}$.

4. Prediction Intervals

From the estimates of the regression parameters, we can obtain an estimate of the expected value of S for some fixed values of the covariates. Given the covariates, it is frequently useful to predict the actual value of S. In a regression setup, the actual value of S is related to its expected value as

$$S = \hat{E}(S|X) + \epsilon,$$

where ϵ is the error term. Since ϵ is unobserved, it is not possible to predict the actual S. In contrast, the prediction interval is a constructed interval that contains the predicted value of actual S. In this section, a method for calculating the PI for S is proposed. Let S_0 denote the response given the covariate $x_0 = (1, x_{01}, \ldots, x_{0p})$. Thus, we have $S_0 = \hat{E}(S_0|x_0) + \epsilon$, where $\hat{E}(S_0|x_0) = \exp(x_0\hat{\delta}) = \hat{\mu}_0$ (say). Assuming $E(\epsilon) = 0$, we get, $E(S_0) = \hat{\mu}_0$. Additionally, we have $V(S_0) = V(\hat{\mu}_0) + V(\epsilon)$. Hence, the $100(1-\alpha)\%$ PI for S_0 is given by $[k_1, k_2]$, such that

$$P[k_1 \leq S_0 \leq k_2] = 1 - \alpha, \tag{23}$$

where $\alpha \in (0,1)$. Here, k_1 and k_2 correspond, respectively, to the lower $\left(\frac{\alpha}{2}\right)^{th}$ and upper $\left(\frac{\alpha}{2}\right)^{th}$ percentiles of the distribution of S_0, which is the compound CMP gamma distribution with mean $E(S_0)$ and variance $V(S_0)$. Since $V(S_0)$ depends on $V(\hat{\mu}_0)$, we proceed as below to obtain an expression for $V(\hat{\mu}_0)$. To begin, consider

$$\hat{\mu}_0 = \exp(x_0\hat{\delta}) \implies \ln(\hat{\mu}_0) = x_0\hat{\delta}. \tag{24}$$

Using the Taylor series expansion of $\ln(A)$ at $E(A)$, we have

$$\ln(A) \approx \ln(E(A)) + (A - E(A))\frac{1}{E(A)}.$$

Thus, we have

$$E(\ln(A)) \approx \ln(E(A)) \tag{25}$$

and

$$V(\ln(A)) \approx \frac{V(A)}{E(A)^2}. \tag{26}$$

Taking A to be $\hat{\mu}_0$ in Equations (25) and (26), we obtain $E(\ln(\hat{\mu}_0)) \approx \ln E(\hat{\mu}_0)$ and $V(\ln(\hat{\mu}_0)) \approx \frac{V(\hat{\mu}_0)}{E(\hat{\mu}_0)^2}$. From Equation (24), we establish that

$$E(\ln(\hat{\mu}_0)) \approx E(x_0\hat{\delta}) = x_0 E(\hat{\delta})$$
$$\implies E(\hat{\mu}_0) \approx \exp(x_0 E(\hat{\delta})) = \exp(x_0\delta) = \mu_0.$$

In a similar manner, we obtain

$$V(\hat{\mu}_0) \approx V(\ln(\hat{\mu}_0))E(\mu_0)^2 = V(x_0\hat{\delta})\mu_0^2 = x_0 V(\hat{\delta})x_0'\mu_0^2$$
$$= x_0\text{diag}(\Sigma_\delta)x_0'\mu_0^2.$$

An estimate of $V(\epsilon)$, namely, $\hat{V}(\epsilon)$, can be obtained by dividing the residual sum of squares (RSS) of the compound CMP regression model by $m - (p+1)$. Using $V(\hat{\mu}_0)$ and $\hat{V}(\epsilon)$, we obtain $V(S_0)$. However, obtaining the values of k_1 and k_2 from Equation (23) is not easy since the cumulative distribution function of the compound CMP gamma distribution is not invertible. One may use bootstrap procedures to identify k_1 and k_2. We propose below a heuristic method to obtain the PI using the two-part GLM methodology given in the previous section.

The PI for S_0 is obtained using the PIs of N_0 and $\bar{Y}_0$, where $N_0 = \hat{E}(N_0|x_0) + \epsilon$ and $\bar{Y}_0 = \hat{E}(\bar{Y}_0|x_0) + \epsilon$. Note that $\hat{E}(N_0|x_0)$ is obtained from the GLM of N on X and $\hat{E}(\bar{Y}_0|x_0)$ is obtained using the GLM of $\bar{Y}$ on X. Denoting $\hat{E}(N_0|x_0) = \hat{\mu}_{01}$ and $\hat{E}(\bar{Y}_0|x_0) = \hat{\mu}_{02}$, we have, $\hat{\mu}_{01} = \exp(x_0\hat{\alpha})$ and $\hat{\mu}_{02} = \exp(x_0\hat{\beta})$. Proceeding along similar lines for obtaining the PI for S_0, the PIs for N_0 and $\bar{Y}_0$ can be obtained, respectively, as $[a_1, a_2]$ and $[b_1, b_2]$, such that

$$P[a_1 \leq N_0 \leq a_2] = 1 - \alpha$$

and

$$P[b_1 \leq \bar{Y}_0 \leq b_2] = 1 - \alpha,$$

where $\alpha \in (0,1)$. Since N_0 has a mean reparameterized CMP distribution given in Equation (3), a_1 and a_2 are respectively, the lower $\left(\frac{\alpha}{2}\right)^{th}$ and upper $\left(\frac{\alpha}{2}\right)^{th}$ percentiles of the CMP distribution with mean $\hat{\mu}_{01}$ and dispersion parameter $\phi = \dfrac{\hat{\mu}_{01}}{V(\hat{\mu}_{01}) + \hat{V}(\epsilon)}$, where $V(\hat{\mu}_{01}) = x_0\mathrm{diag}(\Sigma_\alpha)x_0'\mu_{01}^2$. Likewise, b_1 and b_2 correspond respectively, to the lower $\left(\frac{\alpha}{2}\right)^{th}$ and upper $\left(\frac{\alpha}{2}\right)^{th}$ percentiles of the mean reparameterized gamma distribution given in Equation (15) with mean $\hat{\mu}_{02}$ and dispersion parameter $\psi = \dfrac{V(\hat{\mu}_{02}) + \hat{V}(\epsilon)}{\hat{\mu}_{02}^2}$, where $V(\hat{\mu}_{02}) = x_0\mathrm{diag}(\Sigma_\beta)x_0'\mu_{02}^2$. Supposing Σ_α and Σ_β are not known, the corresponding sample Hessian matrices can be used to compute $V(\hat{\mu}_{01})$ and $V(\hat{\mu}_{02})$. The values of $\hat{V}(\epsilon)$ of the CMP and gamma regression models can be obtained by dividing the RSS of the corresponding regression models by $m - h$ and $t - h$, where h denotes the number of regression parameters in the model.

The PI for S_0 given x_0 can be constructed using the PIs of N_0 and $\bar{Y}_0$. By virtue of equality $S = N\bar{Y}$, a trivial PI for S_0 given x_0 can be taken to be $[k_1, k_2] = [a_1b_1, a_2b_2]$. When N is large, it may be useful to know the PI for S_0. For example, in modeling aggregate claim amounts from insurance data, the company may want to know the PI for the aggregate claim amount for high claim frequencies so that enough funds can be maintained. In this case, the PI for S_0 given x_0 can be defined as $[a_2b_1, a_2b_2]$. This definition of PI is used in the remaining part.

5. Numerical Illustration

5.1. Simulation Study

This section provides a numerical illustration of how to compute the PI for S using simulated data for the independent and dependent compound regression models. To generate random samples from the CMP and gamma regression models with a single covariate $\vec{X}_1 = (x_{11}, x_{21}, \ldots, x_{m1})'$, generated from a standard normal distribution, the following steps are implemented:

1. Generate $n_k, k = 1, 2, \ldots, m$, from the CMP distribution given in Equation (3) with mean $\mu_{1k} = \exp(\alpha_0 + \alpha_1 x_{k1})$ by fixing α_0, α_1 and ϕ. Obtain $\vec{n} = (n_1, n_2, \ldots, n_m)'$.
2. For each $n_k > 0$, generate $y_{kj}, j = 1, 2, \ldots, n_k$ from the gamma distribution given in Equation (5) with mean μ_{2k} by fixing $\psi, \beta_0, \beta_1,$ and θ, where $\mu_{2k} = \exp(\beta_0 + \beta_1 x_{k1})$ for the independent compound regression model and $\exp(\beta_0 + \beta_1 x_{k1} + \theta n_k)$ for the dependent compound regression model. Compute $\bar{y}_k$ and obtain $\vec{\bar{y}} = (\bar{y}_1, \bar{y}_2, \ldots, \bar{y}_m)'$.

For simulation, the values of the regression parameters are taken as $\alpha_0 = 0.5, \alpha_1 = 0.3, \beta_0 = 1, \beta_1 = 0.5$ and $\theta = 0.5$. The dispersion parameter ψ of the gamma distribution is set to 1.5. To accommodate over-, equi- and under-dispersion in N, three choices of the dispersion parameter ϕ, namely, $\phi = -1.6, 0,$ and 1.6, are considered. The CMP and gamma GLMs are fitted to the generated $\vec{n}$ and $\vec{\bar{y}}$ values, using their respective log-link functions for both the independent and dependent compound regression models.

All the computations are carried out in R (version 4.1.1). The `cmp()` function in `cmpreg` package (Ribeiro Jr [18]) and the `glm()` function are used to carry out the CMP and gamma regression, respectively. To compute the value of $M'_N(\hat{\theta})$ in the dependent compound regression model, the `com.expectation()` function in `compoisson` package is employed. `qcom()` function in the `compoisson` package is used to determine the quantile values from the CMP distribution and the function `qgammaAlt()` in the `EnvStats` package is used to determine quantile values from the gamma distribution. For the above choices of the parameters, the 95% PI for S is obtained for the independent and dependent compound regression models under three choices of sample size (m), namely, $m = 25, 50$ and 100. The actual S observations, denoted by $\vec{s} = (s_1, s_2, \ldots, s_m)'$, are computed by $s_k = n_k \bar{y}_k, k = 1, 2, \ldots, m$.

The proportion of $\vec{s}$ lying within its PI is presented in Table 1 for the various choices of m and ϕ. Additionally, the plots of the corresponding prediction bands are displayed in Tables 2 and 3. From Table 1, it can be observed that, for the choices of the covariate and coefficients considered, the proportion is large for $\phi = 1.6$ in the independent compound regression model and for $\phi = -1.6$ in the dependent compound regression model.

Table 1. Proportion of S lying in its respective PIs.

m	ϕ	Independent Model	Dependent Model
	-1.6	0.6667	0.9444
25	0	0.7777	0.8333
	1.6	0.8400	0.8400
	-1.6	0.7353	0.8529
50	0	0.6500	0.7000
	1.6	0.7656	0.8297
	-1.6	0.6615	0.9077
100	0	0.7088	0.9493
	1.6	0.7777	0.9393

Table 2. Prediction bands of independent compound regression model for over-, equi- and under-dispersed data.

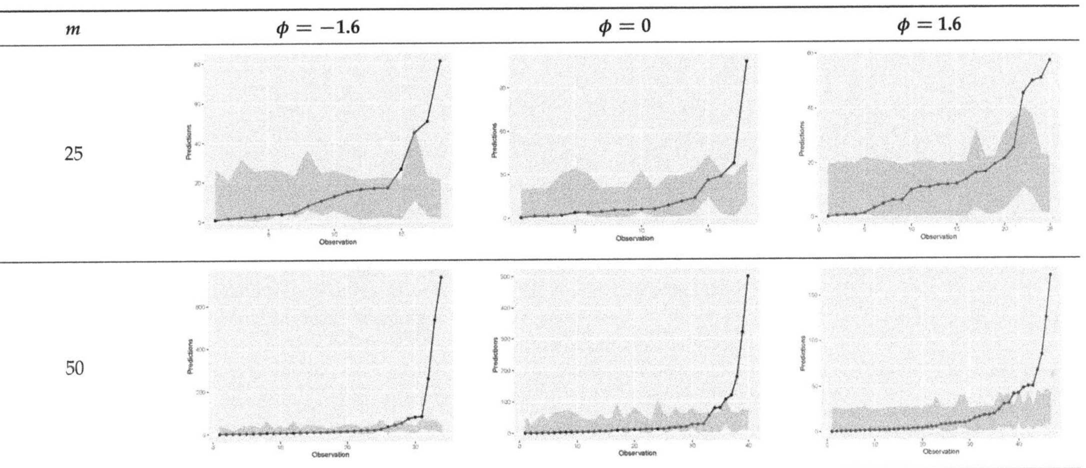

Table 2. *Cont.*

m	$\phi = -1.6$	$\phi = 0$	$\phi = 1.6$
100			

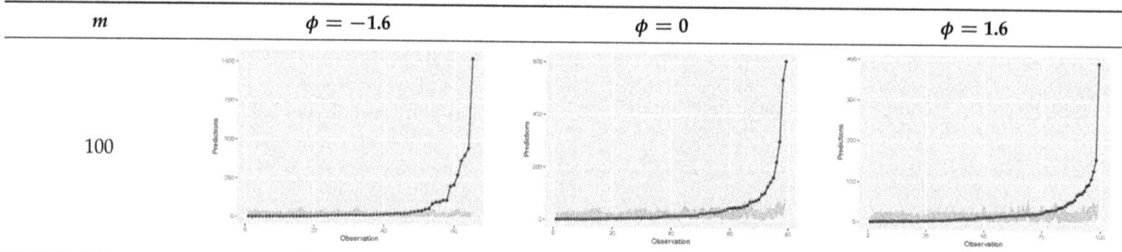

Table 3. Prediction bands of dependent compound regression model for over-, equi- and under-dispersed data.

m	$\phi = -1.6$	$\phi = 0$	$\phi = 1.6$
25			
50			
100			

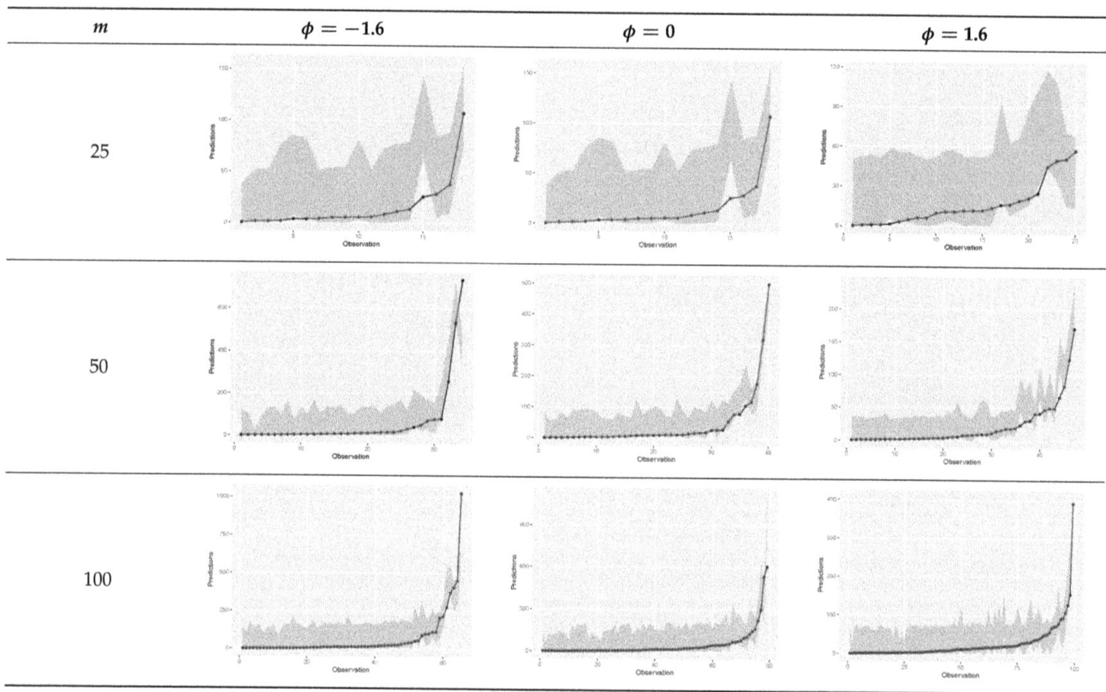

5.2. Real-Life Application

In this section, the proposed two-part methodology to obtain the PI for the compound CMP gamma regression is applied to real-life vehicle insurance claims data. The dataset pertains to the average damage claims for privately owned and insured vehicles in Britain in the year 1975. See Dutang and Charpentier [19]. It consists of 128 observations on five variables, namely, the owner's age (X_1), car age (X_2), model (X_3), number of claims (N) and average claim amount ($\bar{Y}$) in pounds. The variable X_1 consists of eight categories of age group; the variable X_2, four categories of car age; and the variable X_3, four categories of model. The aggregate claim amount (S) for each observation is obtained by multiplying the average claim amount by the number of claims. A dispersion test on N, performed using the function `dispersiontest()` available in R under AER package, resulted in a dispersion index of 119.8246 and a p-value of 2.091×10^{-6}, indicating that N is over-dispersed. Similarly, the Kolmogorov–Smirnov test on $\bar{Y}$ yielded a p-value of 0.7191 to assess the goodness-of-fit of the gamma distribution. As a result, the CMP distribution can be used to model N, whereas the gamma distribution can be used to model $\bar{Y}$. To implement the proposed estimation methodology and validate its performance, 80% of the observations are randomly chosen as training data and the rest 20% as test data. The observations in the training data are used to fit the independent and dependent compound regression models. The owner's age, car age and car model are the considered covariates in the model. The in-built functions `cmp()` function in `cmpreg` package and the `glm()` function are used to obtain the estimates of CMP and gamma regression models, respectively. The estimates of the regression parameters, their corresponding p-values (in parenthesis) and the AIC values are given in Table 4. Using the AIC values for the CMP and gamma regression models, the combined AIC values for the compound regression models are obtained as 2110.31 and 2108.31, respectively. For each observation in the test data, the PI for S is computed using the estimates of the fitted model. The corresponding prediction band of the independent and dependent compound regression model is displayed in Figure 1. From this figure, it can be noted that some observations do not fall within the prediction band. One reason for this is that these observations have large claim frequencies when compared with the other observations, and the corresponding limits of the PI based on the CMP regression are also large. As a result, the limits of the PI of such observations deviate from their observed values. The proportion of observed S in the test data lying within its PI is found to be 0.4782 and 0.6956 for the independent and dependent compound regression models, respectively. Based on the combined AIC values and the proportions, it can be inferred that the dependent compound regression model provides a relatively better fit for modeling the aggregate claim amount.

Table 4. Parameter estimates, p-values and AIC for the CMP and gamma regression models for the real-life data.

Covariates	CMP Regression Model	Gamma Regression Model (Independent Case)	Gamma Regression Model (Dependent Case)
(Intercept)	1.5007 ($< 2 \times 10^{-16}$)	5.7421 ($< 2 \times 10^{-16}$)	5.7754 ($< 2 \times 10^{-16}$)
OwnerAge21–24	1.5885 ($< 2 \times 10^{-16}$)	−0.2010 (0.0670)	−0.1800 (0.0964)
OwnerAge25–29	2.6237 ($< 2 \times 10^{-16}$)	−0.1129 (0.2705)	−0.0497 (0.6357)
OwnerAge30–34	2.7585 ($< 2 \times 10^{-16}$)	−0.3276 (0.0034)	−0.2542 (0.0262)
OwnerAge35–39	2.8854 ($< 2 \times 10^{-16}$)	−0.3150 (0.0047)	−0.2271 (0.0496)
OwnerAge40–49	3.5362 ($< 2 \times 10^{-16}$)	−0.2722 (0.0081)	−0.1140 (0.3528)
OwnerAge50–59	3.3678 ($< 2 \times 10^{-16}$)	−0.1854 (0.0843)	−0.0590 (0.6219)
OwnerAge60+	3.0280 ($< 2 \times 10^{-16}$)	−0.3054 (0.0036)	−0.2120 (0.0553)
ModelB	1.0255 ($< 2 \times 10^{-16}$)	0.0584 (0.4260)	0.1414 (0.0877)
ModelC	0.6930 ($< 2 \times 10^{-16}$)	0.1083 (0.1387)	0.1500 (0.0450)
ModelD	−0.1889 (0.00485)	0.4041 (6.01×10^{-7})	0.3762 (2.40×10^{-6})
CarAge10+	−1.9174 ($< 2 \times 10^{-16}$)	−0.8138 ($< 2 \times 10^{-16}$)	−0.9494 (5.87×10^{-16})
CarAge4–7	−0.1558 (6.65×10^{-5})	−0.0615 (0.3959)	−0.0727 (0.3089)

Table 4. *Cont.*

Covariates	CMP Regression Model	Gamma Regression Model (Independent Case)	Gamma Regression Model (Dependent Case)
CarAge8–9	-1.4876 $(< 2 \times 10^{-16})$	-0.4188 (8.64×10^{-8})	-0.5323 (2.02×10^{-8})
NClaims	-	-	-0.0010 (0.0301)
$\hat{\phi}$	-0.8374 $(< 2 \times 10^{-16})$	-	-
$\hat{\psi}$	-	0.0667	0.0644
AIC	984.7148	1125.6	1123.6

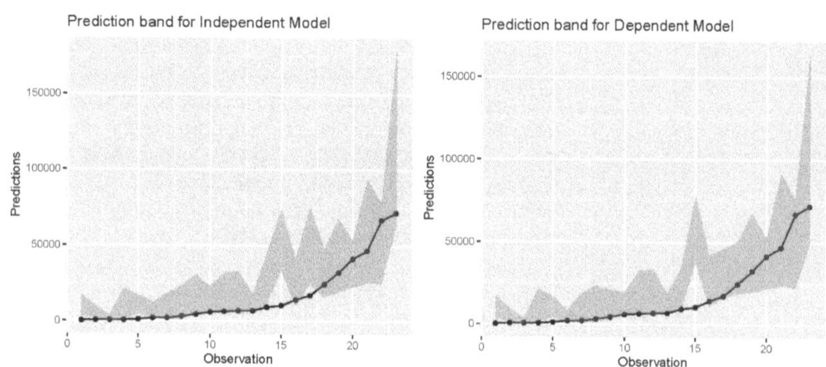

Figure 1. Prediction band for the test data under independent model and dependent model.

6. Conclusions

The Poisson distribution is generally used in compound regression models as the counting distribution. In practice, the Poisson distribution's equi-dispersion assumption is frequently violated. The methodology presented in this paper provided a way to handle non-equi-dispersed count data in the context of compound regression models by using the CMP distribution. The proposed compound regression model can be used when the count data are over- or under-dispersed. The estimation of the parameters was carried out using a two-part GLM approach for the independent and dependent compound regression models. This approach is less complex and provides separate estimates for the count and the continuous distribution involved in the model. Since, in practice, knowledge of the actual value of the response variable rather than its predicted value is more useful, a methodology to obtain the prediction interval of the response variable was proposed. An application of the two-part GLM method to real-life data revealed that the dependent compound regression model performs relatively better than the independent compound regression model. Thus, in practice, one can start with the dependent compound regression model and look for the significance of the count variable in the model. If the count variable is found to be not significant, then the independent compound regression model can be used. To conclude, the proposed compound CMP regression model could be an alternative to modeling a compound random variable when the count data are not equi-dispersed.

Author Contributions: J.M. has contributed to the conceptualization, methodology, mathematical derivation and simulation. V.S.V. and C.C. have contributed equally to mathematical derivation and original draft preparation. All authors have read and agreed to the published version of the manuscript.

Funding: This research received no external funding.

Conflicts of Interest: The authors declare no conflict of interest.

References

1. Gómez-Déniz, E.; Pérez-Rodríguez, J.V. Modelling distribution of aggregate expenditure on tourism. *Econ. Model.* **2019**, *78*, 293–308. [CrossRef]
2. Klugman, S.A.; Panjer, H.H.; Willmot, G.E. *Loss Models: From Data to Decisions*; John Wiley & Sons: New York, NY, USA, 2012; Volume 715.
3. Bahnemann, D. *Distributions for Actuaries*; Casualty Actuarial Society: Arlington, VA, USA, 2015; Volume 2.
4. Jørgensen, B.; Paes De Souza, M.C. Fitting Tweedie's compound Poisson model to insurance claims data. *Scand. Actuar. J.* **1994**, *1994*, 69–93. [CrossRef]
5. Consul, P.C.; Jain, G.C. A generalization of the Poisson distribution. *Technometrics* **1973**, *15*, 791–799. [CrossRef]
6. Shmueli, G.; Minka, T.P.; Kadane, J.B.; Borle, S.; Boatwright, P. A useful distribution for fitting discrete data: Revival of the Conway-Maxwell-Poisson distribution. *J. R. Stat. Soc. Ser. (Appl. Stat.)* **2005**, *54*, 127–142. [CrossRef]
7. Conway, R.W.; Maxwell, W.L. A queuing model with state dependent service rates. *J. Ind. Eng.* **1962**, *12*, 132–136.
8. Sellers, K.F.; Borle, S.; Shmueli, G. The COM-Poisson model for count data: A survey of methods and applications. *Appl. Stoch. Model. Bus. Ind.* **2012**, *28*, 104–116. [CrossRef]
9. Sellers, K.F.; Premeaux, B. Conway-Maxwell-Poisson regression models for dispersed count data. *Wiley Interdiscip. Rev. Comput. Stat.* **2021**, *13*, e1533. [CrossRef]
10. Saavithri, V.; Priyadharshini, J.; Banu, Z.P. Compound COM-Poisson Distribution with Binomial Compounding Distribution. Acailable online: https://www.internationaljournalssrg.org/uploads/specialissuepdf/ICRMIT/2018/MTT/ICRMIT-P122.pdf (accessed on 15 January 2023).
11. Frees, E.W.; Gao, J.; Rosenberg, M.A. Predicting the frequency and amount of health care expenditures. *N. Am. Actuar. J.* **2011**, *15*, 377–392. [CrossRef]
12. Andersen, D.A.; Bonat, W.H. Double generalized linear compound Poisson models to insurance claims data. *Electron. J. Appl. Stat. Anal.* **2017**, *10*, 384–407.
13. Delong, Ł.; Lindholm, M.; Wüthrich, M.V. Making Tweedie's compound Poisson model more accessible. *Eur. Actuar. J.* **2021**, *11*, 185–226. [CrossRef]
14. Ribeiro, E.E., Jr.; Zeviani, W.M.; Bonat, W.H.; Demétrio, C.G.; Hinde, J. Reparametrization of COM-Poisson regression models with applications in the analysis of experimental data. *Stat. Model.* **2020**, *20*, 443–466. [CrossRef]
15. Jorgensen, B. *The Theory of Dispersion Models*; CRC Press: Boca Raton, FL, USA, 1997.
16. De Jong, P.; Heller, G.Z. *Generalized Linear Models for Insurance Data*; Cambridge University Press: Cambridge, UK, 2008.
17. Garrido, J.; Genest, C.; Schulz, J. Generalized linear models for dependent frequency and severity of insurance claims. *Insur. Math. Econ.* **2016**, *70*, 205–215. [CrossRef]
18. Ribeiro, E.E., Jr. *Cmpreg: Reparametrized COM-Poisson Regression Models*; R Package Version 0.0.1. Available online: https://rdrr.io/github/JrEduardo/cmpreg/ (accessed on 15 January 2023).
19. Dutang, C.; Charpentier, A. *CASdatasets: Insurance Datasets*; 2019. R Package Version 1.0-11. Available online: http://cas.uqam.ca/ (accessed on 15 January 2023).

Mathematical and Computational Applications

Article

Computation of the Distribution of the Sum of Independent Negative Binomial Random Variables

Marc Girondot [1,*] and Jon Barry [2]

[1] Laboratoire Écologie, Systématique et Évolution, Université Paris-Saclay, CNRS, AgroParisTech, 91190 Gif-sur-Yvette, France

[2] Lowestoft Laboratory, Centre for Environment, Fisheries and Aquaculture Science, Pakefield Road, Lowestoft, Suffolk NR33 0HT, UK

* Correspondence: marc.girondot@universite-paris-saclay.fr

Abstract: The distribution of the sum of negative binomial random variables has a special role in insurance mathematics, actuarial sciences, and ecology. Two methods to estimate this distribution have been published: a finite-sum exact expression and a series expression by convolution. We compare both methods, as well as a new normalized saddlepoint approximation, and normal and single distribution negative binomial approximations. We show that the exact series expression used lots of memory when the number of random variables was high (>7). The normalized saddlepoint approximation gives an output with a high relative error (around 3–5%), which can be a problem in some situations. The convolution method is a good compromise for applied practitioners, considering the amount of memory used, the computing time, and the precision of the estimates. However, a simplistic implementation of the algorithm could produce incorrect results due to the non-monotony of the convergence rate. The tolerance limit must be chosen depending on the expected magnitude order of the estimate, for which we used the answer generated by the saddlepoint approximation. Finally, the normal and negative binomial approximations should not be used, as they produced outputs with a very low accuracy.

Keywords: negative binomial distribution; computation; R package; sum of negative binomial variables

Citation: Girondot, M.; Barry, J. Computation of the Distribution of the Sum of Independent Negative Binomial Random Variables. *Math. Comput. Appl.* **2023**, *28*, 63. https://doi.org/10.3390/mca28030063

Academic Editor: Sandra Ferreira

Received: 7 February 2023
Revised: 13 April 2023
Accepted: 26 April 2023
Published: 28 April 2023

1. Introduction

The negative binomial (NB) distribution is a discrete probability distribution that models counts [1]. It widely used in statistics, from statistics of accidents [2] to animal counts [3]. The NB distribution can be used to describe the distribution of the number of successes or failures. Suppose that there is a sequence of independent Bernoulli trials, each trial having two potential outcomes called "success" and "failure". The probability of success is p and of failure $q = 1 - p$. We observe this sequence until a predefined number, r, of successes has occurred. Then, the random number of failures has the NB distribution $X \sim NB(r; p)$ with density $P(X = x)$, x being a particular realization of X:

$$P(X = x) = \frac{(x + r - 1)!}{x!(r - 1)!} \, p^r (1 - p)^x \tag{1}$$

with $0 < p < 1$, x and r being integer > 0. The mean is $\mu = r\,(1 - p)/p$.

The moment-generating function of the NB distribution is:

$$M(t) = \left(\frac{q}{1 - p\,e^t}\right)^r \tag{2}$$

An alternative parametrization $X \sim NB(\mu, \theta)$ can also be derived from assuming that the mean parameter of a Poisson distribution has a gamma distribution:

$$P(X = x) = \frac{\Gamma(x + \theta)}{x! \Gamma(\theta)} \left(\frac{\theta}{\mu + \theta} \right)^{\theta} \left(\frac{\mu}{\mu + \theta} \right)^{x} \tag{3}$$

with $\mu > 0$ and $\theta > 0$. Note that θ is not necessarily an integer, contrary to r in (1); hence, the gamma function is used in (3) instead of a factorial, with $\Gamma(x) = (x - 1)!$. The variance of the NB distribution is $\mu (1 + \mu/\theta)$. As θ approaches infinity, the NB distribution tends to follow the Poisson distribution, with the mean μ.

1.1. Sum of Negative Binomials

The sum of independent NB variables is of special interest in different contexts, such as the study of animal distribution [4,5], fecal egg counts in infected goats [6], the number of emergency medical calls [7], empirical distribution of the duration of wet periods in days [8] or insurance risk [9]. When the sum of several independent NB counts is available, determining the distribution of $\sum X_i$ with $X_i \sim NB(r_i; p_i)$ is a problem. When the p_is are all the same and equal to p, a classical result is $\sum X_i \sim NB(\sum r_i; p)$ [10], but more general forms without this constraint are often needed. For example, if counts are available for various spatial or temporal units of the form $X \sim NB(\mu_i; r_i)$, p_i being $r_i/(\mu_i + r_i)$, it implies that the p_is are not all the same, because μ_i varies among the units [4].

With the mean and variance of the $NB(r; p)$ distribution being $r(1 - p)/p$ and $r(1 - p)/p^2$, respectively, it follows that the mean and variance of the sum of n-independent NB variables are respective:

$$mean(S_n) = \sum_{i=1}^{n} (r_i(1 - p_i)/p_i) \text{ and } var(S_n) = \sum_{i=1}^{n} \left(r_i(1 - p_i)/p_i^2 \right) \tag{4}$$

Our paper has developed some novel methods in relation to the practical computation and use of the convolution approach [9]. However, the paper also collects five different methods and presents them in one place, a useful resource for the working data scientist or statistician. We describe and reference these methods and outline the computational difficulties in getting them to work. We also point the reader to the freely available R software that implements each of the methods (plus a sixth method based purely on simulation).

Two methods have been published to estimate the distribution of the sum of NB independent variables using a finite-sum exact expression [11] or the convolution method [9]. However, the computer implementation of both methods was not available, and we have detected potential problems when a practitioner implements them. The finite-sum exact expression computer implementation is relatively straightforward, but memory overflow can occur, and the time of computing will increase as a function of the factorial of the number of observations, x. This precision was not given in the original publication [11]. The convolution method is very complex to implement and has been described as being "cumbersome" [12]; indeed, we found that its implementation was not straightforward and was even counterintuitive. The method uses a sum to infinity, and the condition to stop the recursion was not defined in the original publication.

Our solution for computation of the convolution method, presented here, is novel and has proven to be robust for extensive testing. A naïve tolerance condition has been used by one of the authors of this note (MG) (recursion stops when the change is lower than the tolerance limit) as in [4,5], but the other author (JB) found that outputs can be strongly biased in some conditions. It has been the beginning of a collaboration between the two authors to understand and solve the origin of this bias. We detected two problems: (1) the tolerance check must be applied only when the first-order change of the estimate is negative (convergence criteria being adaptive), and (2) the value of the tolerance must be proportional to the expected estimate. Then, it was necessary to have an estimate of the density to set the tolerance, to better estimate the correct density. To solve this, we used the saddlepoint approximation of the density. We show that the absolute error of this approximation can be on the order of 5%, being too high to be used in many applications, but it is sufficiently low to define a correct tolerance to be used with the convolution method.

1.2. Normal and Negative Binomial Approximations

When working on the sum of variables, the first thought is to use the central limit theorem [13] that establishes that, in many situations, the distribution of the sum-independent random variables tends to go toward a normal distribution. An alternative is to model the distribution of the sum of NB variables, as an NB distribution is based on the observation that the distribution of the sum of NB variables is a mixture NB distribution [9], according to Theorem 2, proposed by Makun, Abdulganiyu, Shaibu, Otaru, Okubanjo, Kudi, and Notter [6].

1.3. Finite-Sum Exact Expression

An exact form for the distribution of the sum of NB is:

$$P(S_n = x) = \sum_{\mu_1 + \cdots + \mu_n = x} \prod_{j=1}^{n} \frac{\Gamma(\mu_j + \theta)}{\mu! \Gamma(r\theta_j)} p_j^{\theta_j} q_j^{\mu_j} \tag{5}$$

The expression (5) is compact and the exact value can be computed [11].

1.4. Approximation by Convolution

When $X_i \sim NB(r_i; p_i)$, with i from 1 to n, the distribution of $S_n = \sum X_i$ is a mixture NB [9], with the probability mass function being approximated by:

$$P(S_n = x) = R \sum_{k=0}^{\infty} \delta_k \frac{\Gamma(r + x + k)}{\Gamma(r + k)x!} M_1^{r+k}(1 - M_1)^x, \ x = 0, \, 1, \, 2, \, \cdots \tag{6}$$

where $r = \sum_{i=1}^{n} r_i$, and $M_1 = max_j(p_j)$.

$$R = \prod_{j=1}^{n} \left(\frac{q_j M_1}{(1 - M_1)p_j} \right)^{-r_j}$$

and $\delta_{k+1} = \frac{1}{k+1} \sum_{i=1}^{k+1} i \, \xi_i \, \delta_{k+1-i}$, $k = 0, \, 1, \, \ldots$ with $\delta_0 = 1$ and

$$\xi_i = \sum_{j=1}^{n} \frac{r_j \left(1 - (1 - M_1)p_j / q_j M_1\right)^i}{i}$$

Expression (6) is used iteratively, with k being the counter of the rank of iterations, but a condition to stop the iterations when a certain level of approximation is reached was not defined in the original publication [9].

1.5. Saddlepoint Approximation

The saddlepoint approximation method provides a highly accurate approximation formula for any probability density function (continuous distribution) or probability mass function (discrete distribution) of a distribution, based on the moment-generating function [14].

Taking the log of the moment-generating function of the NB distribution (2) and summing over n-independent NB variables, the cumulant of sum of NBs is:

$$K(t) = \sum_{i=1}^{n} r_i \left(log(q_i) - log(1 - p_i \, e^t)\right)$$

Or $K(t) = \sum_{i=1}^{n} \theta_i \left(log(\theta_i) - log(\theta_i + \mu_i(1 - e^t))\right)$ \tag{7}

The first and second order of the derivatives of $K(t)$ are:

$$K'(t) = \sum_{i=1}^{n} \frac{\theta_i \, \mu_i \, e^t}{\theta_i + \mu_i(1 - e^t)} \tag{8}$$

$$K''(t) = \sum_{i=1}^{n} \frac{\theta_i \, \mu_i \, (\theta_i + \mu_i) \, e^t}{(\theta_i + \mu_i(1 - e^t))^2} \tag{9}$$

The saddlepoint, s_x, is found by solving $K'(s_x) = x$. Once s_x is found, $P(S_n = x)$ can be approximated by:

$$P(S_n = x) \approx \frac{1}{\sqrt{2\pi \, K''(s_x)}} \, e^{(K(s_x) - x \, s_x)} \tag{10}$$

The value $P(S_n = x)$ is normalized to ensure that $\sum P(S_n) = 1$.

In the remainder of this note, we describe the computational problems that applied statisticians or practitioners face in implementing the distribution of the sum of NB-independent variables using finite-sum exact expression [11], the convolution method [9], saddlepoint approximation, or the approximation by normal and NB distributions. We describe how these have been overcome in the publicly available R package (HelpersMG package version 5.9 and higher (https://CRAN.R-project.org/package=HelpersMG, accessed on 6 February 2023). The code can be checked after loading this package with the command ?dSnbinom.

2. Computations

Figure 1 gives two examples of the sum S of independent NB random variables, and how these distributions are approximated using the four methods (convolution, saddlepoint, single normal, single NB) outlined in this note. In (A), we use $n = 10$, $j = 1 \ldots$ n, $p_j = 0.4 + \frac{j}{10}$ and $r_j = j \times 10$, and in (B) $n = 2$, $p_j = \frac{j}{10}$, and $r_j = j$.

2.1. Normal and Negative Binomial Approximations

When n is large and standard deviation is small as compared to the mean, the normal approximation with $P(S_n = x) = \int_{x-0.5}^{x+0.5} \mathcal{N}(\mu, \sigma)$, where $\mathcal{N}(\mu, \sigma)$ is the normal probability density function with $\mu = mean(S_n)$ and $\sigma = \sqrt{var(S_n)}$ can be used as an approximation for the distribution of the sum of independent NB random variables (Figure 1A). However, for a small n or large standard deviation, as compared to the mean (corresponding to a highly skewed distribution), the approximation can be very poor (Figure 1B). The NB distribution modeled with the probability density function, $NB(\mu, \theta)$, such that $\mu = mean(S_n)$ and $\theta = mean(S_n)^2 / (var(S_n) - mean(S_n))$, better fits the exact distribution of the sum of NB variables, but still with a bias (Figure 1B). This confirms that the distribution of the sum of independent NB variables is not an NB, as wrongly stated in [6]. It is, rather, a mixture NB (see below) [9]. In summary, the normal and NB approximations generate the highest errors (>30% in some cases) and they should not be used, especially as there are better alternatives.

2.2. Finite-Sum Exact Expression

This method permits the calculation of the exact value for $P(S_n = x)$. It will therefore be used as a reference here.

For the finite form exact expression method [11], a table of n columns with all the combinations of integers, from 0 to x, that produce a sum of x ($m_1 + \cdots + m_n = x$), must be first established. The number of different ways to distribute x-indistinguishable objects into n-distinguishable categories is $C(x + n - 1, n - 1)$. This is the memory-consuming part of the Vellaisamy and Upadhye [11] method. The density $P(X = x)$ in Equation (1) is calculated n times for each of these combinations in the final table (the $\prod_{j=1}^{n}$ part of Equation (5)). This is the computationally time-consuming part of the method.

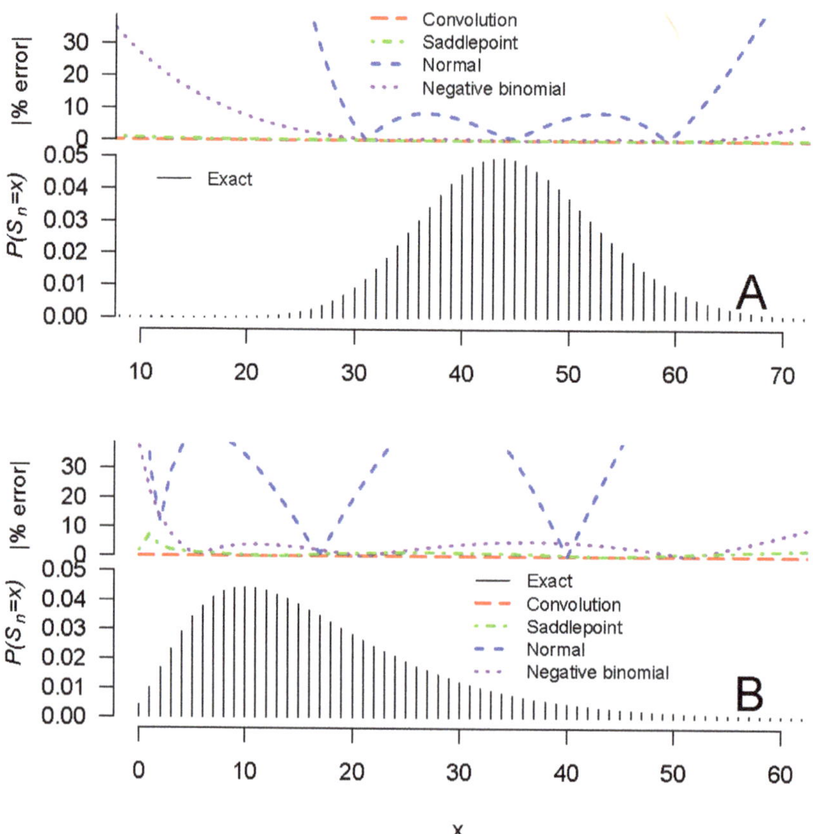

Figure 1. Sum of independent NB distributions approximated with convolution, saddlepoint approximation, normal, and single NB distribution. (**A**) $n = 10$, $j = 1 \ldots n$, $p_j = 0.4 + \frac{j}{10}$ and $r_j = j \times 10$; $mean(S_n) = 183.92$, $var(S_n) = 270.75$. (**B**) $n = 2$, $p_j = \frac{j}{10}$ and $r_j = j$; $mean(S_n) = 17$, $var(S_n) = 130$. The bar plots show the exact distribution, and the top graphs show the absolute % of error of the approximation.

When n and/or x are large, this method requires lot of iterations. For example, there are 1,081,575 different combinations of 17 objects in nine categories. Then, Equation (1) must be applied 9,734,175 times to estimate $P(S_n = x)$ when using Equation (5).

2.3. Approximation by Convolution

The coefficients of Equation (6) are iteratively defined, and we rewrite the published formula to make the computation more efficient using the recursion:

$$W(S_n = x)_0 = \frac{\Gamma(r+x)}{\Gamma(r)x!} M_1^{\,r}(1 - M_1)^x$$

$$W(S_n = x)_{k+1} = W(S_n = x)_k + \delta_{k+1} \frac{\Gamma(r+x+k+1)}{\Gamma(r+k+1)x!} M_1^{\,r+k+1}(1 - M_1)^x$$

$$P(S_n = x)_k = R\, W(S_n = x)_k \tag{11}$$

Intermediate estimates in (11) used log of expressions to prevent a computing overflow. The conditions to stop the iterations were not defined in Furman's original publication.

A typical method in such situations is to stop the recursion when the change in the final output is below a defined tolerance. However, it cannot be used in the context of Equations (6) or (11), because, at the beginning of iterations, the change in $P(S_n = x)$ is sometimes so small that recursions will be stopped immediately and the resulting probability $P(S_n = x)$ will be biased. An example of this is shown in Figure 2A, which shows the value of $P(S_7 = 6)_k$ ($n = 7, j = 1 \dots n$, $p_j = j/10$, and $r_j = j$) as a function of the recursive iterations k from Equation (6). For the first eight iterations, the change in $P(S_7 = 6)_k$ is very small. To alleviate this problem, many iterations can be used, but without being sure that they are sufficient, and it is done at the expense of running time. This solution was chosen with at least 1000 iterations in [5], but this number of iterations is not always large enough to ensure a correct estimate when n or x are large.

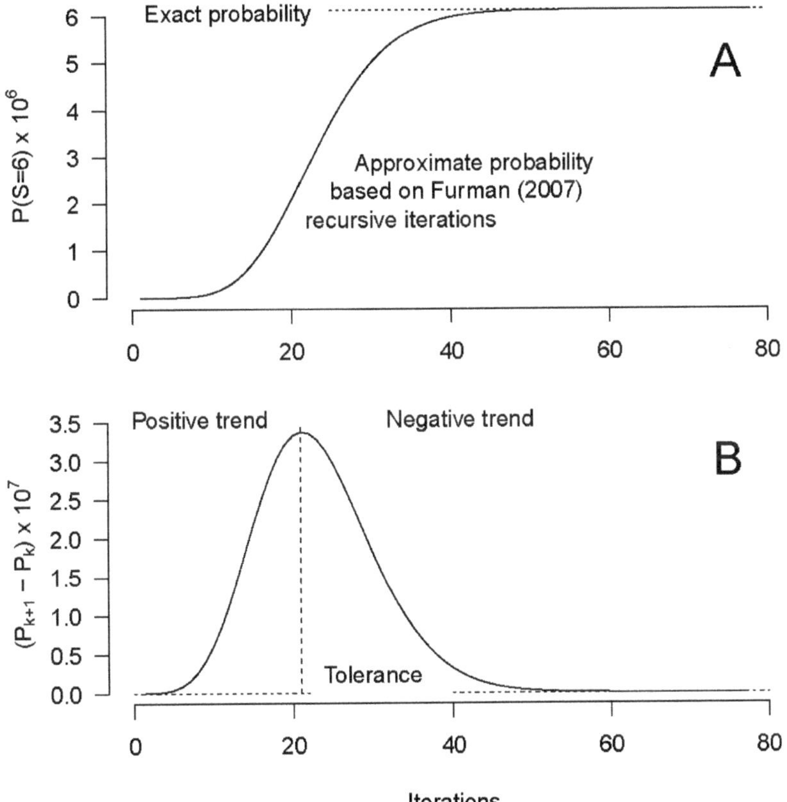

Figure 2. (**A**) Dynamics of the convergence of $P(S_7 = 6)_k$ with $n = 7$, $j = 1 \dots n$, $p_j = j/10$, and $r_j = j$ using Equation (6), with k being the rank of iterations. (**B**) Trend of the changes in P_k with tolerance $= 10^{-12}$.

A better approach came from the study on the trend of the rate of change of $P(S_n = x)$ according to the rank of iteration k: $P_k - P_{k-1}$ vs. $P_{k+1} - P_k$, where P_k denotes the value of $P(S_n = x)$ at iteration k. In its initial phase, the rate of change of P is positive, with $P_k - P_{k-1} > P_{k+1} - P_k$, then it shows a peak and becomes negative, with $P_k - P_{k-1} < P_{k+1} - P_k$ (Figure 2B). The tolerance threshold must be used only after the occurrence of the peak to ensure that the phase of rapid change of P is reached. The number of iterations before the peak is dependent on the values of n, x, r_i and p_i, and cannot be easily anticipated at the beginning of the iterations. We have therefore developed an adaptative strategy to stop the recursion when two conditions are met: the rate of change of $P(S_n = x)$ is negative, and the change of $P(S_n = x)$ is less than the user-defined tolerance. The tolerance value must be lower than $P(S_n = x)$ or the output will be biased. As an example, if μ = (0.01, 0.02, 0.03) and θ = (2, 2, 2), then $P(S_3 = 20) = 7.73139 \times 10^{-35}$ using the exact method. If the Furman method is used with the tolerance set to 10^{-12}, $P(S_3 = 20) = 3.879379 \times 10^{-35}$, which is two times lower than the exact answer. The solution is to define a tolerance much lower than the anticipated results, for example, here, with the tolerance being 10^{-45}, $P(S_3 = 20) = 7.73139 \times 10^{-35}$, which is the correct probability. This can be done using the saddlepoint estimate (see below).

The comparison of the results obtained by Equation (6), with an adaptative stopping of the recursion and tolerance setup, using saddlepoint approximation (see below) and Equation (5), are shown in Table 1 with the corresponding computing time. This table is similar to those used in Tables 1 and 2 in Vellaisamy and Upadhye [11].

Table 1. Comparison of accuracy and computing time of the sum of n numbers x = 3, 5, 8, 10, and 15 obtained from NB distributions with j = 1 … n, p_j = j/10, r_j = j, and n from 2 to 7 based on Equations (5), (11), and (10). For each (n, x) combination in (**A**), the top number is the probability $P(S_n = x)$ and the bottom number is the number of iterations. In (**B**), the number of recursions required to stabilize $P(S_n = x)$ is shown. The $P(S_n = x)$ values are exactly the same as those in (**A**) and are not shown. In (**C**), the $P(S_n = x)$ values for saddlepoint approximation are shown. Computing times for the different methods are shown at the right of each table. The code for Equation (5) was parallelized on an 8-core computer in R 4.2.3 and HelpersMG package version 5.9 (https://CRAN.R-project.org/package=HelpersMG, accessed on 6 February 2023).

A: Vellaisamy and Upadhye [11]: Exact Probabilities						No Parallel	Parallel 8-Core
	$x = 3$	$x = 5$	$x = 8$	$x = 10$	$x = 15$	Time (s)	Time (s)
$n = 2$	0.02320400	0.03403236	0.04283461	0.04425234	0.03856123	0.001	0.011
	16	36	81	121	256		
$n = 3$	0.00273650	0.00730772	0.01724312	0.02421915	0.03607386	0.003	0.011
	40	126	405	726	2176		
$n = 4$	0.00020980	0.00094784	0.00408465	0.00785680	0.02099302	0.014	0.012
	80	336	1485	3146	13,056		
$n = 5$	0.00001503	0.00010490	0.00076597	0.00196540	0.00920145	0.062	0.015
	140	756	4455	11,011	62,016		
$n = 6$	0.00000131	0.00001291	0.00014555	0.00047692	0.00365038	0.249	0.023
	224	1512	11,583	33,033	248,064		
$n = 7$	0.00000017	0.00000218	0.00003427	0.00013604	0.00154413	0.906	0.049
	336	2772	27,027	88,088	868,224		

B: Furman [9]: Convolution					$\mathrm{Tol} = P_{saddlepoint}(S_n = x) \times 10^{-10}$	
	$x = 3$	$x = 5$	$x = 8$	$x = 10$	$x = 15$	Time (s)
$n = 2$	13	14	15	16	18	0.007
$n = 3$	19	20	23	24	27	0.008
$n = 4$	27	29	32	34	38	0.009
$n = 5$	39	42	45	48	54	0.009
$n = 6$	58	62	67	70	79	0.009
$n = 7$	92	97	104	109	122	0.011

C: Normalized Saddlepoint Approximation						
	$x = 3$	$x = 5$	$x = 8$	$x = 10$	$x = 15$	Time (s)
$n = 2$	0.02372254	0.03448835	0.04314218	0.04442429	0.03841261	0.007
$n = 3$	0.00283042	0.00748306	0.01754862	0.02458058	0.03637448	0.007
$n = 4$	0.00021836	0.00097613	0.00418037	0.00802118	0.02132508	0.008
$n = 5$	0.00001571	0.00010840	0.00078653	0.00201341	0.00938611	0.008
$n = 6$	0.00000137	0.00001337	0.00014977	0.00048960	0.00373283	0.008
$n = 7$	0.00000018	0.00000226	0.00003531	0.00013984	0.00158133	0.018

2.4. Saddlepoint Approximation

The saddlepoint approximation (we used the Brent' algorithm [15] for the minimization needed to find the saddlepoint) is computationally fast. However, the estimate must be normalized so that the density function sums to 1 [16]. The normalization used the sum of $P(S_n = x)$, with x from 0 to $mean(S_n) + Max\sqrt{var(S_n)}$, with $mean(S_n)$ and $var(S_n)$ from Equation (4) and $Max = 20$. A test was performed to ensure that $P\left(S_n = mean(S_n) + Max\sqrt{var(S_n)}\right)$ was 0 or that the Max was increased until this condition was reached. The relative difference between the exact value of $P(S_n = x)$ and the saddlepoint approximation can be sometimes of the order of 5% (Figure 1). On the other hand, this approximation is good enough to set the tolerance of the approximation by convolution, using a tolerance equal to $P_{saddlepoint}(S_n = x) \times 10^{-10}$.

The tolerance value to cut the iterations for an approximate Furman [9] method must be of the same order as the value of $P(S_n = x)$, multiplied by the tolerance and set at the value of 10^{-10}, to have an estimate precise up to the 10th digit. The difficulty is that $P(S_n = x)$ needs to be estimated here. The chosen solution was to use a rough estimate of $P(S_n = x)$ from the very fast saddlepoint approximation method first. This approach proved to be very efficient, because the estimates of the approximate Furman [9] method are exactly the same as for the exact method (Table 1A).

Equation (5) has the advantage that it is parallelizable, but for a large n and x (see Table 1A), doing so requires a large number of both iterations and memory. Equations (6) and (11), however, are not disadvantaged by these problems. Vellaisamy and Upadhye [11] indicated that Equation (5) required less computing time than Equation (6), even for n = 7 and n = 15. This would be true only if the authors used a very large number of iterations to stop the iterations in Equation (6), or if their conditions to stop the iterations were sub-optimal.

As a general conclusion, we consider that the approximate form of distribution for the sum of independent NB proposed by Furman [9] should be used in all the contexts, whatever the parameters n, x, p_i or r_i. The tolerance can be approximated by using the value of $P(S_n = x)$, estimated using the saddlepoint approximation method. This solution is used by default in the R package HelpersMG (version > 5.9), available in CRAN: The Comprehensive R Archive Network [17].

Author Contributions: Conceptualization, M.G. and J.B.; software, M.G. and J.B.; validation, M.G. and J.B.; writing, M.G. and J.B. All authors have read and agreed to the published version of the manuscript.

Funding: This research received no external funding.

Data Availability Statement: All methods used in this note, as well as an approximation using the generation of random numbers, are available in the functions dSnbinom, pSnbinom, qSnbinom, and rSnbinom in the R package HelpersMG (version > 5.9), available in CRAN: The Comprehensive R Archive Network [17].

Acknowledgments: We thank all four reviewers for their positive and helpful comments. In particular, one referee pointed us to the saddlepoint approach, of which we were previously unaware.

Conflicts of Interest: The authors declare no conflict of interest.

References

1. Fisher, R.A.; Corbet, A.S.; Williams, C.B. The relation between the number of species and the number of individuals in a random sample of an animal population. *J. Anim. Ecol.* **1943**, *12*, 42–58. [CrossRef]
2. Carlson, T. Negative binomial rationale. *Proc. Casualty Actuar. Soc.* **1962**, *49*, 177–183.
3. Power, J.H.; Moser, E.B. Linear model analysis of net catch data using the negative binomial distribution. *Can. J. Fish. Aquat. Sci.* **1999**, *56*, 191–200. [CrossRef]
4. Girondot, M. Optimizing sampling design to infer marine turtles seasonal nest number for low-and high-density nesting beach using convolution of negative binomial distribution. *Ecol. Indic.* **2017**, *81*, 83–89. [CrossRef]

5. Omeyer, L.C.M.; McKinley, T.J.; Bréheret, N.; Bal, G.; Balchin, G.P.; Bitsindou, A.; Chauvet, E.; Collins, T.; Curran, B.K.; Formia, A.; et al. Missing data in sea turtle population monitoring: A Bayesian statistical framework accounting for incomplete sampling. *Front Mar. Sci.* **2022**, *9*, 817014. [CrossRef]

6. Makun, H.J.; Abdulganiyu, K.A.; Shaibu, S.; Otaru, S.M.; Okubanjo, O.O.; Kudi, C.A.; Notter, D.R. Phenotypic resistance of indigenous goat breeds to infection with *Haemonchus contortus* in northwestern Nigeria. *Trop. Anim. Health Prod.* **2020**, *52*, 79–87. [CrossRef] [PubMed]

7. Lee, H.; Lee, T. Demand modelling for emergency medical service system with multiple casualties cases: K-inflated mixture regression model. *Flex. Serv. Manuf. J.* **2021**, *33*, 1090–1115. [CrossRef]

8. Korolev, V.; Gorshenin, A. Probability models and statistical tests for extreme precipitation based on generalized negative binomial distributions. *Mathematics* **2020**, *8*, 604. [CrossRef]

9. Furman, E. On the convolution of the negative binomial random variables. *Stat. Probab. Lett.* **2007**, *77*, 169–172. [CrossRef]

10. Johnson, N.; Kotz, S.; Kemp, A. *Univariate Discrete Distributions*, 2nd ed.; Wiley: New York, NY, USA, 1992.

11. Vellaisamy, P.; Upadhye, N.S. On the sums of compound negative binomial and gamma random variables. *J. Appl. Probab.* **2009**, *46*, 272–283. [CrossRef]

12. Baena-Mirabete, S.; Puig, P. Computing probabilities of integer-valued random variables by recurrence relations. *Stat. Probab. Lett.* **2020**, *161*, 108719. [CrossRef]

13. Laplace, P.-S. Mémoire sur les approximations des formules qui sont fonctions de très grands nombres, et sur leur application aux probabilités. *Mémoires Cl. Sci. Mathématiques Phys. L'institut Fr.* **1809**, *1809*, 353–415.

14. Daniels, H.E. Saddlepoint approximations in statistics. *Ann. Math. Stat.* **1954**, *25*, 631–650. [CrossRef]

15. Brent, R. *Algorithms for Minimization without Derivatives*; Prentice-Hall: Englewood Cliffs, NJ, USA, 1973.

16. Lugannani, R.; Rice, S. Saddle point approximation for the distribution of the sum of independent random variables. *Adv. Appl. Probab.* **2016**, *12*, 475–490. [CrossRef]

17. Girondot, M. *HelpersMG: Tools for Environmental Analyses, Ecotoxicology and Various R Functions*; The Comprehensive R Archive Network: Indianapolis, IN, USA, 2023.

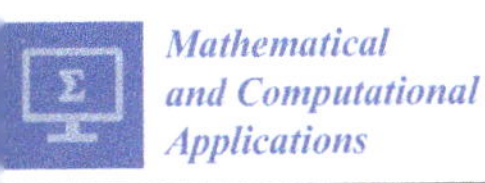

Mathematical and Computational Applications

MDPI

Article

A New Sine Family of Generalized Distributions: Statistical Inference with Applications

SidAhmed Benchiha [1], Laxmi Prasad Sapkota [2], Aned Al Mutairi [3], Vijay Kumar [2], Rana H. Khashab [4,*], Ahmed M. Gemeay [5], Mohammed Elgarhy [6] and Said G. Nassr [7,*]

[1] Laboratory of Statistics and Stochastic Processes, University of Djillali Liabes, BP 89, Sidi Bel Abbes 22000, Algeria; sidahmed.benchiha@univ-sba.dz

[2] Department of Mathematics and Statistics, DDU Gorakhpur University, Gorakhpur 273001, India; laxmisapkota75@gmail.com (L.P.S.); vkgkp@rediffmail.com (V.K.)

[3] Department of Mathematical Sciences, College of Science, Princess Nourah bint Abdulrahman University, Riyadh 11671, Saudi Arabia; aoalmutairi@pnu.edu.sa

[4] Mathematical Sciences Department, College of Applied Sciences, Umm Al-Qura University, Makkah 21961, Saudi Arabia

[5] Department of Mathematics, Faculty of Science, Tanta University, Tanta 31527, Egypt; ahmed.gemeay@science.tanta.edu.eg

[6] Mathematics and Computer Science Department, Faculty of Science, Beni-Suef University, Beni-Suef 62521, Egypt; m_elgarhy85@sva.edu.eg

[7] Department of Statistics and Insurance, Faculty of Commerce, Arish University, Al-Arish 45511, Egypt

* Correspondence: rhkhashab@uqu.edu.sa (R.H.K.); dr.saidstat@gmail.com (S.G.N.)

Abstract: In this article, we extensively study a family of distributions using the trigonometric function. We add an extra parameter to the sine transformation family and name it the alpha-sine-G family of distributions. Some important functional forms and properties of the family are provided in a general form. A specific sub-model alpha-sine Weibull of this family is also introduced using the Weibull distribution as a parent distribution and studied deeply. The statistical properties of this new distribution are investigated and intended parameters are estimated using the maximum likelihood, maximum product of spacings, least square, weighted least square, and minimum distance methods. For further justification of these estimates, a simulation experiment is carried out. Two real data sets are analyzed to show the suggested model's application. The suggested model performed well compares to some existing models considered in the study.

Keywords: sine function; Weibull distribution; moments; estimation methods; hazard function

Citation: Benchiha, S.; Sapkota, L.P.; Al Mutairi, A.; Kumar, V.; Khashab, R.H.; Gemeay, A.M.; Elgarhy, M.; Nassr, S.G. A New Sine Family of Generalized Distributions: Statistical Inference with Applications. *Math. Comput. Appl.* **2023**, *28*, 83. https://doi.org/10.3390/mca28040083

Academic Editor: Sandra Ferreira

Received: 13 June 2023
Revised: 5 July 2023
Accepted: 13 July 2023
Published: 16 July 2023

1. Introduction

Statistical distributions are commonly used to study real-world phenomena. The theory of statistical distributions is extensively studied, as are new developments for their application. Several families of distributions have been developed to describe various real-world phenomena. In reality, this new development in distribution theory is a continuing practice. Most probability distributions proposed in the literature have many parameters to make the model more flexible. According to some authors, these estimates are difficult to obtain using numerical resources (see [1]). For modeling real data, it is preferable to create models with few parameters and a high degree of flexibility. To achieve this goal, a group of researchers decided to look for new distributions using trigonometric functions. In the last several years, researchers have been attracted to trigonometric models due to their flexibility and mathematical tractability. Among the various trigonometric G-families, ref. [2] defined the new class of distribution using the sine trigonometric function and defined the sine-exponential model as its member. The probability density function (PDF) of this family is given by

$$f(x;\varphi) = \frac{\pi}{2}g(x;\varphi)\sin\left\{\frac{\pi}{2}G(x;\varphi)\right\} \quad x \in \Re, \tag{1}$$

where $G(x;\varphi)$ is the cumulative distribution function (CDF) of any baseline model and $g(x;\varphi)$ its PDF.

At the same time, the arctangent function defines the arc-tan-G family of distribution [3]. The authors presented the new family of distribution which was used to model Norwegian fire insurance data. This distribution family was proposed for an underlying Pareto distribution and a new distribution called the Pareto arctan distribution, and it was discovered that this distribution provides a good fit when compared to other well-known distributions. Using a similar technique of sine-G, the cosine-G family of distributions was introduced by [4], who also introduced the cosine-Weibul distribution as a member of cosine-G class. Similarly, [5] introduced another sine-G class and studied the sine inverse Weibull distribution as a particular member. Ref. [6] developed the new sine-G family and analyzed the sine-inverse Weibull model in particular. Ref. [7] defined the sine Kumaraswamy-G family of distributions as having two extra parameters. Ref. [8] defined the exponentiated sine-G family and analyzed the particular distribution as the exponentiated sine-Weibull distribution. Further arcsine-G distributions were introduced by [9], and the arcsine exponential distribution with constant and sharp decreasing hazard functions was defined. Another trigonometric function-related probability model introduced by [10] is called the arctan generalized exponential distribution. Using the sine-G family of distribution, [11] developed a new two-parameter model called the sine Burr XII distribution. Hence, we noticed that the simple functions are associated with trigonometric distribution and are mathematically tractable (see [2,5]). Further, we observed that the sine transformation can remarkably enhance the flexibility of $G(x)$ [7]. A new extended cosine-G family of distributions was proposed by [12]. Truncated Cauchy power family of distributions was studied by [13]. Truncated Cauchy power Weibull-G class of distributions was proposed by [14]. The sine half-logistic inverse Rayleigh and sine inverse exponential distributions were discussed in [15,16]. Due to these pleasant features, we are motivated to conduct research on the sine transformation family.

In this study, we developed a new family of trigonometric models using the sine function by introducing an additional scale parameter α, and we called it the alpha sine-G family (AS-G) of distributions.

The remaining sections of this study are organized as follows. The methodology of model development and some key functions of the family of distributions are introduced in Section 2. Some general properties of the AS-G family of distributions (AS-D FD) are presented in Section 3. In Section 4, a particular member of the AS-G family is introduced. A detailed study and application of this model are presented in Section 5. We discuss parametric estimation and simulation experiments in Sections 6 and 7. The applicability of the suggested model is presented in Section 8. Finally, we present the conclusion in Section 9.

2. The New Sine Family of Distributions

2.1. Methodology

To develop a new family of distributions, [17] defined a relation of $G(x;\varphi)$, the cumulative distribution function (CDF) of any baseline distribution and $r(t)$, the PDF of any arbitrary distribution, to obtain the CDF of the new family as

$$F(x;\varphi) = \int_0^{G(x;\varphi)} r(t)\,dt, \tag{2}$$

where $F(x; \varphi)$ is the CDF of the new class of distributions and φ is the parameter space of baseline distribution. To develop the new sine-G family, Equation (2) can be written as

$$F(x; \varphi) = \int_0^{\frac{\pi}{2} G(x; \varphi)} \cos(t)\, dt = \sin\left\{\frac{\pi}{2} G(x; \varphi)\right\} \quad x \in \Re. \tag{3}$$

Using the structure of Equation (3), we introduce an additional parameter to Equation (3) and the new CDF of AS-G FD can be expressed as

$$F(x; \alpha, \varphi) = \frac{\sin\left\{\frac{\pi\alpha}{2} G(x; \varphi)\right\}}{\sin\left(\frac{\pi\alpha}{2}\right)}; \quad x \in \Re, 0 < \alpha < 1. \tag{4}$$

The PDF corresponding to Equation (4) is

$$f(x; \alpha, \varphi) = \frac{\pi\alpha}{2\sin\left(\frac{\pi\alpha}{2}\right)} g(x; \varphi) \cos\left\{\frac{\pi\alpha}{2} G(x; \varphi)\right\}; \quad x \in \Re, 0 < \alpha < 1, \tag{5}$$

where α is the scale parameter of the AS-G distribution.

Special Case of AS-G FD. When $\alpha = 1$ in the CDF of AS-G FD defined in Equation (4), it is reduced to the sine-G family defined by [2]. Hence, the sine-G family is a special case of AS-G FD.

2.2. Some Important Functional Forms of the New Sine Family of Distributions

In this subsection, we explicitly present some important functions that are necessary for survival analysis, reliability theory, etc.

- Reliability function: In probability theory, the reliability function is a function that offers the probability that a system or device will function correctly for a given amount of time, assuming that it has not failed up to that point. Intuitively, the reliability function offers the probability that the device or system will continue to function beyond time x given that it has not failed up to that point. The reliability function for AS-G FD can be expressed as

$$R(x; \alpha, \varphi) = 1 - \frac{\sin\left\{\frac{\pi\alpha}{2} G(x; \varphi)\right\}}{\sin\left(\frac{\pi\alpha}{2}\right)}; \quad x \in \Re, 0 < \alpha < 1. \tag{6}$$

- Hazard function: In probability theory, the hazard function is a function that describes the rate at which an event occurs given that the event has not yet occurred up to a certain time. The hazard function is often used in survival analysis to model the failure rate of a system over time. The AS-G FD can be defined as

$$h(x) = \frac{\pi\alpha}{2} \frac{g(x; \varphi) \cos\left\{\frac{\pi\alpha}{2} G(x; \varphi)\right\}}{\sin\left(\frac{\pi\alpha}{2}\right) - \sin\left\{\frac{\pi\alpha}{2} G(x; \varphi)\right\}}; \quad x \in \Re. \tag{7}$$

- Odd function: Odd functions are a useful tool in probability theory for describing certain types of distributions and for simplifying calculations involving them. Here, the odd function for AS-G FD can be expressed as

$$O(x) = \frac{\sin\left\{\frac{\pi\alpha}{2} G(x; \varphi)\right\}}{\sin\left(\frac{\pi\alpha}{2}\right) - \sin\left\{\frac{\pi\alpha}{2} G(x; \varphi)\right\}}; \quad x \in \Re. \tag{8}$$

- Failure rate average (FRA): The failure rate average function has important applications in reliability engineering and survival analysis, where it is used to model the behavior of systems and estimate their probability of failure over time. It can also be used to compare different systems' reliability and identify the factors that affect their failure rates.

$$K(x) = -\frac{1}{x}\left[\log\left\{\sin\left(\frac{\pi\alpha}{2}\right) - \sin\left\{\frac{\pi\alpha}{2}G(x;\varphi)\right\}\right\} - \log\left\{\sin\left(\frac{\pi\alpha}{2}\right)\right\}\right]; \quad x \in \Re. \quad (9)$$

3. Properties of the New Sine Family of Distributions

3.1. Linear Representation

One can derive useful linear expansions using exponentiated distributions, specifically the exponentiated-G (Exp-G) distribution with power parameter $z > 0$ which has the CDF:

$$G_z(x;\varphi) = [G(x;\varphi)]^z; x \in \Re, \quad (10)$$

where $x \in \Re$. The corresponding PDF can be expressed as

$$g_z(x;\varphi) = zg(x;\varphi)[G(x;\varphi)]^{(z-1)}, x \in \Re. \quad (11)$$

These notations are used in the following discussion. Exponentiated distributions have well-known properties for a wide range of baseline CDF $G(x;\varphi)$ (for more information, see [5,18,19]). The linear representations of $F(x;\varphi)$ and $f(x;\varphi)$ in terms of Exp-G functions are shown in the following result. Using the Tayler expansion for trigonometric function $Sin(x)$, the CDF of AS-G FD can be expressed as

$$F^*(x;\alpha,\varphi) = \sum_{j=0}^{\infty} \Delta_j G^{2j+1}(x;\varphi), \quad (12)$$

where $\Delta_j = \frac{(-1)^j\left(\frac{\pi\alpha}{2}\right)^{2j+1}}{(2j+1)!\sin\left(\frac{\pi\alpha}{2}\right)}$. The PDF corresponding to Equation (12) can be calculated by differentiating it with respect to x; we obtain

$$f^*(x;\alpha,\varphi) = \sum_{j=0}^{\infty} \Delta_j^* G^{2j}(x;\varphi)g(x;\varphi), \quad (13)$$

where $\Delta_j^* = \Delta_j(2j+1)$.

3.2. Critical Points of the New Sine Family of Distributions

By solving equation $\frac{f(x;\alpha,\varphi)}{dx} = 0$ for x, we can obtain the critical points of $f(x;\alpha,\varphi)$. Let the solution of this equation be x_1, which can be calculated from

$$\sin\left(\frac{\pi\alpha}{2}G(x_1)\right)[g(x_1)]^2 + \cos\left(\frac{\pi\alpha}{2}G(x_1)\right)g'(x_1) = 0. \quad (14)$$

Similarly, the critical points for hazard function $h(x)$ can be obtained by solving the following equation for solution x_2:

$$\left\{\sin\left(\frac{\pi\alpha}{2}\right) - \sin\left(\frac{\pi\alpha}{2}G(x_2)\right)\right\}\left\{\cos\left(\frac{\pi\alpha}{2}G(x_2)\right)g'(x_2) + \frac{\pi\alpha}{2}\sin\left(\frac{\pi\alpha}{2}G(x_2)\right)[g(x_2)]^2\right\} = 0. \quad (15)$$

3.3. Quantile Function

The quantile function is useful in statistical analysis and modeling as it provides a way to estimate percentiles and other summary statistics of a probability distribution. Suppose $Q(p)$ is the smallest value of X for which the probability that X is less than or equal to that value is at least p. The quantile function of CDF $F(x;\alpha,\varphi)$ of AS-G FD can be obtained as

$$Q(p;\alpha,\varphi) = G^{-1}\left(\frac{2}{\pi\alpha}\arcsin\left\{p\sin\left(\frac{\pi\alpha}{2}\right)\right\}\right); \quad p \in (0,1). \quad (16)$$

Using Equation (16), we can calculate the median, upper and lower quartile, quartile deviation (QD), coefficient of QD, skewness, and kurtosis as presented in Table 1.

Table 1. Various measures based on quantiles.

Statistical Measure	Expression
Median	$G^{-1}\left(\dfrac{2}{\pi\alpha}\arcsin\left\{0.5\sin\left(\dfrac{\pi\alpha}{2}\right)\right\}\right)$
Lower Quartile	$G^{-1}\left(\dfrac{2}{\pi\alpha}\arcsin\left\{0.25\sin\left(\dfrac{\pi\alpha}{2}\right)\right\}\right)$
Upper Quartile	$G^{-1}\left(\dfrac{2}{\pi\alpha}\arcsin\left\{0.75\sin\left(\dfrac{\pi\alpha}{2}\right)\right\}\right)$
QD	$\dfrac{1}{2}\left[G^{-1}\left(\dfrac{2}{\pi\alpha}\arcsin\left\{0.75\sin\left(\dfrac{\pi\alpha}{2}\right)\right\}\right) - G^{-1}\left(\dfrac{2}{\pi\alpha}\arcsin\left\{0.25\sin\left(\dfrac{\pi\alpha}{2}\right)\right\}\right)\right]$
Coefficient of QD	$\dfrac{\left[G^{-1}\left(\dfrac{2}{\pi\alpha}\arcsin\left\{0.75\sin\left(\dfrac{\pi\alpha}{2}\right)\right\}\right) - G^{-1}\left(\dfrac{2}{\pi\alpha}\arcsin\left\{0.25\sin\left(\dfrac{\pi\alpha}{2}\right)\right\}\right)\right]}{\left[G^{-1}\left(\dfrac{2}{\pi\alpha}\arcsin\left\{0.75\sin\left(\dfrac{\pi\alpha}{2}\right)\right\}\right) + G^{-1}\left(\dfrac{2}{\pi\alpha}\arcsin\left\{0.25\sin\left(\dfrac{\pi\alpha}{2}\right)\right\}\right)\right]}$
Skewness [20]	$\dfrac{Q\left(\dfrac{3}{4};\alpha,\varphi\right) - 2Q\left(\dfrac{1}{2};\alpha,\varphi\right) + Q\left(\dfrac{1}{4};\alpha,\varphi\right)}{Q\left(\dfrac{3}{4};\alpha,\varphi\right) - Q\left(\dfrac{1}{4};\alpha,\varphi\right)}$
Kurtosis [21]	$\dfrac{Q\left(\dfrac{7}{8};\alpha,\varphi\right) - Q\left(\dfrac{5}{8};\alpha,\varphi\right) - Q\left(\dfrac{1}{8};\alpha,\varphi\right) + Q\left(\dfrac{3}{8};\alpha,\varphi\right)}{Q\left(\dfrac{3}{4};\alpha,\varphi\right) - Q\left(\dfrac{1}{4};\alpha,\varphi\right)}$

3.4. Moments

In probability theory and statistics, moments of a random variable X are numerical quantities that measure various aspects of its probability distribution. The moments of X are calculated using the values of X and the PDF of X. The Kth moment about the origin can be calculated as

$$\mu'_k = \int_{-\infty}^{\infty} x^k f(x)dx. \tag{17}$$

Now, considering the integral and summation terms exist and are interchangeable, using the PDF defined in Equation (5), we can calculate the Kth moment as

$$\mu'_k = \sum_{j=0}^{\infty} \Delta_j^* \int_{-\infty}^{\infty} x^k G^{2j}(x;\varphi)g(x;\varphi)dx. \tag{18}$$

Further, the Kth moment can also be calculated using the quantile function (for more detail, see [22]) as

$$\mu'_k = \sum_{j=0}^{\infty} \Delta_j^* \int_{0}^{1} x^{2j}\{Q(x;\alpha,\varphi)\}^k dx. \tag{19}$$

3.5. Moment Generating Function

Let $M_X(t)$ be the MGF of X. Using Equation (18), MGF can be defined as

$$M_X(t) = \sum_{j=0}^{\infty} \sum_{m=0}^{\infty} \frac{t^m \Delta_j^*}{m!} \int_{-\infty}^{\infty} x^k G^{2j}(x; \varphi) g(x; \varphi) dx. \tag{20}$$

Similarly, using Equation (19), MGF can be expressed as

$$M_X(t) = \sum_{j=0}^{\infty} \sum_{m=0}^{\infty} \frac{t^m \Delta_j^*}{m!} \int_{0}^{1} x^{2j} \{Q(x; \alpha, \varphi)\}^k dx. \tag{21}$$

3.6. Mean Residual Life Function

Suppose t is the lifetime of a component or item; then, MRF can be obtained as

$$\mu_{MRF}(t) = \frac{1}{R(t)} \left[E(t) - \sum_{j=0}^{\infty} \Delta_j^* \int_{0}^{t} x \, G^{2j}(x; \varphi) g(x; \varphi) dx \right], \tag{22}$$

where $R(t)$ is the reliability function.

4. Alpha-Sine Weibull Distribution

Model Presentation

In this section, a particular model of AS-G FD is introduced, and we analyze this model briefly. To define the new member, we select $G(x; \varphi)$ as the CDF of Weibull distribution as

$$G(x; \delta, \lambda) = 1 - \exp\left[-\left(\frac{x}{\delta}\right)^{\lambda}\right]; \quad x \geqslant 0, \quad \delta, \lambda > 0. \tag{23}$$

The PDF corresponding to CDF (23) can be written as

$$g(x; \delta, \lambda) = \lambda \delta^{-\lambda} x^{\lambda-1} \exp\left[-\left(\frac{x}{\delta}\right)^{\lambda}\right]; \quad x \geqslant 0, \quad \delta, \lambda > 0. \tag{24}$$

Substituting Equation (23) in the CDF of AS-G FD defined in Equation (4), we obtain the new member distribution called the AS-Weibull (AS-W) distribution with CDF:

$$F(x; \alpha, \delta, \lambda) = \frac{1}{\sin\left(\frac{\pi\alpha}{2}\right)} \sin\left\{ \frac{\pi\alpha}{2} \left[1 - \exp\left(-\left(\frac{x}{\delta}\right)^{\lambda}\right) \right] \right\}; \quad x \geqslant 0, \quad 0 < \alpha < 1, \delta, \lambda > 0. \tag{25}$$

The PDF of the AS-W distribution can be obtained by differentiating Equation (25) and can be expressed as

$$f(x; \alpha, \delta, \lambda) = \frac{\pi\alpha\lambda\delta^{-\lambda}}{2\sin\left(\frac{\pi\alpha}{2}\right)} x^{\lambda-1} \exp\left[-\left(\frac{x}{\delta}\right)^{\lambda}\right] \cos\left\{ \frac{\pi\alpha}{2} \left[1 - \exp\left(-\left(\frac{x}{\delta}\right)^{\lambda}\right) \right] \right\}. \tag{26}$$

Similarly, the HRF of the AS-W distribution is given by

$$h(x; \alpha, \delta, \lambda) = \frac{\pi}{2} \alpha\lambda\delta^{-\lambda} x^{\lambda-1} \exp\left[-\left(\frac{x}{\delta}\right)^{\lambda}\right] \frac{\cos\left\{ \frac{\pi\alpha}{2} \left[1 - \exp\left(-\left(\frac{x}{\delta}\right)^{\lambda}\right) \right] \right\}}{\sin\left(\frac{\pi\alpha}{2}\right) - \sin\left\{ \frac{\pi\alpha}{2} \left[1 - \exp\left(-\left(\frac{x}{\delta}\right)^{\lambda}\right) \right] \right\}}. \tag{27}$$

We demonstrate the various shapes of PDF and HRF for varying two parameters keeping δ and λ constant, respectively; Figures 1 and 2. In Figure 3, we use all three parameters with different combinations. From all these graphical investigations, we find that the suggested model is versatile regarding skewness and kurtosis. Both PDF and HRF can have either increasing or decreasing or bathtub or inverted bathtub or -j- or reverse-j-shaped curves according to parameter values. Hence, the AS-W model is capable of fitting highly skewed heterogeneous data sets.

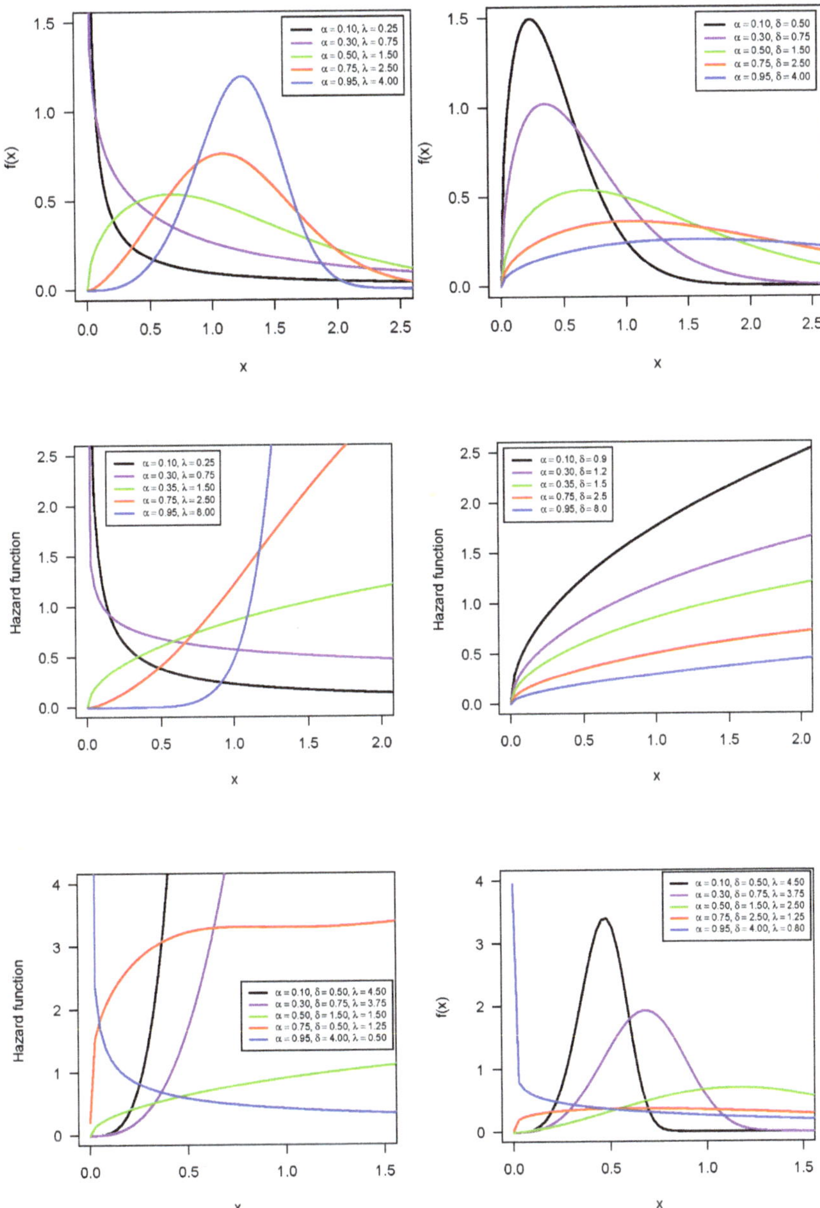

Figure 3. The plots of PDF and HRF with a variation of all three parameters.

5. Properties of the Alpha-Sine Weibull Distribution

5.1. Quantile Function

The QF can be used in statistical analysis and modeling to estimate probability distribution percentiles and other summary statistics. The QF for the AS-W distribution can be expressed as

$$Q(p; \alpha, \delta, \lambda) = \left[-\delta^\lambda \log \left\{ 1 - \frac{2}{\pi\alpha} \arcsin \left\{ p \sin \left(\frac{\pi\alpha}{2} \right) \right\} \right\} \right]^{\frac{1}{\lambda}}; \quad p \in (0, 1). \tag{28}$$

Using Equation (28), we can obtain various statistical measures provided in Table 1. Also, for generating random numbers to the distribution AS-W, we can use the following expression:

$$x = \left[-\delta^\lambda \log \left\{ 1 - \frac{2}{\pi\alpha} \arcsin \left\{ u \sin \left(\frac{\pi\alpha}{2} \right) \right\} \right\} \right]^{\frac{1}{\lambda}}; \quad u \in (0, 1). \tag{29}$$

Using the formulae defined by [20,21] for skewness and kurtosis using quantiles, we plotted the graphs of skewness and kurtosis with various combinations of parameter values presented in Figures 4 and 5.

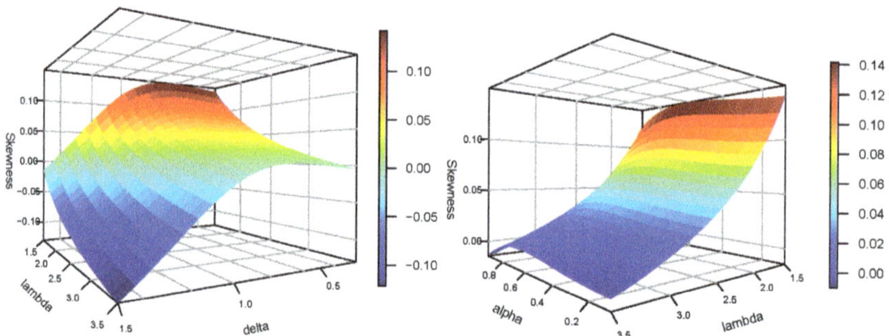

Figure 4. The plots of skewness with constant $\alpha = 0.5$ (**left**) and constant $\delta = 0.75$ (**right**).

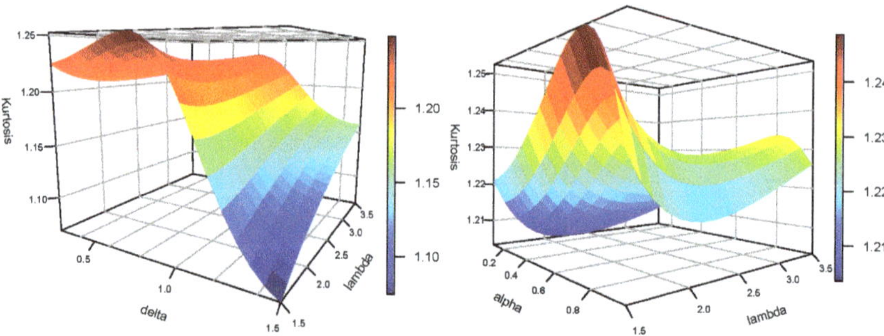

Figure 5. The plots of kurtosis with constant $\alpha = 0.5$ (**left**) and constant $\delta = 0.75$ (**right**).

5.2. Linear Expansion of Alpha-Sine Weibull Distribution

Using Equation (12), the expansion of the CDF defined in Equation (25) is given by

$$F^*(x; \alpha, \delta, \lambda) = \sum_{j=0}^{\infty} \Delta_j \left\{ 1 - \exp \left[-\left(\frac{x}{\delta} \right)^\lambda \right] \right\}^{(2j+1)}; \quad x \geqslant 0, \tag{30}$$

where $\Delta_j = \dfrac{(-1)^j \left(\frac{\pi\alpha}{2}\right)^{2j+1}}{(2j+1)!\sin\left(\frac{\pi\alpha}{2}\right)}$. Further using the binomial expansion, Equation (30) can be expressed as

$$F^*(x; \alpha, \delta, \lambda) = \sum_{j=0}^{\infty} \sum_{m=0}^{\infty} \Delta_{jm}^* \exp\left[-m\left(\frac{x}{\delta}\right)^{\lambda}\right]; \quad x \geqslant 0, \tag{31}$$

where $\Delta_{jm}^* = \Delta_j(-1)^m \dbinom{2j+1}{m}$. The PDF corresponding to Equation (31) can be written as

$$f^*(x; \alpha, \delta, \lambda) = \sum_{j=0}^{\infty} \sum_{m=0}^{\infty} \Delta_{jm}^{**} x^{\lambda-1} \exp\left[-(1+m)\left(\frac{x}{\delta}\right)^{\lambda}\right]; \quad x \geqslant 0, \tag{32}$$

where $\Delta_{jm}^{**} = \Delta_j(-1)^m \dbinom{2j}{m} \lambda \delta^{-\lambda}$.

5.3. Moments

The Kth moment of random variable $X \sim AS - W(\alpha, \delta, \lambda)$ can be obtained by using the following expression:

$$\mu_k' = \sum_{j=0}^{\infty} \sum_{m=0}^{\infty} \Delta_{jm}^{**} \frac{\Gamma\left(\frac{k+\lambda}{\lambda}\right)}{\left(\frac{1+m}{\delta^{\lambda}}\right)^{\frac{k+\lambda}{\lambda}}}. \tag{33}$$

5.4. Moment Generating Function of Alpha-Sine Weibull Distribution

The MGF of AS-W for any real number t can be expressed as

$$M_X(t) = \sum_{j=0}^{\infty} \sum_{m=0}^{\infty} \sum_{n=0}^{\infty} \Delta_{jm}^{**} \frac{t^n}{n!} \frac{\Gamma\left(\frac{k+\lambda}{\lambda}\right)}{\left(\frac{1+m}{\delta^{\lambda}}\right)^{\frac{k+\lambda}{\lambda}}}, \tag{34}$$

where $\Gamma(.)$ is the gamma function.

5.5. Mean Waiting Time Function

Let t denote the waiting time or time to failure of an item or event; then, the MWT function can be defined as

$$\mu(t) = t - \frac{1}{F(t; \alpha, \delta, \lambda)} \sum_{j=0}^{\infty} \sum_{m=0}^{\infty} \lambda^{-1} \Delta_{jm}^{**} \frac{\gamma\left(\frac{1}{\lambda}, \frac{1+m}{\delta^{\lambda}} t^{\lambda}\right)}{\left(\frac{1+m}{\delta^{\lambda}}\right)^{\frac{1}{\lambda}}}, \tag{35}$$

where $\gamma(.)$ is the lower incomplete gamma function.

6. Estimation Methods

In this part of the work, we consider different methods for estimating the parameters of the AS-W distribution.

6.1. Maximum Likelihood Method

Consider a simple random sample $x = (x_1, x_2, \ldots, x_m)$ of size m following the AS-W distribution; then, the likelihood function can be presented as

$$
\begin{aligned}
L(x;\alpha,\delta,\lambda) &= \prod_{j=1}^{m}\left(\frac{\pi\alpha\lambda\delta^{-\lambda}}{2\sin\left(\frac{\pi\alpha}{2}\right)}\right)e^{-\left(\frac{x_j}{\delta}\right)^{\lambda}}(x_j)^{\lambda-1}\cos\left(\frac{\pi\alpha\left(1-e^{-\left(\frac{x_j}{\delta}\right)^{\lambda}}\right)}{2}\right) \\
&= \left(\frac{\pi\alpha\lambda\delta^{-\lambda}}{2\sin\left(\frac{\pi\alpha}{2}\right)}\right)^{m}e^{\left(-\sum_{j=1}^{m}\left(\frac{x_j}{\delta}\right)^{\lambda}\right)}\prod_{j=1}^{m}(x_j)^{\lambda-1}\cos\left(\frac{\pi\alpha\left(1-e^{-\left(\frac{x_j}{\delta}\right)^{\lambda}}\right)}{2}\right).
\end{aligned}
$$

$$(36)$$

Hence, the corresponding log-likelihood function is given as

$$
\begin{aligned}
\log L\,(x;\alpha,\delta,\lambda) &= m\log(\pi)+m\log(\alpha)+m\log(\lambda)-m\lambda\log(\delta)-m\log\left(2\sin\left(\frac{\pi\alpha}{2}\right)\right) \\
&\quad -\sum_{j=1}^{m}\left(\frac{x_j}{\delta}\right)^{\lambda}+(\lambda-1)\sum_{j=1}^{m}\log(x_j)+\sum_{j=1}^{m}\log\left[\cos\left(\frac{\pi\alpha\left(1-e^{-\left(\frac{x_j}{\delta}\right)^{\lambda}}\right)}{2}\right)\right].
\end{aligned}
$$

The MLEs $\hat{\Theta}=(\hat{\alpha},\hat{\delta},\hat{\lambda})$ of $\Theta=(\alpha,\delta,\lambda)$ are obtained, respectively, using numerical methods.

6.2. Maximum Product of Spacings Method

Cheng and Amin [23] present this technique as a different method to MLE. It relies on the geometric mean of the spacings, which is

$$
v_j(\alpha,\delta,\lambda)=F(x_{(j)}|\alpha,\delta,\lambda)-F(x_{(j-1)}|\alpha,\delta,\lambda),\, j=1,\ldots,m+1,
$$

where $F(t_{(0)}|\alpha,\delta,\lambda)=0$ and $F(t_{(m+1)}|\alpha,\delta,\lambda)=1$. We can consider that $\sum_{j=1}^{m+1}v_j(\alpha,\delta,\lambda)=1$.
The MPS estimators of $\Theta=(\alpha,\delta,\lambda)$ can be solved by increasing the geometric mean of the spacing,

$$
\vartheta(\alpha,\delta,\lambda|x)=\left[\prod_{j=1}^{m+1}v_j(\alpha,\delta,\lambda)\right]^{\frac{1}{m+1}},
$$

$$(37)$$

or similarly by increasing the natural logarithm of the product spacing function of (37) given by

$$
\psi(\alpha,\delta,\lambda|x)=\frac{1}{m+1}\sum_{j=1}^{m+1}\log v_j(\alpha,\delta,\lambda).
$$

6.3. Least Squares Methods

Our study proposes two variants of least squares, Ordinary Least Squares (OLS) and Weighted Least Squares (WLS).

The OLS estimators can be determined by minimizing

$$
\Delta(\alpha,\delta,\lambda|x) = \sum_{j=1}^{m}\left[F(x_{(j)}|\alpha,\delta,\lambda) - \frac{j}{m+1}\right]^2
$$

$$
= \sum_{j=1}^{m}\left[\frac{\sin\left(\frac{\pi\alpha\left(1-e^{-\left(\frac{x_{(j)}}{\delta}\right)^{\lambda}}\right)}{2}\right)}{\sin\left(\frac{\pi\alpha}{2}\right)} - \frac{j}{m+1}\right]^2 .
$$

However, the WLS estimators can be obtained by minimizing

$$
\Delta_W(\alpha,\delta,\lambda|x) = \sum_{j=1}^{m}\frac{(m+1)^2(m+2)}{j(m-j+1)}\left[F(x_{(j)}|\alpha,\delta,\lambda) - \frac{j}{m+1}\right]^2
$$

$$
= \sum_{j=1}^{m}\frac{(m+1)^2(m+2)}{j(m-j+1)}\left[\frac{\sin\left(\frac{\pi\alpha\left(1-e^{-\left(\frac{x_{(j)}}{\delta}\right)^{\lambda}}\right)}{2}\right)}{\sin\left(\frac{\pi\alpha}{2}\right)} - \frac{j}{m+1}\right]^2 .
$$

6.4. Minimum Distance Methods

Various methods have been proposed based on the minimization of empirical distribution functions and estimated distribution functions. This work uses the Cramer–Von–Mises (CV) and Anderson–Darling (AD) methods. We start with a CV estimator, and we can derive these estimators by minimizing the following functions:

$$
\zeta(\alpha,\delta,\lambda|x) = \frac{1}{12m} + \sum_{j=1}^{m}\left[F(x_{(j)};\alpha,\delta,\lambda) - \frac{2j-1}{2m}\right]^2
$$

$$
= \frac{1}{12m} + \sum_{j=1}^{m}\left[\frac{\sin\left(\frac{\pi\alpha\left(1-e^{-\left(\frac{x_{(j)}}{\delta}\right)^{\lambda}}\right)}{2}\right)}{\sin\left(\frac{\pi\alpha}{2}\right)} - \frac{2j-1}{2m}\right]^2 ,
$$

moreover, the AD estimators are determined by minimizing

$$\zeta(\alpha, \delta, \lambda|x) = -m - \frac{1}{m}\sum_{j=1}^{m}(2j-1)\left\{ \log\left[\frac{\sin\left(\frac{\pi\alpha\left(1-e^{-\left(\frac{x_{(j)}}{\delta}\right)^{\lambda}}\right)}{2}\right)}{\sin\left(\frac{\pi\alpha}{2}\right)}\right]\right.$$

$$\left. + \log\left[1 - \frac{\sin\left(\frac{\pi\alpha\left(1-e^{-\left(\frac{t_{(m+1-j)}}{\delta}\right)^{\lambda}}\right)}{2}\right)}{\sin\left(\frac{\pi\alpha}{2}\right)}\right]\right\}. \tag{38}$$

7. Numerical Simulation

To compare unknown parameter estimates of the (AS-W) distribution, a simulation study is conducted with different parameters, and several sample sizes $m = 30, 60, 100, 150, 200$, and 500 are presented. Based on 1000 runs, we compute the average estimate (AE) and mean square error (MSE), which are considered to be the optimality criteria.

From Tables 2–5, the following is clear from the numerical experiments:

- Based on all estimation methods, the average estimate converges to the true values, which shows that these estimators are consistent.
- The AE tends to its initial values as the sample size increase, so we can say that our estimates are unbiased.
- For all methods, whenever the MSEs decrease, the sample size m increases.
- The MLE estimators perform better than all the other methods considered in this work.

Table 2. The AES and MSEs of $\alpha = 0.4, \delta = 2, \lambda = 3$.

Sample Size		MLE		MPS		LSE		WLS		CVE		ADE	
		AE	MSE	AE	MSE	AE	MSE	AE	MSE	AE	MSE	AE	MSE
30	$\hat{\alpha}$	0.4282	0.2229	0.5727	0.2056	0.4468	0.2089	0.4707	0.2165	0.3965	0.2279	0.4661	0.2144
	$\hat{\delta}$	2.0909	0.3833	2.2162	0.4673	2.1140	0.2179	2.1152	0.1477	2.0552	0.0475	2.1046	0.1377
	$\hat{\lambda}$	3.1459	0.3697	2.7929	0.2730	2.9844	0.3670	3.0017	0.3144	3.2108	0.4036	3.0294	0.2770
60	$\hat{\alpha}$	0.5516	0.0155	0.5333	0.1673	0.4501	0.1589	0.4570	0.1585	0.3957	0.1626	0.4503	0.1587
	$\hat{\delta}$	2.0105	0.0031	2.1136	0.0490	2.0668	0.0376	2.0637	0.0282	2.0285	0.0191	2.0594	0.0284
	$\hat{\lambda}$	3.0861	0.0258	2.8456	0.1225	2.9515	0.1485	2.9718	0.1242	3.0802	0.1362	2.9820	0.1158
100	$\hat{\alpha}$	0.5429	0.0129	0.5160	0.1593	0.4173	0.1547	0.4282	0.1496	0.3918	0.1458	0.4378	0.1519
	$\hat{\delta}$	2.0054	0.0016	2.0937	0.0366	2.0448	0.0194	2.0446	0.0189	2.0211	0.0134	2.0490	0.0206
	$\hat{\lambda}$	3.0741	0.0222	2.8854	0.0665	2.9601	0.0819	2.9766	0.0678	3.0436	0.0701	2.9799	0.0655

Table 2. *Cont.*

Sample Size		MLE		MPS		LSE		WLS		CVE		ADE	
		AE	MSE	AE	MSE	AE	MSE	AE	MSE	AE	MSE	AE	MSE
150	$\hat{\alpha}$	0.5333	0.0100	0.5197	0.1487	0.4428	0.1360	0.4476	0.1369	0.4175	0.1318	0.4463	0.1386
	$\hat{\delta}$	2.0021	0.0006	2.0819	0.0287	2.0388	0.0154	2.0411	0.0155	2.0230	0.0125	2.0409	0.0157
	$\hat{\lambda}$	3.0633	0.0190	2.9047	0.0439	2.9641	0.0539	2.9732	0.0442	3.0188	0.0451	2.9764	0.0427
200	$\hat{\alpha}$	0.5237	0.0071	0.5464	0.1465	0.4247	0.1372	0.4404	0.1344	0.4234	0.1303	0.4439	0.1322
	$\hat{\delta}$	2.0003	0.0001	2.0910	0.0290	2.0323	0.0129	2.0366	0.0137	2.0256	0.0121	2.0366	0.0135
	$\hat{\lambda}$	3.0531	0.0159	2.9242	0.0342	2.9802	0.0411	2.9868	0.0325	3.0209	0.0342	2.9892	0.0327
500	$\hat{\alpha}$	0.5027	0.0008	0.4988	0.1292	0.4153	0.1220	0.4110	0.1198	0.3952	0.1173	0.4115	0.1188
	$\hat{\delta}$	2.0000	0.0000	2.0624	0.0207	2.0232	0.0096	2.0205	0.0095	2.0122	0.0081	2.0200	0.0094
	$\hat{\lambda}$	3.0297	0.0089	2.9533	0.0145	2.9794	0.0166	2.9861	0.0138	3.0008	0.0137	2.9871	0.0136

Table 3. The AES and MSEs of $\alpha = 0.4, \delta = 1, \lambda = 1.5$.

Sample Size		MLE		MPS		LSE		WLS		CVE		ADE	
		AE	MSE	AE	MSE	AE	MSE	AE	MSE	AE	MSE	AE	MSE
30	$\hat{\alpha}$	0.2849	0.1535	0.4189	0.1723	0.3544	0.1660	0.3813	0.1785	0.3276	0.1644	0.3751	0.1642
	$\hat{\delta}$	1.0271	0.0322	1.1228	0.2186	1.0844	0.3253	1.0866	0.0735	1.0361	0.0235	1.0658	0.0312
	$\hat{\lambda}$	1.5828	0.0631	1.4216	0.0535	1.4915	0.0732	1.5027	0.0642	1.6071	0.0870	1.5224	0.0585
60	$\hat{\alpha}$	0.4456	0.0091	0.4559	0.1658	0.3692	0.1467	0.3912	0.1513	0.3620	0.1387	0.3953	0.1506
	$\hat{\delta}$	1.0211	0.0049	1.1065	0.0401	1.0575	0.0192	1.0652	0.0222	1.0406	0.0167	1.0647	0.0230
	$\hat{\lambda}$	1.5492	0.0098	1.4396	0.0255	1.4911	0.0357	1.4999	0.0295	1.5527	0.0348	1.5045	0.0271
100	$\hat{\alpha}$	0.4443	0.0066	0.4371	0.1590	0.3933	0.1348	0.3974	0.1376	0.3694	0.1267	0.4026	0.1380
	$\hat{\delta}$	1.0214	0.0032	1.0922	0.0333	1.0584	0.0173	1.0603	0.0187	1.0403	0.0139	1.0611	0.0192
	$\hat{\lambda}$	1.5381	0.0057	1.4635	0.0143	1.4995	0.0197	1.5059	0.0160	1.5389	0.0182	1.5087	0.0157
150	$\hat{\alpha}$	0.4456	0.0068	0.4645	0.1582	0.3779	0.1307	0.3865	0.1320	0.3649	0.1234	0.3895	0.1330
	$\hat{\delta}$	1.0136	0.0020	1.0971	0.0329	1.0486	0.0138	1.0509	0.0149	1.0363	0.0116	1.0520	0.0157
	$\hat{\lambda}$	1.5297	0.0045	1.4626	0.0106	1.4918	0.0140	1.4976	0.0114	1.5198	0.0122	1.4988	0.0110
200	$\hat{\alpha}$	0.4410	0.0061	0.4412	0.1528	0.3754	0.1280	0.3748	0.1271	0.3583	0.1200	0.3764	0.1296
	$\hat{\delta}$	1.0118	0.0018	1.0898	0.0296	1.0505	0.0129	1.0492	0.0137	1.0379	0.0111	1.0508	0.0146
	$\hat{\lambda}$	1.5245	0.0037	1.4695	0.0080	1.4901	0.0101	1.4969	0.0082	1.5137	0.0086	1.4972	0.0080
500	$\hat{\alpha}$	0.4380	0.0057	0.4484	0.1420	0.3764	0.1161	0.3778	0.1151	0.3629	0.1094	0.3789	0.1153
	$\hat{\delta}$	1.0012	0.0002	1.0815	0.0264	1.0413	0.0110	1.0416	0.0121	1.0330	0.0098	1.0418	0.0122
	$\hat{\lambda}$	1.5115	0.0017	1.4802	0.0034	1.4932	0.0039	1.4960	0.0032	1.5034	0.0032	1.4964	0.0032

Table 4. The AES and MSEs of $\alpha = 0.6, \delta = 2, \lambda = 1$.

Sample Size		MLE		MPS		LSE		WLS		CVE		ADE	
		AE	MSE	AE	MSE	AE	MSE	AE	MSE	AE	MSE	AE	MSE
30	$\hat{\alpha}$	0.6597	0.0195	0.5390	0.2054	0.5081	0.1784	0.5212	0.1858	0.4882	0.1821	0.5191	0.1895
	$\hat{\delta}$	2.0642	0.1340	2.6170	8.4422	2.3168	2.3140	2.3242	1.8946	2.2091	2.7088	2.2942	1.1570
	$\hat{\lambda}$	1.0537	0.0140	0.9369	0.0276	0.9900	0.0380	0.9979	0.0330	1.0665	0.0433	1.0081	0.0285
60	$\hat{\alpha}$	0.6494	0.0099	0.5570	0.1611	0.4988	0.1622	0.5099	0.1605	0.4692	0.1636	0.5016	0.1645
	$\hat{\delta}$	2.0634	0.0127	2.2948	0.4264	2.1652	0.3995	2.1645	0.2667	2.0678	0.1926	2.1487	0.2479
	$\hat{\lambda}$	1.0276	0.0055	0.9435	0.0135	0.9803	0.0164	0.9865	0.0139	1.0219	0.0152	0.9883	0.0125
100	$\hat{\alpha}$	0.6442	0.0088	0.5808	0.1450	0.5099	0.1456	0.5104	0.1474	0.4700	0.1500	0.5034	0.1487
	$\hat{\delta}$	2.0510	0.0102	2.2622	0.3124	2.1219	0.2023	2.1178	0.1835	2.0400	0.1417	2.1039	0.1735
	$\hat{\lambda}$	1.0250	0.0050	0.9625	0.0075	0.9858	0.0086	0.9913	0.0073	1.0135	0.0078	0.9944	0.0070

Table 4. *Cont.*

Sample Size		MLE		MPS		LSE		WLS		CVE		ADE	
		AE	MSE	AE	MSE	AE	MSE	AE	MSE	AE	MSE	AE	MSE
150	$\hat{\alpha}$	0.6378	0.0076	0.5631	0.1385	0.4894	0.1464	0.4867	0.1480	0.4572	0.1500	0.4871	0.1483
	$\hat{\delta}$	2.0552	0.0110	2.2113	0.2558	2.0807	0.1505	2.0728	0.1324	2.0186	0.1090	2.0729	0.1341
	$\hat{\lambda}$	1.0140	0.0028	0.9649	0.0054	0.9829	0.0063	0.9873	0.0051	1.0024	0.0051	0.9883	0.0050
200	$\hat{\alpha}$	0.6272	0.0054	0.5662	0.1418	0.4872	0.1449	0.4817	0.1453	0.4546	0.1477	0.4877	0.1450
	$\hat{\delta}$	2.0500	0.0100	2.2230	0.2529	2.0804	0.1423	2.0684	0.1301	2.0218	0.1096	2.0768	0.1367
	$\hat{\lambda}$	1.0108	0.0022	0.9718	0.0041	0.9895	0.0048	0.9929	0.0039	1.0045	0.0040	0.9931	0.0038
500	$\hat{\alpha}$	0.608	0.0016	0.5878	0.1189	0.4732	0.1353	0.4847	0.1241	0.4688	0.1234	0.4885	0.1239
	$\hat{\delta}$	2.000	0.0000	2.2199	0.2269	2.0446	0.0999	2.0459	0.0965	2.0175	0.0843	2.0511	0.1001
	$\hat{\lambda}$	1.000	0.0000	0.9808	0.0018	0.9936	0.0018	0.9949	0.0015	1.0000	0.0015	0.9949	0.0015

Table 5. The AES and MSEs of $\alpha = 0.7, \delta = 3, \lambda = 2.5$.

Sample Size		MLE		MPS		LSE		WLS		CVE		ADE	
		AE	MSE	AE	MSE	AE	MSE	AE	MSE	AE	MSE	AE	MSE
30	$\hat{\alpha}$	0.7483	0.0274	0.7345	0.1700	0.6787	0.1822	0.6691	0.1730	0.5878	0.2083	0.6681	0.1844
	$\hat{\delta}$	3.0782	0.0368	3.5572	3.0963	3.4365	2.7615	3.3212	1.4609	3.1723	0.7521	3.3279	1.4390
	$\hat{\lambda}$	2.5959	0.0441	2.2631	0.2335	2.4062	0.2695	2.4325	0.2375	2.6046	0.2761	2.4583	0.2202
60	$\hat{\alpha}$	0.7210	0.0063	0.6983	0.1259	0.6762	0.1393	0.6420	0.1358	0.5789	0.1566	0.6587	0.1363
	$\hat{\delta}$	3.0507	0.0152	3.2034	0.1991	3.1919	0.3088	3.1130	0.1402	3.0399	0.1163	3.1341	0.1548
	$\hat{\lambda}$	2.5813	0.0244	2.3468	0.1024	2.4330	0.1317	2.4604	0.1028	2.5513	0.1109	2.4620	0.0987
100	$\hat{\alpha}$	0.7054	0.0016	0.6886	0.1179	0.6692	0.1197	0.6293	0.1264	0.5736	0.1445	0.6494	0.1217
	$\hat{\delta}$	3.0354	0.0106	3.1517	0.1021	3.1244	0.1366	3.0739	0.0787	3.0172	0.0689	3.0901	0.0833
	$\hat{\lambda}$	2.5756	0.0227	2.3837	0.0631	2.4531	0.0757	2.4717	0.0618	2.5291	0.0638	2.4711	0.0590
150	$\hat{\alpha}$	0.7027	0.0008	0.6884	0.1119	0.6650	0.1186	0.6020	0.1345	0.5654	0.1444	0.6330	0.1253
	$\hat{\delta}$	3.0246	0.0074	3.1333	0.0846	3.1112	0.1146	3.0444	0.0621	3.0061	0.0571	3.0696	0.0679
	$\hat{\lambda}$	2.5552	0.0166	2.3985	0.0393	2.4481	0.0500	2.4685	0.0377	2.5075	0.0375	2.4649	0.0371
200	$\hat{\alpha}$	0.7006	0.0002	0.6686	0.1123	0.6532	0.1147	0.5924	0.1309	0.5587	0.1392	0.6294	0.1191
	$\hat{\delta}$	3.0168	0.0050	3.1048	0.0724	3.0849	0.0863	3.0256	0.0514	2.9926	0.0492	3.0551	0.0578
	$\hat{\lambda}$	2.5450	0.0135	2.4154	0.0282	2.4574	0.0366	2.4756	0.0275	2.5059	0.0279	2.4706	0.0272
500	$\hat{\alpha}$	0.7000	0.0000	0.6829	0.0887	0.6505	0.0995	0.5927	0.1110	0.5671	0.1176	0.6345	0.0976
	$\hat{\delta}$	3.0000	0.0000	3.0887	0.0561	3.0578	0.0586	3.0039	0.0359	2.9809	0.0339	3.0369	0.0428
	$\hat{\lambda}$	2.5126	0.0038	2.4351	0.0132	2.4620	0.0144	2.4740	0.0107	2.4876	0.0101	2.4687	0.0112

8. Applications

In this part of the work, we provide two application datasets to show the effectiveness and flexibility of the AS-W distribution. Different statistic measures for the two data sets are presented in Table 6.

Table 6. Summary statistics for the selected datasets.

Datasets	Minimum	One Quntile	Median	Mean	Three Quntile	Maximum	Skew	Kurt
Dataset 1	0.070	1.170	2.490	3.494	5.840	13.300	1.152	3.890
Dataset 2	2.998	21.187	51.385	55.123	75.435	138.500	0.555	2.108

8.1. First Data Set

The first data set represents the total annual rainfall (in inches) during the month of January from 1880 to 1916 recorded at Los Angeles Civic Center; ref. [24] provided the values. The data are reported in Table 7.

Table 7. The total annual rainfall.

1.33	1.43	1.01	1.62	3.15	1.05	7.72	0.2	6.03	0.25	7.83	0.25	0.88	6.29	0.94
5.84	3.23	3.7	1.26	2.64	1.17	2.49	1.62	2.1	0.14	2.57	3.85	7.02	5.04	7.27
1.53	6.7	0.07	2.01	10.35	5.42	13.3								

8.2. Second DataSet

The second data set is the failure times of eight components at three different temperatures, 100, 120, 140, introduced by [25]. The value data are described Table 8.

Table 8. The values of the failure times of eight components at three different temperatures.

14.712	32.644	61.979	65.521	105.50	114.60	120.40
138.50	8.610	11.741	54.535	55.047	58.928	63.391
105.18	113.02	2.998	5.016	15.628	23.040	27.851
37.843	38.050	48.226				

The AS-W distribution is fitted to these two datasets and compared with the following:

- Sine-inverse Weibull [4]:

$$F(x, \alpha, \theta) = \sin\left\{\frac{\pi}{2} e^{\left(-\alpha x^{-\theta}\right)}\right\}.$$

$$f(x; \alpha, \theta) = \frac{\pi}{2} \alpha \theta x^{-\theta-1} e^{\left(-\alpha x^{-\theta}\right)} \cos\left\{\frac{\pi}{2} e^{\left(-\alpha x^{-\theta}\right)}\right\} \quad x > 0, \alpha, \theta > 0.$$

- The inverse Weibull distribution [26]:

$$F(x, \tau, \theta) = e^{-\left(\frac{\theta}{x}\right)^{\tau}}.$$

$$f(x; \tau, \theta) = f(x) = \frac{\tau(\theta/x)^{\tau} e^{-(\theta/x)^{\tau}}}{x} \quad x > 0, \tau, \theta > 0.$$

- Weighted generalized quasi Lindley distribution (WGQLD) [27]:

$$F(x, \alpha, \theta) = 1 - \frac{\begin{array}{c} 24 + 6\alpha^2[2 + x\theta(2 + x\theta)] \\ +6\alpha[6 + x\theta(6 + x\theta(3 + x\theta))] \\ +x\theta[24 + x\theta(12 + x\theta(4 + x\theta))] \end{array}}{12(1 + \alpha)(2 + \alpha)} e^{-\theta x}.$$

$$f(x; \alpha, \theta) = \frac{\theta^3 x^2 \cdot \left(\theta^2 x^2 + 6\alpha\theta x + 6\alpha^2\right) e^{-\theta x}}{12(\alpha + 1)(\alpha + 2)} \quad x > 0, \alpha, \theta > 0.$$

- Sine Burr XII distribution [11]:

$$F(x) = \sin\left\{\frac{\pi}{2}\left[1 - \frac{1}{(1 + x^a)^b}\right]\right\} : a, b, x > 0.$$

$$f(x) = \frac{\pi}{2} \frac{abx^{a+1}}{(1 + x^a)^{b+1}} \cos\left\{\frac{\pi}{2}\left[1 - \frac{1}{(1 + x^a)^b}\right]\right\}, a, b, x > 0.$$

The MLEs, SEs and corresponding log-likelihood $l(.)$ values for AS-G FD model for both datasets are provided in Table 9. For the decision about the best fitting of the competing model, we computed several criteria measures such as the Akaike information criteria (AIc), the consistent Akaike information criteria (CAIc), the Bayesian information criteria (BIc), and the Hannan–Quinn information criteria (HQIc).

Table 9. The MLEs, SEs and corresponding log-likelihood $l(.)$ values for the AS-G FD model.

Datasets	Estimate	SE	$l(x; \cdot)$
Dataset 1	$\hat{\alpha} = 0.0003$	1.1977	-83.265
	$\hat{\lambda} = 1.0495$	0.1381	
	$\hat{\delta} = 3.55905$	0.5862	
Dataset 2	$\hat{\alpha} = 0.002$	1.083	-119.119
	$\hat{\lambda} = 59.518$	9.820	
	$\hat{\delta} = 1.300$	0.216	

From the results given in Tables 10 and 11, we noted that the AS-W model provides a better fit with the minimum value of AIC, AICc, BIC, HQIC, and KS and the largest p-values compared with other models considered in this work. Figures 6 and 7 support this assertion. Box plot and TTT plot along with the PP-Plot for the two real data sets are, respectively, presented in Figures 8 and 9.

Table 10. The goodness of fit tests for Dataset 1.

Model	AIC	AICc	BIC	HQIC	K-S	p-Value
AS-W	172.5304	173.2577	177.3632	174.2342	0.0907	0.9212
Sine-inverse Weibull	184.3137	184.6666	187.5355	185.4495	0.15862	0.3096
Inverse Weibull	190.8537	191.2066	194.0755	191.9896	0.1897	0.1394
WGQLD	206.7907	207.1436	210.0125	207.9265	0.2682	0.0097
Sine Burr XII	181.3963	181.7493	184.6181	182.5322	0.1423	0.4417

Table 11. The goodness of fit tests for Dataset 2.

Model	AIC	AICc	BIC	HQIC	K-S	p-Value
AS-W	244.239	245.439	247.7732	245.1767	0.1271	0.7875
Sine-inverse Weibull	251.187	251.7585	253.5431	251.8121	0.1546	0.5622
Inverse Weibull	255.0592	255.6306	257.4153	255.6843	0.1778	0.3884
WGQLD	252.8124	253.3839	255.1686	253.4375	0.1950	0.2824
Sine Burr XII	284.8518	285.4232	287.2079	285.4768	0.3609	0.0026

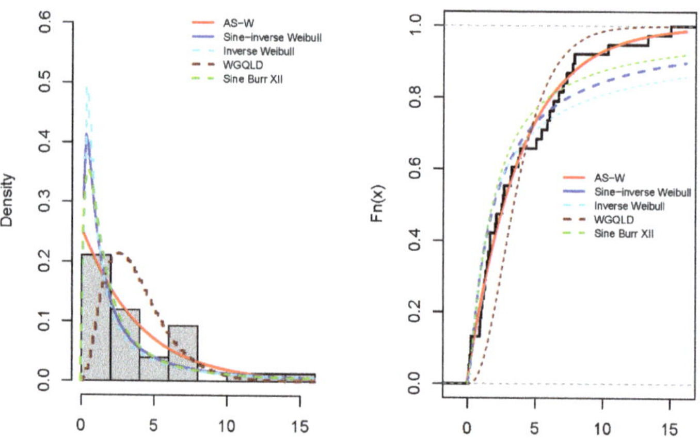

Figure 6. Plots of estimated probability density functions and cumulative distribution functions for Dataset 1.

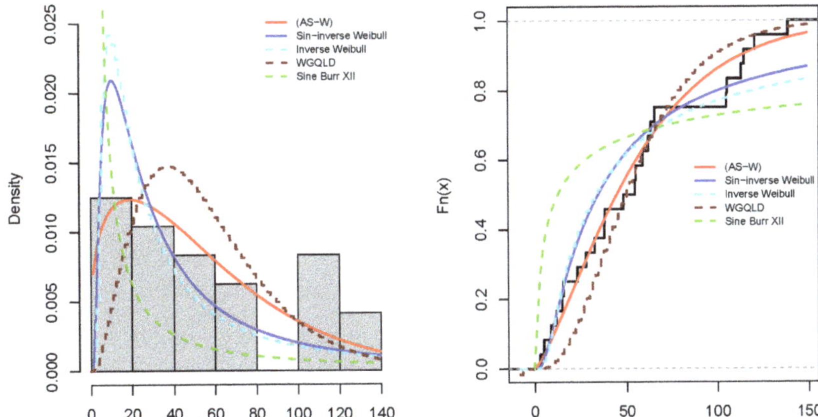

Figure 7. Plots of estimated probability density functions and cumulative distribution functions for Dataset 2.

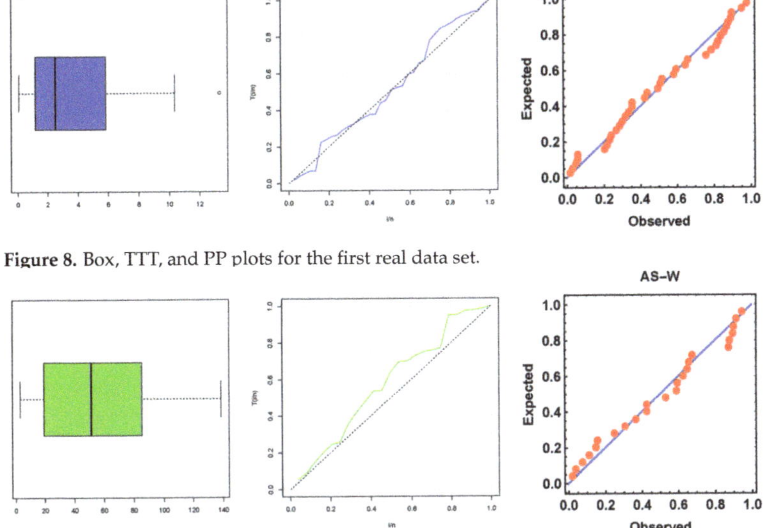

Figure 8. Box, TTT, and PP plots for the first real data set.

Figure 9. Box, TTT, and PP plots for the second real data set.

9. Conclusions

We intensely study a new family of distributions with a trigonometric function. We introduce an extra parameter to the sine transformation family and name it the alpha-sine-G family of distributions. Some important functional forms and properties of the family are provided in a general form. A specific three-parameter sub-model alpha-sine Weibull of this family is also introduced using Weibull distribution as a parent distribution; it is studied deeply. The statistical properties of this new distribution are investigated. From the graphical investigations of the PDF and HRF shapes, we find that the suggested model is versatile regarding skewness and kurtosis. Both the PDF and HRF can have either increasing or decreasing or bathtub or inverted bathtub or -j- or reverse-j-shaped curves according to the parameter values. Hence, the AS-W model is also capable of fitting highly skewed heterogeneous data sets. We obtain the estimates of AS-W parameters using several methods, including MLE, MPS, OLS, WLS, CV, and AD. A simulation experiment is carried out to justify these estimates further and finds that AEs nearly converge to the true values of the parameter, and MSEs are approaching zero as the sample size increases. We study two real data applications and demonstrate that the AS-W distribution is consistently the best model among all its competitors. Hence, we expect that the suggested family of distributions can be used to generate new flexible models for modeling real data, even heterogeneous data from different fields of application. For future works, many authors can use the new suggested family of distributions to generate new continuous statistical models, such as alpha-sine-power Lomax, alpha-sine-power Topp Leone and alpha-sine-power Lindley distributions.

Author Contributions: Conceptualization, S.B., L.P.S., A.A.M., V.K., R.H.K., A.M.G., M.E. and S.G.N.; methodology, S.B., L.P.S., A.A.M., V.K., R.H.K., A.M.G., M.E. and S.G.N.; software, S.B., L.P.S., A.A.M., V.K., R.H.K., A.M.G., M.E. and S.G.N.; validation, S.B., L.P.S., A.A.M., V.K., R.H.K., A.M.G., M.E. and S.G.N.; formal analysis, S.B., L.P.S., A.A.M., V.K., R.H.K., A.M.G., M.E. and S.G.N.; investigation, S.B., L.P.S., A.A.M., V.K., R.H.K., A.M.G., M.E. and S.G.N.; resources, S.B., L.P.S., A.A.M., V.K., R.H.K., A.M.G., M.E. and S.G.N.; data curation, S.B., L.P.S., A.A.M., V.K., R.H.K., A.M.G., M.E. and S.G.N.; writing—original draft preparation, S.B., L.P.S., A.A.M., V.K., R.H.K., A.M.G., M.E. and S.G.N.; writing—review and editing, S.B., L.P.S., A.A.M., V.K., R.H.K., A.M.G., M.E. and S.G.N.; visualization, S.B., L.P.S., A.A.M., V.K., R.H.K., A.M.G., M.E. and S.G.N.; supervision, S.B., L.P.S., A.A.M., V.K., R.H.K., A.M.G., M.E. and S.G.N.; project administration, S.B., L.P.S., A.A.M., V.K., R.H.K., A.M.G., M.E. and S.G.N.; funding acquisition, S.B., L.P.S., A.A.M., V.K., R.H.K., A.M.G., M.E. and S.G.N. All authors have read and agreed to the published version of the manuscript.

Funding: This research has no funds.

Conflicts of Interest: The authors declare no conflict of interest.

References

1. Tarantola, A. *Inverse Problem Theory and Methods for Model Parameter Estimation*; SIAM: Philadelphia, PA, USA, 2005.
2. Kumar, D.; Singh, U.; Singh, S.K. A new distribution using sine function-its application to bladder cancer patients data. *J. Stat. Appl. Probab.* **2015**, *4*, 417.
3. Gómez-Déniz, E.; Calderín-Ojeda, E. Modelling insurance data with the Pareto ArcTan distribution. *ASTIN Bull. J. IAA* **2015**, *45*, 639–660. [CrossRef]
4. Souza, L.; Junior, W.R.O.; de Brito, C.C.R.; Ferreira, T.A.E.; Soares, L.G.M. General properties for the Cos-G class of distributions with applications. *Eurasian Bull. Math.* **2019**, *2*, 63–79.
5. Souza, L.; Junior, W.; De Brito, C.; Chesneau, C.; Ferreira, T.; Soares, L. On the Sin-G class of distributions: Theory, model and application. *J. Math. Model.* **2019**, *7*, 357–379.
6. Mahmood, Z.; Chesneau, C.; Tahir, M.H. A new sine-G family of distributions: Properties and applications. *Bull. Comput. Appl. Math.* **2019**, *7*, 53–81.
7. Chesneau, C.; Jamal, F. The sine Kumaraswamy-G family of distributions. *J. Math. Ext.* **2020**, *15*, 1–26.
8. Muhammad, M.; Alshanbari, H.M.; Alanzi, A.R.A.; Liu, L.; Sami, W.; Chesneau, C.; Jamal, F. A new generator of probability models: The exponentiated sine-G family for lifetime studies. *Entropy* **2021**, *23*, 1394. [CrossRef]
9. Rahman, M.M. Arcsine-G Family of Distributions. *J. Stat. Appl. Probab. Lett.* **2021**, *8*, 169–179.
10. Chaudhary, A.K.; Sapkota, L.P.; Kumar, V. Some properties and applications of arctan generalized exponential distribution. *Int. J. Innov. Res. Sci. Eng. Technol. IJIRSET* **2021**, *10*, 456–468.
11. Isa, A.M.; Ali, B.A.; Zannah, U. Sine Burr XII Distribution: Properties and Application to Real Data Sets. *Arid. Zone J. Basic Appl. Res.* **2022**, *1*, 48–58. [CrossRef]
12. Muhammad, M.; Bantan, R.; Liu, L.; Chesneau, C.; Tahir, M.; Jamal, F.; Elgarhy, M. A New Extended Cosine-G Distributions for Lifetime Studies. *Mathematics* **2021**, *9*, 2758. [CrossRef]
13. Aldahlan, M.; Jamal, F.; Chesneau, C.; Elgarhy, M.; Elbatal, I. The truncated Cauchy power family of distributions with inference and applications. *Entropy* **2020**, *22*, 346. [CrossRef]
14. Alotaibi, N.; Elbatal, I.; Almetwally, E.; Alyami, S.; Al-Moisheer, A.; Elgarhy, M. Truncated Cauchy Power Weibull-G Class of Distributions: Bayesian and Non-Bayesian Inference Modelling for COVID-19 and Carbon Fiber Data. *Mathematics* **2022**, *10*, 1565. [CrossRef]
15. Shrahili, M.; Elbatal, I.; Elgarhy, M. Sine Half-Logistic Inverse Rayleigh Distribution: Properties, Estimation, and Applications in Biomedical Data. *J. Math.* **2021**, *2021*, 4220479. [CrossRef]
16. Shrahili, M.; Elbatal, I.; Almutiry, W.; Elgarhy, M. Estimation of Sine Inverse Exponential Model under Censored Schemes. *J. Math.* **2021**, *2021*, 7330385. [CrossRef]

17. Eugene, N.; Lee, C.; Famoye, F. Beta-normal distribution and its applications. *Commun. Stat. Theory Methods* **2002**, *31*, 497–512. [CrossRef]
18. Nadarajah, S.; Gupta, A.K. The exponentiated gamma distribution with application to drought data. *Calcutta Stat. Assoc. Bull.* **2007**, *59*, 29–54. [CrossRef]
19. Lemonte, A.J.; Barreto-Souza, W.; Cordeiro, G.M. The exponentiated Kumaraswamy distribution and its log-transform. *Braz. J. Probab. Stat.* **2013**, *27*, 31–53. [CrossRef]
20. Kenney, J.F.; Keeping, E.S. *Mathematics of Statistics*; D. Van Nostrand Company, Inc.: New York, NY, USA, 1962.
21. Moors, J.J.A. A quantile alternative for kurtosis. *J. R. Stat. Soc. Ser. D Stat.* **1988**, *37*, 25–32. [CrossRef]
22. Balakrishnan, N.; Cohen, A.C. *Order Statistics and Inference: Estimation Methods*; Academic Press: San Diego, CA, USA, 1991.
23. Cheng, R.C.H.; Amin, N.A.K. Estimating parameters in continuous univariate distributions with a shifted origin. *J. R. Stat. Soc. Ser. B Methodol.* **1983**, *45*, 394–403. [CrossRef]
24. Selim, M.A. Estimation and prediction for Nadarajah-Haghighi distribution based on record values. *Pak. J. Stat.* **2018**, *34*, 77–90. [CrossRef]
25. Murthy, D.P.; Xie, M.; Jiang, R. *Weibull Models*; John Wiley & Sons: Hoboken, NJ, USA, 2004.
26. Keller, A.Z.; Goblin, M.T.; Farnworth, N.R. Reliability analysis of commercial vehicle engines. *Reliab. Eng.* **1985**, *10*, 15–25. [CrossRef]
27. Benchiha, S.; Al-Omari, A.I.; Alotaibi, N.; Shrahili, M. Weighted generalized quasi Lindley distribution: Different methods of estimation, applications for COVID-19 and engineering data. *AIMS Math.* **2021**, *6*, 11850–11878. [CrossRef]

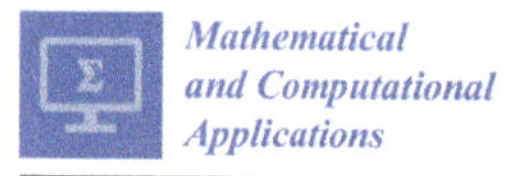
Mathematical and Computational Applications

Article

A Quantile Functions-Based Investigation on the Characteristics of Southern African Solar Irradiation Data

Daniel Maposa [1,*], Amon Masache [2] and Precious Mdlongwa [2]

1   Department of Statistics and Operations Research, University of Limpopo, Private Bag X1106, Sovenga 0727, South Africa
2   Department of Statistics and Operations Research, National University of Science and Technology, Ascot, Bulawayo P.O. Box AC 939, Zimbabwe; amon.masache@nust.ac.zw (A.M.); precious.mdlongwa@nust.ac.zw (P.M.)
*   Correspondence: daniel.maposa@ul.ac.za

Abstract: Exploration of solar irradiance can greatly assist in understanding how renewable energy can be better harnessed. It helps in establishing the solar irradiance climate in a particular region for effective and efficient harvesting of solar energy. Understanding the climate provides planners, designers and investors in the solar power generation sector with critical information. However, a detailed exploration of these climatic characteristics has not yet been studied for the Southern African data. Very little exploration is being done through the use of measures of centrality only. These descriptive statistics may be misleading. As a result, we overcome limitations in the currently used deterministic models through the application of distributional modelling through quantile functions. Deterministic and stochastic elements in the data were combined and analysed simultaneously when fitting quantile distributional function models. The fitted models were then used to find population means as explorative parameters that consist of both deterministic and stochastic properties of the data. The application of QFs has been shown to be a practical tool and gives more information than approaches that focus separately on either measures of central tendency or empirical distributions. Seasonal effects were detected in the data from the whole region and can be attributed to the cyclical behaviour exhibited. Daily maximum solar irradiation is taking place within two hours of midday and monthly accumulates in summer months. Windhoek is receiving the best daily total mean, while the maximum monthly accumulated total mean is taking place in Durban. Developing separate solar irradiation models for summer and winter is highly recommended. Though robust and rigorous, quantile distributional function modelling enables exploration and understanding of all components of the behaviour of the data being studied. Therefore, a starting base for understanding Southern Africa's solar climate was developed in this study.

Keywords: solar irradiation; quantile; quantile function; median rankit; population mean

Citation: Maposa, D.; Masache, A.; Mdlongwa, P. A Quantile Functions-Based Investigation on the Characteristics of Southern African Solar Irradiation Data. *Math. Comput. Appl.* **2023**, *28*, 86. https://doi.org/10.3390/mca28040086

Academic Editor: Sandra Ferreira

Received: 19 May 2023
Revised: 4 July 2023
Accepted: 19 July 2023
Published: 24 July 2023

1. Introduction

With ample sunshine in the Southern African region, an exploratory study of solar irradiation (SI) data can play a significant role in better understanding how this enormous source of energy can be harnessed in a bid to satisfy the energy demands within regional countries. However, solar irradiation is significantly affected by weather elements. In addition, most, if not all, meteorological features have error distributions with finite limits such that assuming normality of the distributions is not appropriate. As a result, deterministic models have intrinsic limitations when dealing with weather data that is characterised by such rapid-fluctuating uncertainties. Therefore, using the measures of central tendency (such as the mean) only to describe the characteristics of solar irradiation data is not enough. Exploring meteorological features using the statistics of the mean can be a misleading summary of a distribution.

As a result, to overcome these limitations, solar irradiation data can be modelled using quantile functions. We can learn the data's skewness, tails and outliers by plotting quantile function graphs. The application of quantile functions to exploratory data analysis considers the data's deterministic and stochastic elements.

1.1. Rationale of the Study

The Southern African region's solar irradiation data characteristics have not yet been studied according to the best of our knowledge. Most authors have been interested in forecasting solar irradiation, and they have been using locational data of at most three sites from the region within the same country. Very little exploration of this data has been done. The little exploratory analysis conducted has focused on measures of central tendency or the statistics of the mean per se. In addition, of course, with the interpretation of measures of dispersion, the standard error and kurtosis are the commonly used explorative statistics to describe the variability of the data. However, data exploration that ends with measures of central tendency and dispersion can be a misleading analysis [1]. The big challenge comes with efforts to explore solar irradiation data in the Southern African region with a minimum error of misleading results. A complete understanding of this data is desired. Therefore, an approach that satisfies this completeness can be the introduction of quantile functions in the exploratory analysis. In addition to the deterministic element, quantile functions model the stochastic element of the data which cannot be done using the statistics of the mean. The two elements of the solar irradiation data can be developed with a common construction kit approach [2]. In addition, the use of quantile functions is part of distributional modelling which cannot be done when exploring data using the statistics of the mean. Moreover, the analysis of empirical distributions tends to focus on only the stochastic element of the variable. Empirical distributions are much more suitable than exploratory analysis for forecasting modelling.

1.2. Contribution of the Study

This explorative investigation helps with the establishment of the solar irradiance climate in the Southern African region. Instead of exploring the deterministic component only, and separately (by applying the statistics of central tendency) and then again exploring separately the stochastic element through a simple analysis of empirical distributions, a complete exploration can be done through quantile distributional function models (QDFMs). In addition, some approaches to solar irradiance modelling are non-parametric like the complete-history persistence ensemble (CH-PeEn) developed by [3]. They lack inferences of statistic(s) like population mean that can be used to describe the behaviour of SI, especially the physical characteristics inherent in the stochastic component. The statistical characteristics and climate of solar irradiation that are explored help planners and designers in the solar panels manufacturing industry and solar power generation sector. They can understand better the factors that affect the efficient generation of solar power. The study may help investors to appreciate how investing in solar power generation can be profitable financially and socio-economically by exposing the characteristics of Southern African solar irradiation into the finance world. Experts in meteorological services will be made aware of how solar irradiation weather studies can be enhanced. Researchers and academics can be made aware of the new data exploratory technique of QDFMs which completes the description of data characteristics by combining deterministic and stochastic elements of variables.

1.3. Review of Literature

Several previous studies on solar irradiation in the Southern African region have been conducted dating back to as early as 1983 by Jain. Unfortunately, only a few have included study of the characteristics of radiation. The majority of the studies were concentrating on measuring and/or predicting (forecasting) solar irradiation in the different countries of the region. The earliest study that included an analysis of the characteristics of solar irradiation in the region, according to the best of our knowledge, was done by J. Andringa in 1988. They used monthly averages to establish the SI pattern in Botswana. Another early study was done by [4], and they concluded that SI data from Botswana showed weak non-seasonal effects while moving average parameters showed strong seasonal effects. Later, [5] reached the same conclusions as [6] after observing that Malawi SI data had average daily maximums in October and minimums in January. This highly significant seasonality characteristic in SI made [7] split the Ritchersveld training data set into two samples, one from January to May and the other one from June to December. Ref. [7] are the only authors so far, according to the best of our knowledge, who have done a periodogram analysis of SI in the region. They identified the largest ordinate periods and produced the harmonic frequencies of the ordinate periods. All of the ordinates they identified were highly significant at a 1% level of significance when using a Fisher's G-test. One of the latest studies to confirm seasonality in SI data was done by [8] using the University of Pretoria data. The interpretation of constructed box plots was used to deduce seasonality in the data. They also came up with a monthly pattern of the data. Earlier on, [9] had already produced a detailed daily SI pattern for Sebele data. They concluded that solar conditions during the summer and winter months tend to be uniform over consecutive months (i.e., the SI series had a memory of two months). Therefore, the data had a persistent pattern. The same conclusion was also made by [10]. Ref. [11] discovered that the introduction of this persistent pattern improved their model performance when predicting distillate production while monitoring meteorological conditions at Malawi Polytechnic. On the other hand, shortly before, their solar distilled water project [12] concentrated on the relationship between SI and the sky clearness index. Their results confirmed that the SI pattern is associated with sky clearness (sunshine duration) or cloud cover. Ref. [13] concurred by deducing that the SI pattern depends on sunshine duration. Probably that was the rationale [14] that applied the K-means algorithm when classifying sunshine duration into four classes. Previously, [15] had already improved the quality of this classification by cutting the hierarchical tree and further produced a fifth class of 'good weather' throughout the day with intermittent clouds passing over.

Other researchers like [16–19] described the distribution of SI in different parts of SA using the measures of skewness and kurtosis. They all found their data to be positively skewed and platykurtic, that is, SI did not follow a normal probability distribution. The non-normality of the data was confirmed by the constructed Q-Q plots which exhibited non-linear relationships between the theoretical and sample quantiles. Refs. [16–19] went further to extract non-linear trends from their respective data sets by fitting penalised cubic smoothing spline functions. They also constructed time plots as well as density plots; however, the time plots constructed by [19] exhibited some dominant cycles. In addition to the various plots they constructed, they computed some measures like the minimum, mean, median and quartiles to describe the SI. Though the data were from different parts of South Africa, the different researchers reached the same conclusions regarding SI characteristics.

However, none of the previous studies reviewed in this study fitted a probability distribution and used it to describe SI. They all concentrated on the statistics of the mean. In contrast, we extend the property description of SI through application of the statistics of quantiles. This includes analysing a fitted QDFM which has never been done in previous studies when investigating the characteristics of SI in the Southern African region and beyond.

2. Materials and Methods

Expressing statistical ideas in terms of quantile functions gives a new perspective on data exploration which is simpler and clearer. Quantile functions enable distributional model development with a common construction kit approach including both the deterministic and stochastic elements in the process. This implies that QDFM can present both deterministic and stochastic components of SI. If we denote a quantile function $Q(p)$ as a function that gives quantile values for all probabilities p, $0 \leq p \leq 1$ then a quantile can be defined as the observation that corresponds to a specified proportion of an ordered sample. That is, if x lies on a proportion p of the way through the data set of n observations, then $x_{(r)}$ lies a proportion p_r of the way through the data set. Therefore, $(x_{(r)}, p_r)$ describes the data where $x_{(r)}$ is the rth observation in the data set and $p_r = \frac{r}{n}$.

2.1. Quantile Functions

If we let X be the random variable and $p = P(X \leq x)$ then we can formally define a quantile function (QF) as follows:

$$x_p = Q(p), \tag{1}$$

where x_p is the p-quantile of the population and $p = F(x)$ is the cumulative distribution function (CDF) such that,

$$Q(p) = F^{-1}(p) \quad \text{and} \quad F(x) = F^{-1}(x). \tag{2}$$

That is, the plot of $Q(p)$ against p corresponds to the plot of x against p. It has to be noted that an empirical distribution replaces the cumulative distribution in practice. According to [20], the p-quantile can be written as

$$x_p = \underbrace{\text{argmin}}_{x} E\left[\rho_p(X - x)\right],$$

for each $p \in (0, 1)$ and ρ_p is the quantile loss function given by

$$\rho_p = \begin{cases} up, & \text{if} \quad u \geq 0 \\ u(p-1), & \text{if} \quad u < 0. \end{cases}$$

Since this quantile loss function is not differentiable, then the statistics of central tendency cannot be applied in a quantile analysis context. The estimate of the p-quantile is computed as a sample quantile, and we consider Theorem 1 (the result of Linderberg's central limit theorem) when finding its asymptotic distribution.

Theorem 1. *Given a random variable X with associated cumulative distribution function F(x), that is continuous in a neighbourhood of the p-th quantile of interest, with $f(x_p) > 0$. Then, the asymptotic distribution of the sample quantile, x'_p, is given by*

$$\sqrt{n}(x'_p - x_p) \xrightarrow{d} N(0, \sigma^2),$$

where $\sigma^2 = \frac{p(1-p)}{f^2(x_p)}$ and $N(0, \sigma^2)$ represents the Gaussian distribution with zero mean and variance σ^2.

If we introduce $S(p)$ as the QF of the basic form of a probability distribution, then

$$Q(p) = \lambda + \eta S(p, \alpha), \tag{3}$$

where λ and η are the position and scale parameters, respectively, and α has components that give the shape parameter of the 'basic distribution'. We assume that:

1.   the uniform transformation rule applies and
2.   ordered U_r leads to the corresponding ordered X_r such that

$$X_r = Q(U_r).$$

We also introduce the statistics of the median and the median rankit, where percentiles are applicable. So, we treat quantile basic forms as QDFM components to provide a flexible and effective means of constructing distributions that mimic observed data properties. The most important property of quantile basic forms is that we can compute the population mean by evaluating the integral of the QDFM overall percentiles [21,22],

$$\mu = \int_0^1 Q(p)dp. \tag{4}$$

This population mean describes simultaneously both the deterministic and stochastic components of a variable. In addition, [18] listed the following two main properties of quantile functions.

1. If X has a quantile distribution, $R(p)$, on the positive axis, $0 \le x < 1$, then the distribution $-R(1-p)$ is the quantile distribution that is its reflection in the axis at $x = 0$, called the reflected distribution on $-1 < x \le 0$.
2. The reciprocal $1/X$ has the reciprocal distribution $1/R(1-p)$ also on $0 \le x < 1$.

2.2. Method of Percentiles

The method involves equating population and sample quantiles (percentiles) on distributions defined by their quantile functions. Percentiles are descriptive statistics of positions (the centrality) of ordered data. These positions are the expected values of the observations in the data set. Letting $p(r)$, $r = 1, 2, 3, \ldots, n$ to be the corresponding ordered sequence probabilities of $X_{(1)}, X_{(2)}, X_{(3)}, \ldots, X_{(n)}$, then any quantile distribution $X = Q(p)$ can be generated from a uniform distribution U on the domain $(0, 1)$ by $X = Q(U)$. That is, ordering X corresponds to ordering U as in (5) here under:

$$X_{(r)} = Q(U_{(r)}). \tag{5}$$

We now obtain the mean of the distribution of the rth order statistic from the uniform distribution as,

$$\overline{p}(r) = \frac{1}{n+1}, \tag{6}$$

and the median is given by:

$$p_M(r) = IIB(0.5, r, n+1-r). \tag{7}$$

IIB in (7) is the acronym for the inverse of the incomplete beta function. $IIB(p, r, n + 1 - r)$ generally gives the quantile distribution for the ordered statistics. Thus, the median for $X_{(r)}$, technically called the median rankit is defined as

$$\text{Median}\left(X_{(r)}\right) = Q\left(\text{Median}\left(U_{(r)}\right)\right) = Q(p_M(r)). \tag{8}$$

Therefore, we analyse the centrality of ordered data, which is ignored by most statistical estimation methods.

2.3. Parameter Estimation

The natural approach to estimating parameters using quantile-based models is the method based on minimising the differences between ordered observations and their predictions. That can be done using either the distributional least squares (DLS) technique (which uses the mean rankit) and/or the distributional least absolute (DLA) technique. The techniques are based on developing some measure of lack of fit (LoF), i.e., fitting a distribution based on deviations between ordered observations and some measure of position derived from the fitted model. In some cases, the mean rankit does not exist; as a result, we extend the parameter estimation procedure by using the median rankit. Thus, we introduce the DLA technique in the parameter estimation exercise. When applying the DLA technique, the best QDFM fit is obtained from parameters that minimise,

$$D_A = \sum \left| x_{(r)} - M_{(r)} \right|, \tag{9}$$

such that the measure of the best fit is the distributional mean absolute error (DMAE), where

$$DMAE = \frac{D_A}{n}.$$ (10)

In Equation (9), $M(r)$ is the median of the distribution of $X(r)$ obtained from the median rankit. The DLA technique is associated with the least absolute deviation (LAD) technique in linear regression. LAD supersedes the ordinary least squares (OLS) technique in that it is resilient to outliers and more accurate as the sample size gets larger. However, LAD is computationally extensive.

2.4. Model Validation

2.4.1. Graphical Analysis

Ref. [22] recommended the use of graphical inspection of suitable plots for testing the adequacy of quantile functions as shown in Table 1.

Table 1. QDFM validation plots.

Name of Plot	y	Against	Comment
Fit observation	$x_{(r)}$	$Q'(p_r)$	Points to exhibit an approximately linear pattern
Distributional plots	$f_r = x_{(r)} - Q'(p_r)$	$Q'(p_r)$	Points to be randomly distributed

2.4.2. Chi-Square Goodness of Fit Test

Hosmer and Leme use a chi-square test statistic on the null hypothesis that the model is a good fit for the data. An insignificant p-value indicates that we fail to reject the null hypothesis.

3. Results and Discussions

3.1. Ground-Based Data

Ground-based data from the Southern African Universities Radiometric Association Network (SAURAN) website was used, and the radiometric stations have geographical locations as shown in Table 2. Some of the stations are currently inactive as shown on the map in Figure 1.

Table 2. SAURAN stations.

Station	Latitude	Longitude	Location	Period
University of Venda (UV)	−23.13100052	30.42399979	Venda	April 2015–April 2022
University of Pretoria (UP)	−25.75308037	28.22859001	Pretoria	July 2017–June 2021
University of KwaZulu-Natal Howard College (UKZNH)	−29.87097931	30.97694969	Durban	December 2015–September 2022
Stellenbosch University (SUN)	−33.92810059	18.86540031	Cape Town	July 2017–June 2021
Namibian University of Science and Technology (NUST)	−22.56500053	17.07500076	Windhoek	July 2017–June 2021
University of Gaborone (UG)	−24.6609993	25.93400002	Gaborone	January 2015–November 2020

3.2. Hourly Solar Irradiance Distributional Modelling

Solar irradiance (SI) for a particular day is significantly affected by the time horizon. This is supported by the time plots from all of the locations which have a general pattern shown in Figure 2. When measured in hours starting from midnight to midnight, [23] demonstrated that ignoring sidebands in the data causes overshoots just before sunrise and after sunset. As a result, we use up to 3 cycles per day which consider the sidebands.

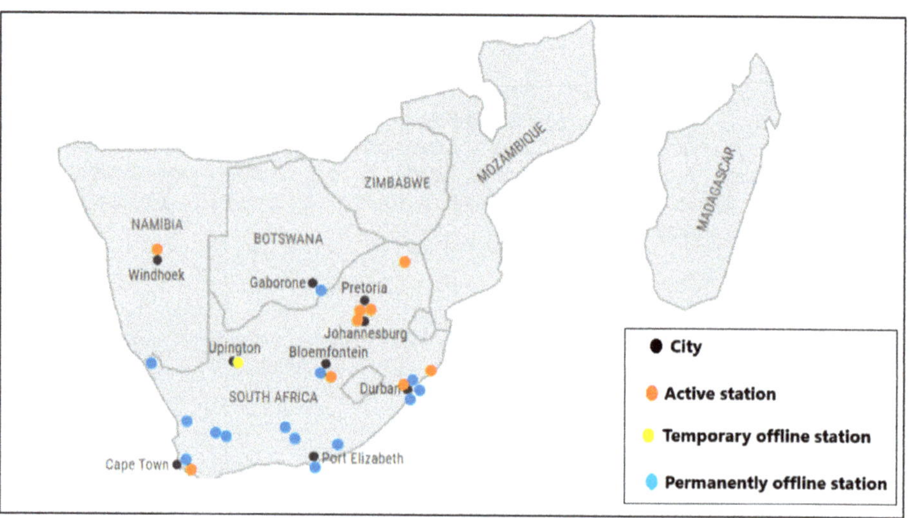

Figure 1. Radiometric Stations in Southern Africa (Source: www.sauran.ac.zw, accessed on 12 June 2022).

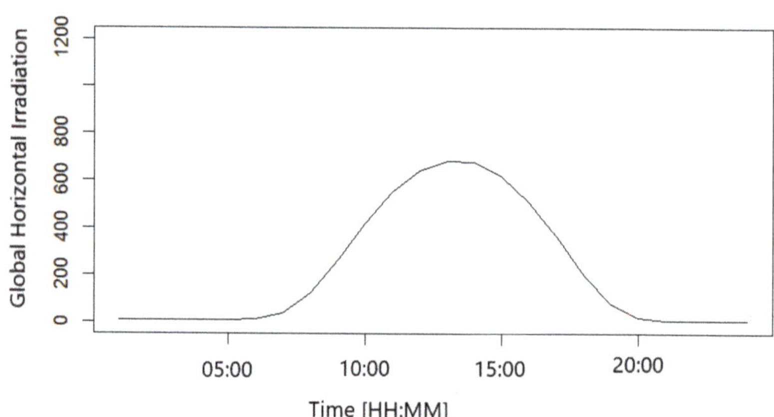

Figure 2. Day's hourly profile.

Ref. [23] modelled this hourly profile for a particular day through a Fourier series. Thus, the mean function of SI in an hour for the three cycles in a day can be modelled as follows:

$$y_t = \beta_0 + \beta_1 Cos\left(\frac{\pi}{12}t\right) + \beta_2 Sin\left(\frac{\pi}{12}t\right) + \beta_3 Cos\left(\frac{2\pi}{12}t\right) + \beta_4 Sin\left(\frac{2\pi}{12}t\right) + \beta_5 Cos\left(\frac{3\pi}{12}t\right) + \beta_6 Sin\left(\frac{3\pi}{12}t\right) + \varepsilon \qquad (11)$$

The Fourier series expansion model should satisfy the following constraints:

- $y_{sunrise} = y_{sunset} = 0.$
- $y_{sunrise-1hr} = y_{sunset+1hr} = 0.$

As a result, this profile is considered on the QDFM of the SI hourly distribution such that we apply the following regression quantile distributional model as suggested by [2]:

$$Q_y(p|t) = y_t + \eta S(p, \alpha, \gamma, \delta, \tau) \qquad t = 1, 2, 3, \ldots, 24. \qquad (12)$$

where $S(p, \alpha, \gamma, \delta, \tau)$ is the basic quantile distribution function of the residuals (from the Fourier series expansion model in (11)) described by α, γ, δ and τ, the respective shape, scale, skewness and kurtosis parameters. We assume that $E(\varepsilon) = 0$ and $S(0.5) = 0$. That is, the deterministic part of the distributional model in (12) becomes Galton's median regression line. This means that

$$M[S(U_r)] \; = \; S(p*) \; = \; M_r \tag{13}$$

which is called the median rankit for $p^* = IIB(0.5, r, n + 1 - r)$.

3.2.1. Venda and Gaborone Hourly Quantile Profiles

The 'fitdistrplus' R package developed by [24] automatically selects the best distribution that particular data follows. The package estimates the distribution parameters through a default maximum likelihood optimisation algorithm. As a result, the residuals on fitting the SI Fourier series for the Venda and Gaborone hourly profile followed a skew normal type 2 (SN2) distribution with the probability distribution parameters as estimated in Table 3. The 'gamlss.dist' R package developed by [25] was used to fit the distributions as shown in Figure A1. That is, the fitted QDFM is as shown in (14),

$$Q_y(p|t) = \beta_0 + \beta_1 Cos\left(\tfrac{\pi}{12}t\right) + \beta_2 Sin\left(\tfrac{\pi}{12}t\right) + \beta_3 Cos\left(\tfrac{2\pi}{12}t\right) + \beta_4 Sin\left(\tfrac{2\pi}{12}t\right) + \beta_5 Cos\left(\tfrac{3\pi}{12}t\right)$$
$$+ \beta_6 Cos\left(\tfrac{3\pi}{12}t\right) + \eta \begin{cases} \alpha + \tfrac{\gamma}{\delta}\Phi^{-1}\left(\tfrac{p(1+\delta^2)}{2}\right), & p \le (1+\delta^2)^{-1} \\ \alpha + \gamma\delta\Phi^{-1}\left(\tfrac{p(1+\delta^2)-1+\delta^2}{2\delta^2}\right), & p > (1+\delta^2)^{-1}. \end{cases} \tag{14}$$

so that the model parameters are as shown in Table 4.

Table 3. Venda and Gaborone distributional parameters.

Location	Shape	Scale	Skewness
Venda	22.676906	-2.308079	-5.612271
Gaborone	23.233404	2.127659	-1.204687

Table 4. Venda and Gaborone model parameters.

Location	$\hat{\beta}_0$	$\hat{\beta}_1$	$\hat{\beta}_2$	$\hat{\beta}_3$	$\hat{\beta}_4$	$\hat{\beta}_5$	$\hat{\beta}_6$	$\hat{\eta}$
Venda	143.24	-327.52	-55.60	148.90	57.37	-17.33	18.81	2.02
Gaborone	422.36	-372.09	-92.34	163.73	71.42	-16.13	-6.71	-8.17

3.2.2. Durban, Pretoria, Cape Town and Windhoek Hourly Quantile Profiles

The residuals on Durban followed a skew exponential power type 3 distribution and the Cape Town and Windhoek profiles followed a sinh-arcsinh distribution. However, the skew exponential power type 3 and sinh-arcsinh probability distributions do not have corresponding quantile functions as yet. As a result, the closest alternative probability distribution is a normal or Cauchy distribution. The results in Table 5 show that the normal distribution better fits the residuals for the three locations than the Cauchy distribution. Thus, the fitted normal distributions (as second best fits) using the 'fitdistrplus' R package are shown in Figure A1.

The Durban and Cape Town residuals from the Fourier series model had means of -2.3122×10^{-16} and 1.1102×10^{-16} and standard deviations of 11.0653 and 13.4113 respectively. The residuals had also respective skewness of 0.051 and -0.055. As a result, the fitted QDFM is

$$Q_y(p|t) = \beta_0 + \beta_1 Cos\left(\tfrac{\pi}{12}t\right) + \beta_2 Sin\left(\tfrac{\pi}{12}t\right) + \beta_3 Cos\left(\tfrac{2\pi}{12}t\right) + \beta_4 Sin\left(\tfrac{2\pi}{12}t\right) + \beta_5 Cos\left(\tfrac{3\pi}{12}t\right)$$
$$+ \beta_6 Cos\left(\tfrac{3\pi}{12}t\right) + \eta\left[\mu + \sigma\Phi^{-1}(p)\right]. \tag{15}$$

Table 5. Residual fitted distribution comparisons.

Location	Metric	Normal	Cauchy
Durban	AIC	187.4920	199.3287
	BIC	189.8481	201.6848
Cape Town	AIC	196.7216	211.7815
	BIC	199.077	214.1376
Windhoek	AIC	218.9350	222.8473
	BIC	221.2911	225.2034

The residuals from the Windhoek and Pretoria deterministic models had a mean ($\mu_{\text{NUST}} = 0.2567696$, $\mu_{\text{UP}} = -1.15597$) and standard deviation of ($\sigma_{\text{NUST}} = 21.3035529$, $\sigma_{\text{UP}} = 2.77733$). However, the residuals from the Windhoek and Pretoria deterministic models have respective skewness of 0.162308 and -0.1442648, which cannot be ignored (that is, the skewness cannot be approximated to zero). That is, the residuals are suggesting some skewness, so considering a skewed lambda quantile distribution (in Equation (16)) for the residuals will give better results [21]. Therefore, we fit the following QDFM for the Pretoria and Windhoek hourly profiles. Thus, the estimated parameters are shown in Table 6.

$$Q_y(p|t) = \beta_0 + \beta_1 Cos\left(\frac{\pi}{12}t\right) + \beta_2 Sin\left(\frac{\pi}{12}t\right) + \beta_3 Cos\left(\frac{2\pi}{12}t\right) + \beta_4 Sin\left(\frac{2\pi}{12}t\right) + \beta_5 Cos\left(\frac{3\pi}{12}t\right) \\ + \beta_6 Cos\left(\frac{3\pi}{12}t\right) + \frac{\eta}{2\sigma}\left[(1-\delta)p^\sigma - (1+\delta)(1-p)^\sigma\right]. \tag{16}$$

Table 6. Pretoria, Cape Town and Windhoek model parameters.

Location	$\hat{\beta}_0$	$\hat{\beta}_1$	$\hat{\beta}_2$	$\hat{\beta}_3$	$\hat{\beta}_4$	$\hat{\beta}_5$	$\hat{\beta}_6$	$\hat{\eta}$
Durban	186.88	−300.05	−27.46	145.53	28.01	−26.56	−11.13	1.089
Cape Town	220.88	−309.44	−111.00	110.03	91.52	−6.93	−11.83	1.034
Windhoek	267.82	−400.60	−137.34	159.85	114.20	−27.07	−29.39	3676.63
Pretoria	247.62	−362.25	−54.47	163.33	51.28	−23.25	−10.66	−312.92

3.2.3. Hourly Population Means

On average, the daily maximum irradiance was observed at 13:00 on all the stations considered, with either the second or third maximum taking place at 12:00 or 14:00. Using the hourly profile QDFMs fitted for each location, we can then estimate the population means at 12:00 up to 14:00 as follows:

$$\mu_t = \int_0^1 Q(p|t)dp, \qquad t = 12, 13, 14. \tag{17}$$

Now, some QDFMs discussed in previous sections include the inverse cumulative distribution function (CDF) of the standard normal distribution, $\Phi^{-1}(p)$. We adopt the method suggested by [26] of probabilistic polynomial approximations to evaluate the inverse. Researchers like [27,28] and the latest [29] concentrated on approximating the CDF. Ref. [29] are claiming to have the most accurate approximation using both the MATLAB Global Optimization Toolbox and BARON, but they did not document evaluating the inverse of the CDF. The approximation developed by [26] is explicit and has an acceptable maximum absolute percentage relative error (APRE) of 1.4×10^{-2}. We find their approximation function simple and very accurate for the purposes of estimating the population mean SI in any time interval of interest. Therefore, Table 7 shows the estimated population mean of the average SI for 12:00, 13:00 and 14:00 time hours at each location.

Table 7. 12:00–14:00 population means (Wh/m^2).

Location	12:00	13:00	14:00
Venda	704.5501	724.3324	664.2824
Pretoria	792.3848	798.1858	720.3530
Durban	653.7334	646.0031	566.3265
Cape Town	647.2710	702.8115	690.4624
Windhoek	856.5969	927.0284	892.8881
Gaborone	789.5647	814.5785	756.4473

That is, for a period of 13:00 $\pm$ 2 h we can have an accumulative radiation of at least 3000 Wh/m^2 which is the amount of energy required to fully charge a 12 Volt and 250 Amp solar battery. This means that given the correct solar panel capacity such a solar battery can be fully charged in at least five hours i.e., a period from 11:00 up to 15:00 at any of the locations in the Southern Africa region.

3.3. Daily Total SI Distributional Modelling

The daily total SI distribution is not that significantly influenced so much by other variables in such a way that it is not necessary to consider other meteorological features when modelling its quantile distribution. That is, a day's total SI distribution for a particular month is presumed identical. The basic quantile functions $S(p,\alpha)$, considered on each month's daily total fitted QDFMs at the locations under study are shown in Table 8. If we look at the population mean daily totals in Table 9, location by location then the maximums in a year were all received in summer (i.e., either November, December or January), except for Windhoek which has its maximum in autumn. The maximum population mean daily totals are shown in bold for each location. All locations receive their population mean daily total minimums in winter. Our results contrast with the conclusion drawn by [6] who had a maximum taking place in October and a minimum in January, though they analysed daily averages for Malawi.

Table 8. Probability distributions' quantile functions.

Probability Distribution	Quantile Function
Normal	$\mu + \sigma\Phi^{-1}(p)$
Lognormal	$\mathrm{Exp}(\mu + \sigma\Phi^{-1}(p))$
Skewed Lambda	$\frac{1}{2\sigma}((1-\delta)p^\sigma - (1+\delta)(1-p)^\sigma)$
Weibull	$\alpha(-\log(1-p))^{1/\gamma}$
Gumbel	$\alpha + \gamma\log(-\log(1-p))$
Reverse Gumbel	$\alpha - \gamma\log(-\log(1-p))$
Logistic	$\alpha + \gamma\log\left(\frac{p}{1-p}\right)$
Cauchy	$\alpha + \gamma Tan(\pi(p-0.5))$
Weibull Type 3	$\beta(-\log(1-p))^{1/\gamma}$

We see it as not a proper descriptive analysis to consider the daily average because the minimum SI on every single day is always zero. In addition, SI is always approximately equal to zero from sunset progressing through the night up to sunrise. However, on some clear nights, we may have significant but very low SI readings. As a result, meaningful daily average analysis has to exclude readings from sunset up to sunrise when targeting the solar power generation industry. On the other hand, comparing the mean daily totals across the locations on each month Windhoek receives the maximum (daily population mean totals with an asterisk) in 75% of the year except for January, February and October. It is Cape Town, instead, which receives maximums in those other three months.

Table 9. Daily total population means (Wh/m^2).

Month	Venda	Pretoria	Durban	Cape Town	Windhoek	Gaborone
January	5808.48	6570.46	**7419.84**	8350.78 *	7966.67	7045.33
February	5118.63	5796.38	5569.62	7339.92 *	6655.05	6741.43
March	5328.46	5549.78	5727.71	5478.89	6969.69 *	5847.43
April	4218.16	4563.87	3869.33	4241.18	5855.68 *	5143.91
May	4189.18	4626.59	2832.39	3321.19	5183.17 *	4593.42
June	4207.39	4002.05	3543.30	2380.00	4946.30 *	4292.30
July	4463.09	4554.78	3146.75	3077.00	5109.11 *	4522.42
August	4338.57	5237.01	4393.84	3331.33	10,342.86 *	3966.38
September	5820.81	6381.69	4684.33	4937.00	**10,678.41 ***	6310.75
October	5441.11	6508.65	5773.34	7396.06 *	7342.81	6881.60
November	**5992.28**	7045.96	5197.02	7909.29	8022.61 *	**7370.91**
December	5786.87	**7165.13**	7118.95	8392.25	8799.95 *	6856.38
Maximum	5992.28	7165.13	7419.84	8350.78	10,678.41	7370.91
Minimum	4189.79	4002.05	2832.39	2379.96	4946.30	4292.30

* means a monthly maximum and bold means a locational maximum.

3.4. Monthly Total SI Distribution Modelling

The monthly total SI for a particular year is significantly affected by the month. The deterministic component of monthly totals is suspected to be affected by the seasons of summer and winter because from Table 9 we can conclude that the daily population mean totals are affected by seasonal variation. This agrees with the results of [30], which showed that SI greatly changed its pattern according to seasonal variation. Figure 3 exhibits some cyclical variations in the monthly totals at all locations. As a result, we can attribute these cyclical variations to seasonal effects that were also discovered by [5–7] from different countries in Southern Africa. Thus, our cycle must have a period of 12 months. Therefore, we can fit the deterministic component of the monthly totals as the following trigonometric regression model:

$$y_t = \beta_0 + \beta_1 Cos\left(\frac{\pi}{12}t\right) + \beta_2 Sin\left(\frac{\pi}{12}t\right) + \varepsilon \tag{18}$$

If a trend is observed on the time series plot of the monthly totals, then a trend component can be added to the deterministic model as follows:

$$y_t = \beta_0 + \beta_1 t + \beta_2 Cos\left(\frac{\pi}{12}t\right) + \beta_3 Sin\left(\frac{\pi}{12}t\right) + \varepsilon. \tag{19}$$

Thus, the quantile distribution of the monthly totals can now be modelled as

$$Q_y(p|t) = y_t + \eta S(p, \alpha, \gamma), \tag{20}$$

where $S(p, \alpha, \gamma)$ is the quantile distribution function of the residuals, ε, from the trigonometric regression model. However, the time series plots exhibited in Figure 3 show that we can suspect a trend in the Pretoria and Venda monthly totals' time series, but fitting both the trigonometric regression models with and without a trend gave the results in Table 10. We can conclude that monthly total solar irradiance in the Southern African region is neither increasing nor decreasing. There is no significant trend in SI monthly totals from year to year. However, it is evident that due to global warming, atmospheric temperatures are increasing [31–33]. In contrast, our time series plots and model comparisons do not show that. Thus, the effects of global warming may not be influencing SI in the Southern African region. Rather, in variable selection concepts, the temperature is a significant explanatory variable for SI as demonstrated by researchers like [8,16,34,35] who had the meteorological feature as one of the important predictors of SI in their forecasting models. As a result, all of the QDFMs for the monthly totals are fitted without considering trend regression being part of the deterministic component.

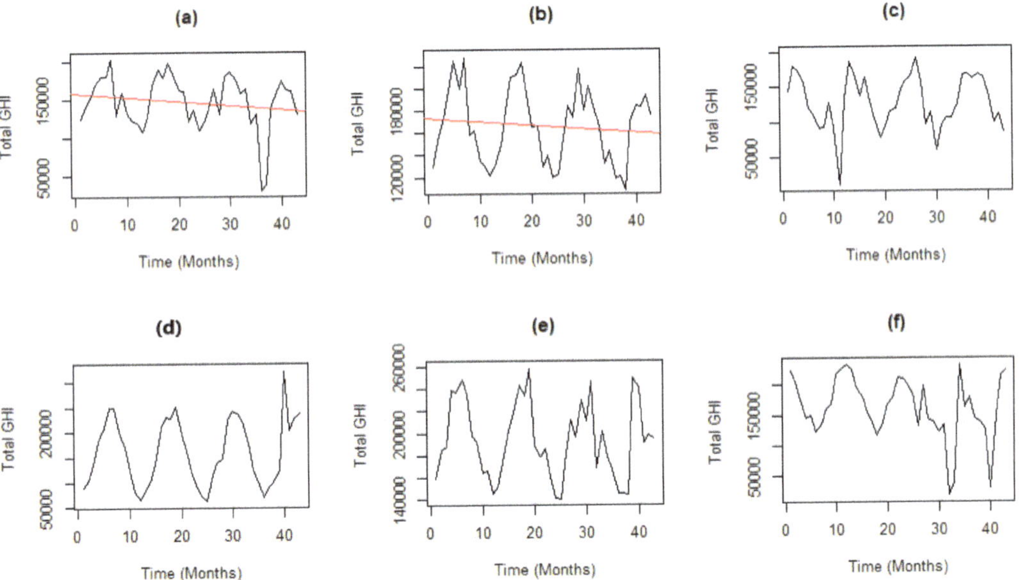

Figure 3. Monthly total solar irradiation for (**a**) Venda; (**b**) Pretoria; (**c**) Durban; (**d**) Cape Town; (**e**) Windhoek; (**f**) Gaborone.

Table 10. Trend model AIC comparison.

Location	With	Without
Venda	266.7684	265.613
Pretoria	256.3586	255.4424

The residuals for Cape Town and Durban followed sinh-arcsinh and skew exponential power type 2 distributions, respectively. Like the sinh-arcsinh distribution, the skew exponential power type 2 distribution does not have an existing quantile function. Likewise, we compare the closest two distributions to them as shown in Table 11. As a result, the better distribution was the normal distribution. Figure A4 shows the fitted normal distributions.

Table 11. Comparisons of residual distributions on Cape Town.

Location	Metric	Normal	Cauchy
Cape Town	AIC	187.4920	199.3287
	BIC	189.8481	201.6848
Durban	AIC	268.5895	271.3327
	BIC	269.5593	272.3025

The residuals in the other locations were best fitted by the distributions shown in Table 12 and are also shown graphically as in Figure A4. Our results are in tandem with the results from [36]. The original residual distributions are different over the year and the day. However, because some distributions do not have existing quantile functions, Durban and Cape Town had the same second-best-fitted distribution over the day and the year. The fitted QDFMs for the monthly totals have the estimated parameters as shown in Table 12. All stations received maximum total population mean solar irradiation during summer and minimum in winter. These results agree with the seasonality in SI observed by researchers who studied the meteorological feature in Southern Africa. Durban is receiving the maximum total population mean all year round of all the locations considered, while the minimum is received in Cape Town (Figure 4). Therefore, Durban is the best location to set up a solar farm in the region when considering the monthly accumulated solar irradiation.

Table 12. Monthly total SI model parameters.

Location	Probability Distribution	$\hat{\beta}_0$	$\hat{\beta}_1$	$\hat{\beta}_2$	$\hat{\eta}$	$\hat{\alpha}$	$\hat{\gamma}$
Venda	R. Gumbel	1,678,882.00	−8767.19	40,937.26	2013.06	−768.98	9.11
Pretoria	R. Gumbel	3,692,969.00	−9175.68	20,756.98	4163.51	−852.62	8.72
Windhoek	SN2	−24,798,121	−5434.35	36,610.50	2870.26	8700.05	−0.69

Location	Probability Distribution	$\hat{\beta}_0$	$\hat{\beta}_1$	$\hat{\beta}_2$	$\hat{\eta}$	$\hat{\mu}$	$\hat{\sigma}$
Cape Town	Normal	155,245.11	12,380.08	82,328.01	−39.04	-2.31×10^{-16}	11.06526
Durban	Normal	197,409.84	3445.95	37,525.34	2536.12	2488.44	9834.54
Gaborone	Normal	148,521.33	22,150.61	41,991.97	2372.42	2863	23,670.46

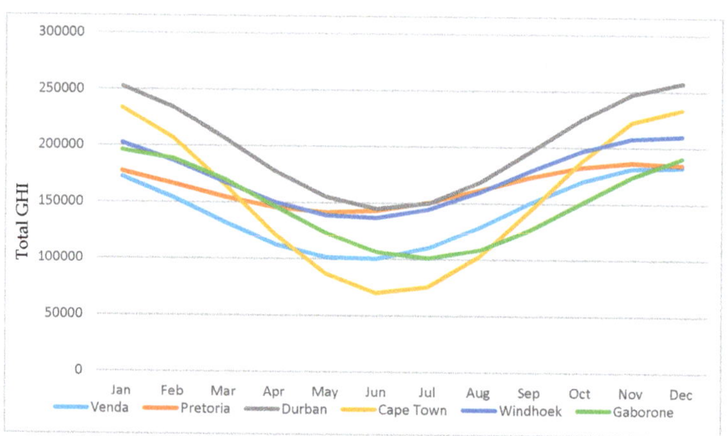

Figure 4. Monthly population mean totals (Wh/m^2).

3.5. Model Validations

The Hosmer and Lemeshow (HL) goodness of fit test done on all of the fitted QDFMs had a p-value greater than 0.05 to indicate that all of the QDFMs were good fits to the respective data. In addition, a runs test on all the fitted models showed that the QDFMs were generating random fitted values except for the Venda and Gaborone monthly QDFMs. The Hosmer and Lemeshow *p*-values as well as those for the runs test are shown in Table 13.

All of the fit-observation plots were approximately linear as shown in Figures A2 and A5. All of the distributional residual plots did not exhibit any pattern. The points on the plots were haphazardly distributed on the scatter plots as shown in Figures A3 and A6. Therefore, all of the fitted models are valid to use in describing the characteristics of solar irradiation in the locations studied.

Table 13. Goodness of fit test *p*-values.

Location	Hourly QDFM		Monthly QDFM	
	HL	Runs test	HL	Runs test
Venda	1	0.09498	1	0.0154
Pretoria	1	1	1	0.2259
Durban	1	0.4038	1	0.5431
Cape Town	1	0.4038	1	0.2154
Windhoek	1	0.4038	1	0.2259
Gaborone	1	0.2105	1	0.0154

4. Conclusions

The main objective of this study was not to predict but to explore the behaviour of SI using the unpopular quantile distributional functions modelling approach. The application of QFs has been shown to be a practical tool and gives more information than the use of only empirical distributions when exploring data. Both the deterministic and stochastic elements inherent in SI could be modelled on par to give a complete description of data characteristics. Application of the Fourier series in our residual analysis gave a direct physical interpretation of the deterministic component while QFs modelled the stochastic element. It enabled the representation of seasonality in the data when we considered different seasons. However, the seasonal modelling could be done over the year at once like the study from [37]. Therefore, the QDFM structure was developed by combining the two modelling components.

Although QDFMs are comprehensive and powerful data exploration tools, some probability distributions do not have existing QFs. This emerges as a drawback in accurately estimating the stochastic properties inherent in the data that follow such probability distributions. Therefore, further studies can be done on developing QFs of such probability distributions. Another challenge is approximating the inverse of the cumulative standardised normal distribution function. The approximations developed so far are complex. More studies can be done on simplifying the approximation process as well as increasing its accuracy.

Daily SI recorded on an hourly time horizon is cyclical, and that pattern can be modelled using a Fourier series. In the Southern African region, the meteorological feature is received on the earth's surface at a maximum between 12:00 and 14:00 depending on seasonal variations, but on average the maximum is experienced during the 13th hour of the day throughout the whole year. Therefore, maximum solar power generation can be done within two hours of midday at any location in Southern Africa regardless of any weather conditions. Maximum daily totals are generally being received during summer (November, December and January) across the region except at Windhoek where the maximum true mean daily total is being received in autumn. We also conclude that Windhoek can be the best solar power generation location in the region when considering daily accumulated solar irradiation because it had the maximum daily population mean total in 9 months of the year, then followed by Cape Town. However, if we consider the monthly accumulated solar irradiance, then Durban is the best location to set up a solar farm in the region. All maximum monthly population mean totals are received at that location in the region. The monthly total SI across the region is a maximum in summer and a minimum in winter. This shows that SI is highly seasonal in the region. Therefore, we suggest that when forecasting SI in the region the modelling process should be split into summer models and winter models. Though seasonal in nature, we can also conclude that Southern Africa's solar irradiance is not being influenced by global warming yet. With such solar irradiance climatic information, then, planners, designers and investors in the solar power generation industry can use this research when identifying where, when and how effective and efficient electricity generation can be operationalised in this region.

Finally, we acknowledge the availability of some meteorology approaches that can be used to further describe the climate of solar irradiation. Therefore, this research creates a starting platform for understanding solar irradiance climate in Southern Africa.

Author Contributions: Conceptualisation, A.M., D.M. and P.M.; methodology, A.M., D.M. and P.M.; software, A.M.; validation, A.M., D.M. and P.M.; formal analysis, A.M., D.M. and P.M.; investigation, A.M., D.M. and P.M.; resources, A.M.; data curation, A.M.; writing—original draft preparation, A.M.; writing—review and editing, A.M., D.M. and P.M.; visualisation, A.M., D.M. and P.M.; supervision, D.M. and P.M.; project administration, A.M. All authors have read and agreed to the published version of the manuscript.

Funding: This research received no external funding.

Data Availability Statement: The data used in this study are from Southern African Universities Radiometric Network (SAURAN), website (https://sauran.ac.za, accessed on 12 June 2022).

Conflicts of Interest: The authors declare no conflict of interest.

Appendix A

Appendix A.1. Fitted Probability Distributions on Modelling Residuals from Trigonometric Regression of the Hourly Profiles

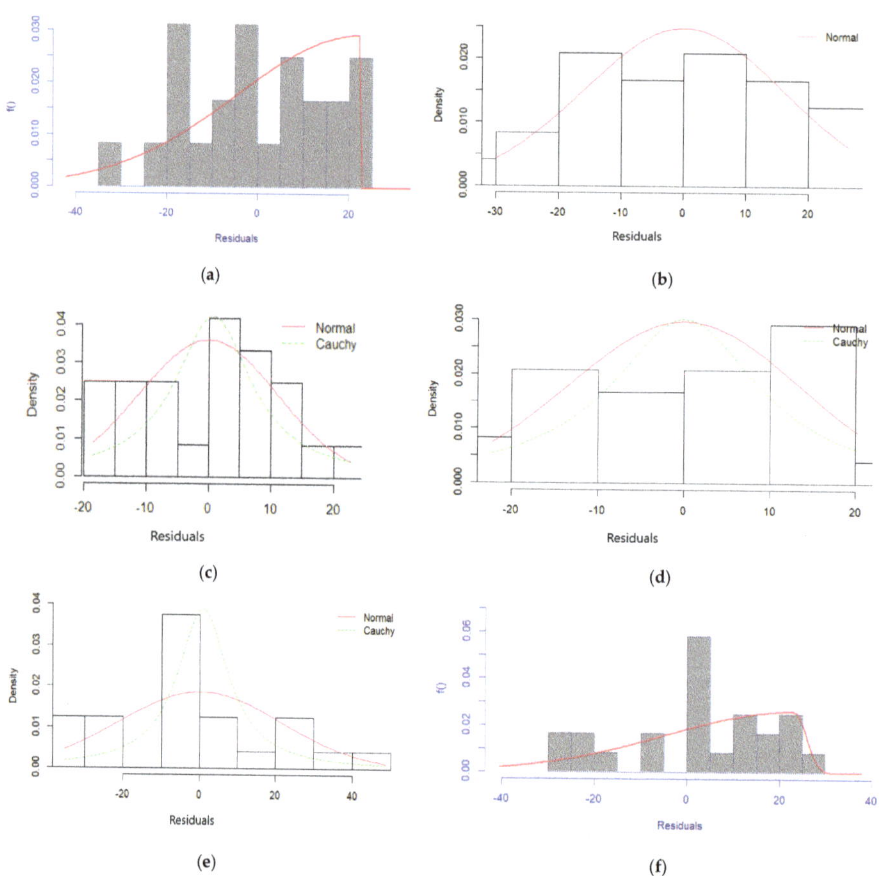

Figure A1. Fitted residual distribution plot for (**a**) Venda; (**b**) Pretoria; (**c**) Durban; (**d**) Cape Town; (**e**) Windhoek; (**f**) Gaborone.

Appendix A.2. Hourly Profile QDFM Validation Plots

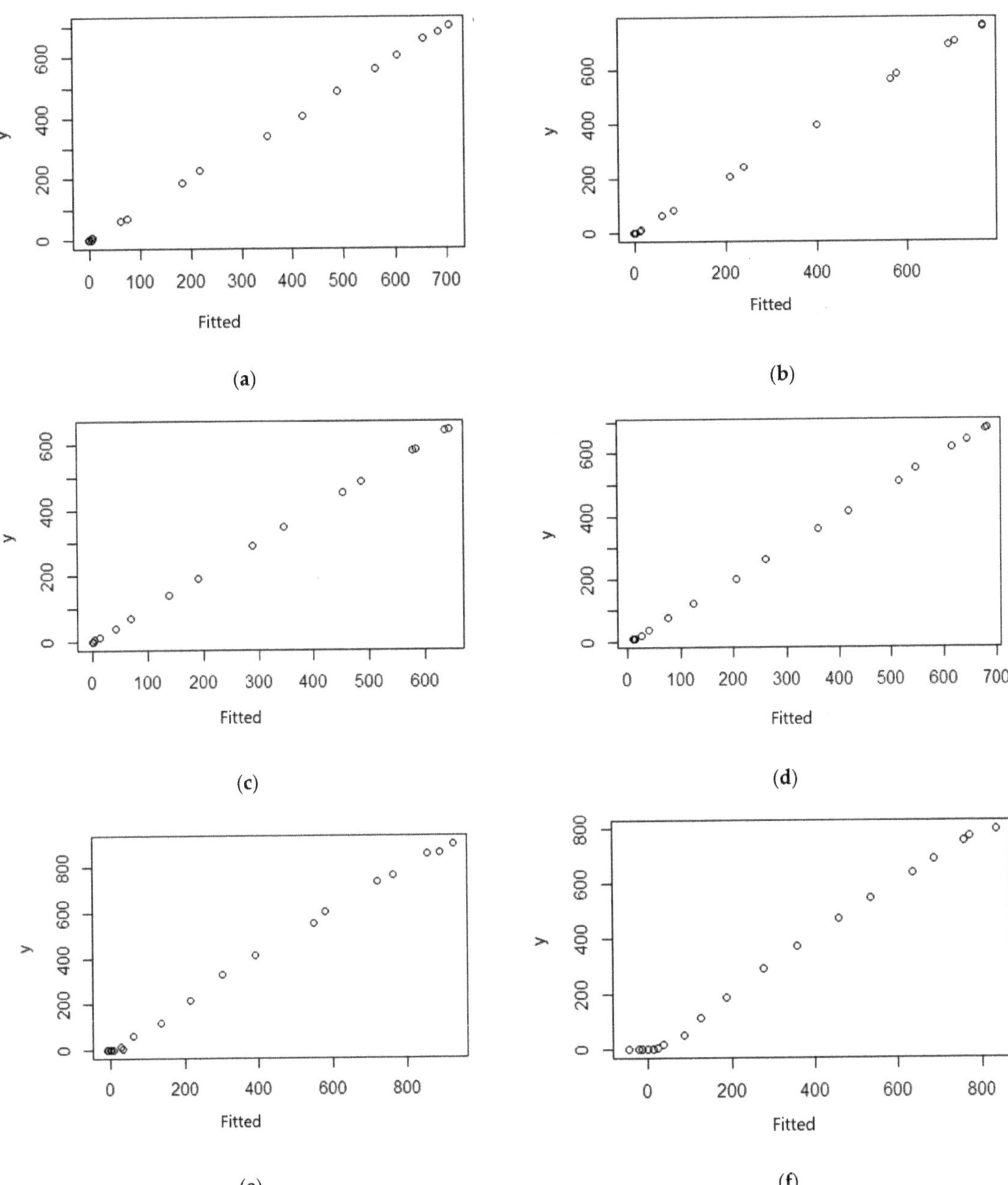

Figure A2. Fit-observation plot for (**a**) Venda; (**b**) Pretoria; (**c**) Durban; (**d**) Cape Town; (**e**) Windhoek; (**f**) Gaborone.

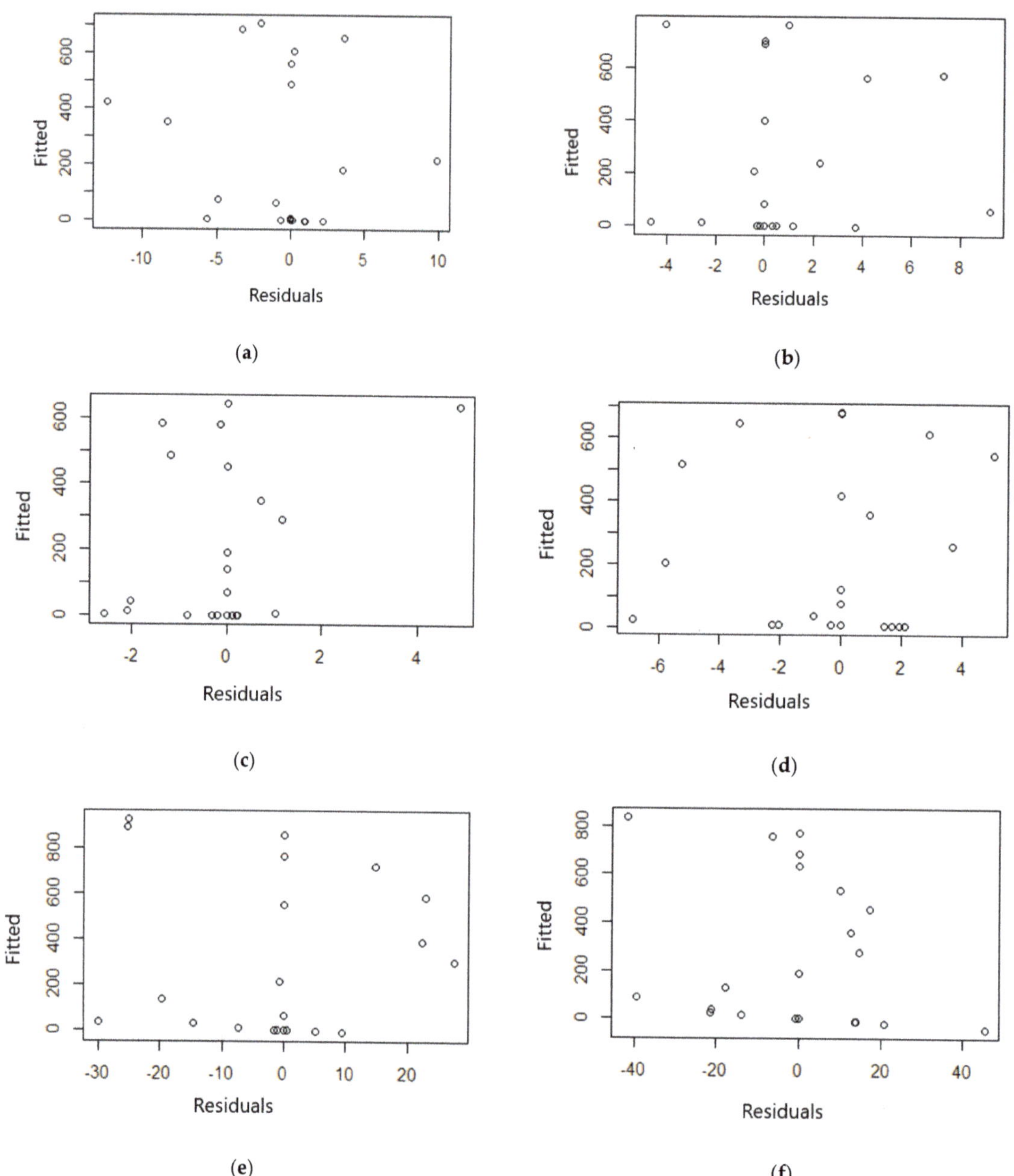

Figure A3. Distributional residual plots (**a**) Venda; (**b**) Pretoria; (**c**) Durban; (**d**) Cape Town; (**e**) Windhoek; (**f**) Gaborone.

Appendix B

Appendix B.1. Fitted Probability Distributions on Modelling Residuals from Trigonometric Regression of Monthly Totals

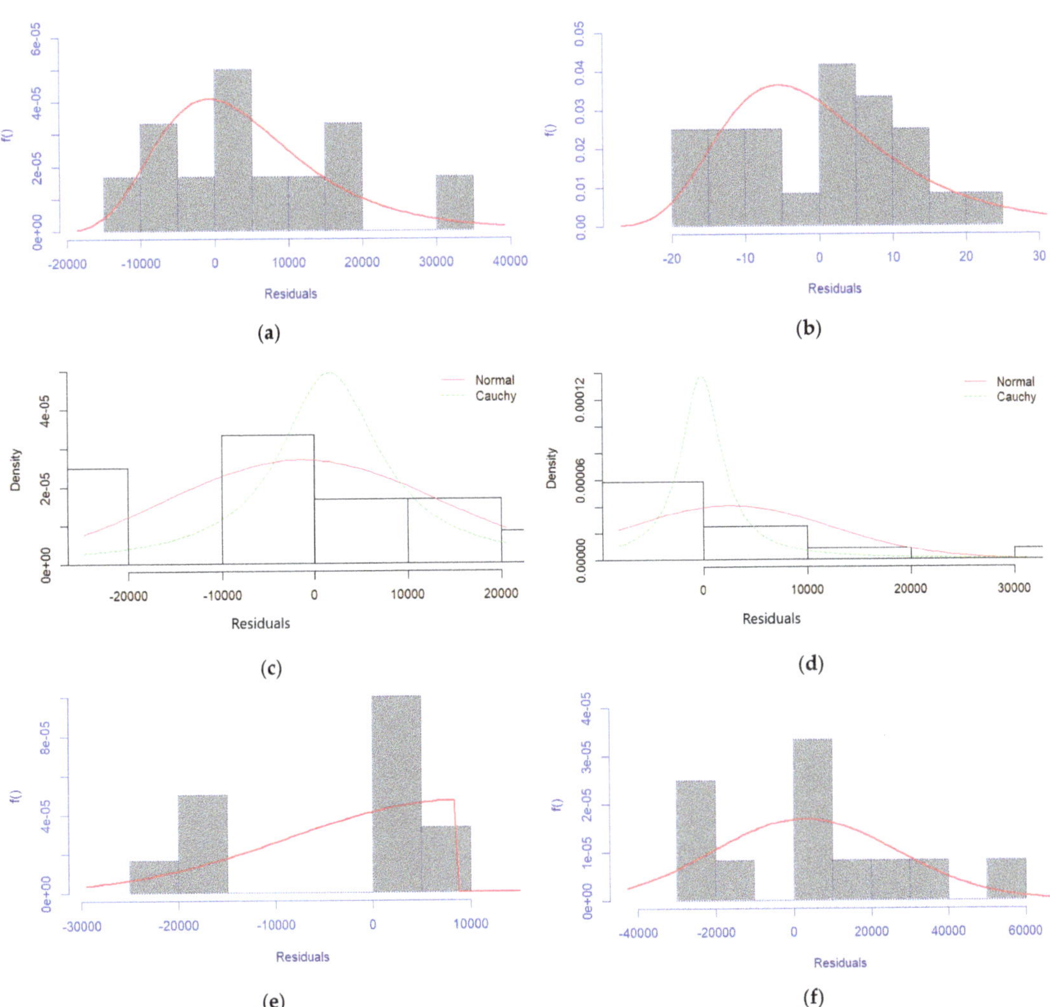

Figure A4. Fitted residual distribution plot for (**a**) Venda; (**b**) Pretoria; (**c**) Durban; (**d**) Cape Town; (**e**) Windhoek; (**f**) Gaborone.

Appendix B.2. Monthly Total Profile QDFMS Validation Plots

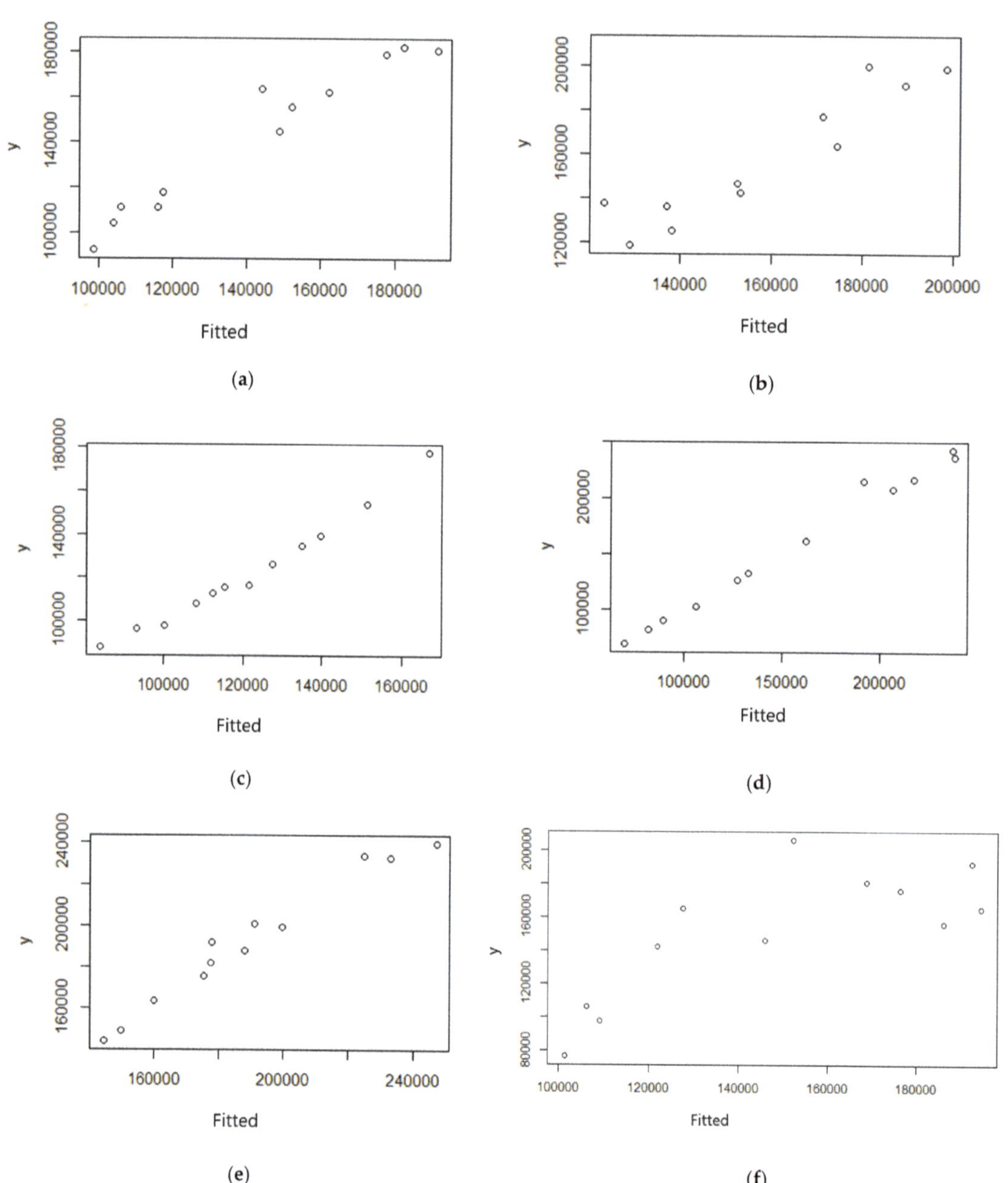

Figure A5. Fit-observation plots (**a**) Venda; (**b**) Pretoria; (**c**) Durban; (**d**) Cape Town; (**e**) Windhoek; (**f**) Gaborone.

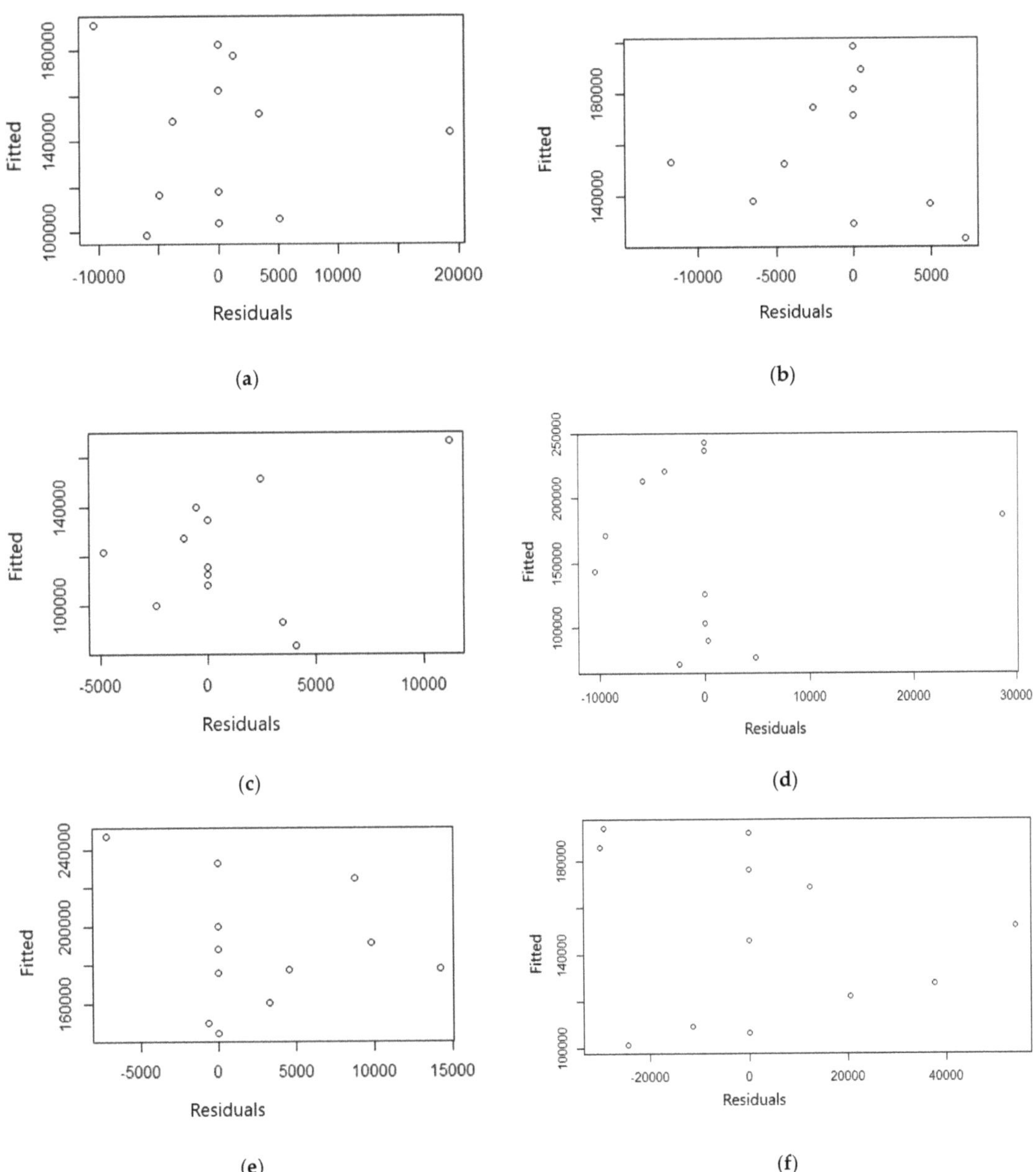

Figure A6. Distributional residual plots (**a**) Venda; (**b**) Pretoria; (**c**) Durban; (**d**) Cape Town; (**e**) Windhoek; (**f**) Gaborone.

References

1. Parzen, E. Quantile probability and statistical modelling. *Stat. Sci.* **2004**, *19*, 652–662. [CrossRef]
2. Gilchrist, W.G. Regression Revisited. *Int. Stat. Rev.* **2008**, *76*, 401–439. [CrossRef]
3. Yang, D. A universal benchmarking method for probabilistic solar irradiance forecasting. *Sol. Energy* **2019**, *184*, 410–416. [CrossRef]

4. Jain, P.K.; Lungu, E.M.; Prakash, J. Stochastic characteristics of solar irradiation—Extremum temperatures processes. In Proceedings of the World Renewable Energy Congress VII (WREC 2002), Cologne, Germany, 29 June–5 July 2002.

5. Jain, P.K.; Prakash, J.; Lungu, E.M. Correlation between temperature and solar irradiation in Botswana: Bivariate model. In Proceedings of the 2nd IASTED Africa Conference Modelling and Simulation (Africa MS 2008), Gaborone, Botswana, 8–10 September 2008.

6. Salima, G.; Chavuka, G.M.S. Determining Angstrom constants for estimating solar radiation in Malawi. *Int. J. Geosci.* **2012**, *3*, 391–397. [CrossRef]

7. Sivhugwana, K.S.; Ranganai, E. Intelligent techniques, harmonically coupled and SARIMA models in forecasting solar radiation data: A hybridisation approach. *J. Energy South. Afr.* **2020**, *31*, 14–37. [CrossRef]

8. Mutavhatsindi, T.; Sigauke, C.; Mbuvha, R. Forecasting Hourly Global Horizontal Solar Irradiance in South Africa. *IEEE Access* **2020**, *8*, 19887. [CrossRef]

9. Jain, P.K.; Lungu, E.M. Stochastic models for sunshine duration and solar irradiation. *Renew. Energy* **2002**, *27*, 197–209. [CrossRef]

10. Jain, P.K.; Prakash, J.; Lungu, E.M. Climate characteristics of Botswana. In Proceedings of the Sixth IASTED International Conference, Gaborone, Botswana, 11–13 September 2006.

11. Madhlopa, A. Study of diurnal production of distilled water by using solar irradiation distribution about solar noon. In Proceedings of the EuroSun 2006 Conference, Glasgow, Scotland, 27–30 June 2006.

12. Madhlopa, A. Solar radiation climate in Malawi. *Sol. Energy* **2006**, *80*, 1055–1057. [CrossRef]

13. Jain, P.K.; Lungu, E.M.; Prakash, J. Bivariate models: Relationships between solar irradiation and either sunshine or extremum temperatures. *Renew. Energy* **2003**, *28*, 1211–1223. [CrossRef]

14. Govender, P.; Brooks, M.J.; Mathews, A.P. Cluster analysis for classification and forecasting of solar irradiance in Durban, South Africa. *J. Energy South. Afr.* **2018**, *29*, 1–6. [CrossRef]

15. Bessafi, M.; Delage, O.; Jeanty, P.; Heintz, A.; Cazal, J.-D.; Delsaut, M.; Gangat, Y.; Partal, L.; Lan-Sun-Luk, J.-D.; Chabriat, J.-P.; et al. Research collaboration in solar radiometry between the University of Reunion Island and the University of Kwazulu-Natal. In Proceedings of the Third Southern African Solar Energy Conference, Mpumalanga, South Africa, 11–13 May 2015.

16. Mpfumali, P.; Sigauke, C.; Bere, A.; Mlaudzi, S. Day Ahead Hourly Global Horizontal Irradiance Forecasting-Application to South African Data. *Energies* **2019**, *12*, 3569. [CrossRef]

17. Ranganai, E.; Sigauke, C. Capturing Long-Range Dependence and Harmonic Phenomena in 24-Hour Solar Irradiance Forecasting. *IEEE Access* **2020**, *8*, 172204–172218. [CrossRef]

18. Ratshilengo, M.; Sigauke, C.; Bere, A. Short-Term Solar Power Forecasting Using Genetic Algorithms: An Application Using South African Data. *Appl. Sci.* **2021**, *11*, 4214. [CrossRef]

19. Chandiwana, E.; Sigauke, C.; Bere, A. Twenty-four-hour ahead probabilistic global horizontal irradiation forecasting using Gaussian process regression. *Algorithms* **2021**, *14*, 177. [CrossRef]

20. Conde-Amboage, M.; Gonzalez-Manteiga, W.; Sanchez-Sellero, C. Quantile regression: Estimation and lack-of-fit tests. *Bol. De Estad. E Investig. Oper.* **2018**, *34*, 97–116.

21. Gilchrist, W.G. *Statistical Modelling with Quantile Functions*; Chapman and Hall/CRC: Boca Raton, FL, USA, 2007.

22. Karian, Z.A.; Dudewicz, E.J. *Handbook of Fitting Statistical Distributions with R.*; Chapman and Hall/CRC: Boca Raton, FL, USA, 2010.

23. Boland, J. Time series modelling of solar radiation. In *Modelling Solar Radiation at the Earth's Surface: Recent Advances*; Badescu, V., Ed.; Springer-Verlag: Berlin/Heidelberg, Germany, 2008; Chapter 11; pp. 283–312.

24. Delignette-Muller, M.-L.; Dutang, C.; Pouillot, R.; Denis, J.-B.; Siberchicot, A. Package 'fitdistrplus'. *J. Stat. Softw.* **2015**, *24*, 1–14.

25. Stasinopoulos, D.M.; Rigby, A. Generalized additive models for location scale and shape (GAMLSS) in R. *J. Stat. Softw.* **2007**, *23*, 507–554. [CrossRef]

26. Richards, W.A.; Antoine, R.; Sahai, A.; Acharya, M.R. An Efficient Polynomial Approximation to the Normal Distribution Function and Its Inverse Function. *J. Math. Res.* **2010**, *2*, 47–51. [CrossRef]

27. Aludaat, K.M.; Alodat, M.T. A note on approximating the normal distribution function. *Appl. Math. Sci.* **2008**, *2*, 425–429.

28. Soranzo, A.; Epure, E. Very Simply Explicitly Invertible Approximations of Normal Cumulative and Normal Quantile Function. *Appl. Math. Sci.* **2014**, *8*, 4323–4341. [CrossRef]

29. Lipoth, J.; Tereda, Y.; Papalexiou, S.N.; Spiteri, R.J. A new very simply explicitly invertible approximation for the standard normal cumulative distribution function. *AIMS Math.* **2022**, *7*, 11635–11646. [CrossRef]

30. Yan, K.; Shen, H.; Wang, L.; Zhou, H.; Xu, M.; Mo, Y. Short-Term Solar Irradiance Forecasting Based on a Hybrid Deep Learning Methodology. *Information* **2020**, *11*, 32. [CrossRef]

31. Crowley, T.J. Causes of Climate Change Over the Past 1000 Years. *Science* **2000**, *289*, 270–277. [CrossRef]

32. Argueso, D.; Evans, J.P.; Fita, L.; Kathryn, J. Temperature response to future urbanization and climate change. *Clim. Dyn.* **2014**, *42*, 2183–2199. [CrossRef]

33. Chapman, S.; Watson, J.E.M.; Salazar, A.; Thatcher, M.; McAlpine, C.A. The impact of urbanization and climate change on urban temperatures: A systematic review. *Landsc. Ecol.* **2017**, *32*, 1921–1935. [CrossRef]

34. Paulescu, M.; Tulcan-Paulescu, E.; Sudhansu, S.S. A temperature-based model for global solar irradiance and its application to estimate daily irradiation values. *Int. J. Energy Res.* **2011**, *35*, 520–529. [CrossRef]

35. Mohanty, S.; Patra, P.K.; Sahoo, S.S. Prediction of global solar radiation using nonlinear autoregressive network with exogenous inputs (narx). In Proceedings of the 2015 39th National Systems Conference (NSC), IEEE, Greater Noida, India, 14–16 December 2015.
36. Grantham, A.; Gel, Y.R.; Boland, J. Nonparametric short-term probabilistic forecasting for solar radiation. *Sol. Energy* **2016**, *133*, 465–475. [CrossRef]
37. Boland, J. Characterising seasonality of solar radiation and solar farm output. *Energies* **2020**, *13*, 471. [CrossRef]

MDPI

St. Alban-Anlage 66

4052 Basel

Switzerland

www.mdpi.com

Mathematical and Computational Applications Editorial Office

E-mail: mca@mdpi.com

www.mdpi.com/journal/mca